D0207919

MEDICINE, MIND, AND THE DOUBLE BRAIN

ANNE HARRINGTON

Medicine, Mind, and the Double Brain

A Study in Nineteenth-Century Thought

PRINCETON UNIVERSITY PRESS
PRINCETON, NEW JERSEY

Published by Princeton University Press, 41 William Street,
Princeton, New Jersey 08540

In the United Kingdom: Princeton University Press, Guildford, Surrey

Library of Congress Cataloging in Publication Data will be
found on the last printed page of this book

ISBN 0-691-08465-3

Publication of this book has been aided by a grant from the
Whitney Darrow Fund of Princeton University Press

This book has been composed in Linotron Sabon

Clothbound editions of Princeton University Press books
are printed on acid-free paper, and binding materials are
chosen for strength and durability. Paperbacks, although satisfactory
for personal collections, are not usually suitable for library rebinding

Printed in the United States of America by Princeton University Press,
Princeton, New Jersey

To my mother, the hand I can always reach for; the friend who sees and knows (all too well) the child in me, but is no less proud of the woman I have become and am becoming.

Contents

List of Figures and Tables

Acknowledgments

THROUGHOUT THE research and composition of this study, I have been the recipient of great riches in the way of support, encouragement, and critical guidance, and I am very happy to have the chance finally to acknowledge some of the many people to whom I am indebted.

This book is a revised version of a 1985 Oxford D.Phil. thesis, "Hemisphere Differences and 'Duality of Mind' in Nineteenth Century Medical Science, c. 1860–1900." Charles Whitty, assigned to me as a "neurological" supervisor when I came to Oxford in the fall of 1982, helped to smooth some of the bumpier technical bits of my research, and served as an ever-patient, perceptive sounding board for my developing ideas. Early on, I also had the opportunity to be supervised by the psychiatrist G. E. Berrios, at Cambridge University, whose encyclopediac knowledge of the broad area in which I was working guided me to useful background material and generally helped me get my bearings. John Durant of the Oxford External Studies Department provided exemplary supervision—and no inconsiderable friendship—during my third year, closely following the actual writing of the thesis, enriching my views with his own sensitivity to important historical links and signposts and invariably helping me to remember why I fell in love with the history of science in the first place. As I settled down to the task of revamping the thesis for publication, William Bynum at the London Wellcome Institute for the History of Medicine diligently read through all the revisions and made numerous perceptive suggestions. The final chapter of this study has especially benefited from his scrutiny.

I am grateful to Charles Webster and the rest of the staff at the Oxford Wellcome Unit for opening their doors to me in 1982, and then facilitating the progress of my research in all sorts of quiet, critical ways. I am also very indebted to the Marshall Aid Commemoration Commission, which made it possible for me to come to Oxford in the first place and then financed my research for three years. The final revision of the dissertation was made possible by a grant from the Wellcome Trust, which enabled me to spend a year as a post-doctoral fellow at the London Wellcome Institute for the History of Medicine. I am glad to have a chance to express my appreciation for the assistance, stimulation, and friendship of all the people there.

Julian Jaynes at Princeton University knows how much I owe to him for encouraging me to produce a preliminary summary of this work for the journal *The Behavioral and Brain Sciences*. The effect of that project on this book has been quite significant, both for what it taught me in its own right and for the range of interdisciplinary feedback it generated. Other people I would like to single out for special mention include the late Kenneth Dewhurst; John Forrester; Ruth Harris; the late Henry Hécaen; Georges Lanteri-Laura; Gert-Jan Lokhorst; John Marshall; Pierre Morel; Rodney Needham; Aaron Smith; Roger Smith; Peter Swales; Paul Weindling; and Oliver Zangwill.

I am grateful to my two readers from Princeton University Press: Francis Schiller, for his vote of confidence, and Lorraine Daston, for her perceptive suggestions on how I might improve the manuscript. I am most indebted to Rod Andrews for his beautiful diagrams and drawings. I also am pleased to express my thanks to Judith May, Science Editor at Princeton University Press, for her faith that all my labors would finally materialize into a proper book, for her skillful handling of all the practical aspects of this venture, and for her ready and friendly accessibility. Finally, I would like to note my debt to the Oxford University Psychology Department, which granted me access to its word processing machines and then tolerated my insistent presence during various hours of the day and night.

A Note on References and Style

IN THIS STUDY, I have adopted the system of references used in all modern scientific literature: author's name, year of publication, and page number (if necessary) placed in parenthesis after the relevant passage of text. All sources are then listed by author and year of publication in the bibliography. In spite of the slight aesthetic disadvantages of this system, it is finding increasing favor with historians of science (e.g., Sulloway 1979; Gould 1981), since it cuts down appreciably on the bulky apparatus of scholarly argumentation and gives the reader immediate access to the most pertinent information he or she requires. The relatively small number of notes that elaborate on or clarify a point made in the text have been referenced in the conventional manner, using footnotes placed at the bottom of the page.

I will add that in composing this text—and I do not think it a trivial point—I have often chosen to use the conventional terms *he, man, mankind*, etc. when referring generally to the human species. Almost without exception, this study is concerned with male scientists, addressing themselves to male colleagues, and viewing the male half of humanity as a norm against which all deviants (including women) might be measured. For this reason, I came to the conclusion that my adoption of such self-consciously androgynous or liberal terms as *humankind, she and he*, etc. might easily have distorted or obscured both the underlying spirit and (quite often) the deeper ideological significance of the different scientific and medical ideas these scientists were propounding.

MEDICINE, MIND, AND THE DOUBLE BRAIN

Introduction

The effect of physiology was to put mind back into nature.
—A. N. Whitehead, 1926

THE HUMAN BRAIN consists of two apparently identical half-spheres, or hemispheres, joined at the median line by great tracts of fibers. Curiously, the two hemispheres as a rule are not endowed with the same functional capacities. Most strikingly, only the left seems able to "speak." What are the implications of this duality and functional asymmetry in the organ of human consciousness and personality? In this study, I am concerned with nineteenth-century medico-scientific responses to that question. The focus is chiefly on France, where most of the interest centred, but attention is paid as well to England, the German-speaking countries, and to a lesser degree, the United States. The study concentrates on, without being strictly limited to, the years 1860–1900 and encompasses explorations into the concepts of symmetry and asymmetry in early nineteenth-century neurology; French aphasiology under Paul Broca, and the discovery of functional asymmetry; the first efforts to conceptualize perceived dichotomies of the human mind in terms of the differential functioning of the left and right sides of the brain; the motif of mental duality and the post-Broca upsurge of interest in the duality of the brain; the late nineteenth-century hypnosis and hysteria theory and research in Paris under Jean-Martin Charcot and his followers; the philosophical physiology of John Hughlings Jackson in England; the possibility of a Freudian indebtedness to Jackson's views on hemisphere specialization and the "duality of mental operations"; and "the fate of the double brain" early in the twentieth century.

This is a study in conceptual history, an attempt to examine critically the ideas and arguments being put forward in the medico-scientific literature of the last century, the basis upon which they rested, and the problems they were intended to help resolve. A fair portion of the literature with which I am concerned has received very little previous attention by modern commentators, historical or otherwise. First and foremost, then, this study aims to serve as a reliable map to a strikingly neglected dimension of the nineteenth century intellectual terrain. Like every map, it is not, and cannot be, exhaustive but has a necessary bias or point of view. The bias in this case is toward what, to my mind, is one of the most interesting and important themes in the history of science: how the human animal has variously struggled to apply the categories of scientific understanding to himself—a brain and mind poised between a universe of social and moral realities, and a universe that seems to stand outside such realities and that he calls "natural." In a variety of ways, then, I have chosen in the present study to focus on the role that the duality and asymmetry of the human brain seem to have played in the nineteenth century's often uneasy and ambivalent efforts to naturalize the human mind, in both health and disease.

Although this study is devoted largely to detailed analysis of scientific and medical texts, this does not mean that I am unaware, as an historian, of the importance of practical human events—so-called contextual factors—in the evolution of scientific and medical thought. I thoroughly agree with Lacapra (1984, 310) on the need in intellectual history for a "noncanonical reading of canonical texts, open to their contestatory dimensions and alert to the problem of how to relate them to artifacts and issues excluded from established canons." At the same time, I am rather wary of that class of currently fashionable studies in the history of science and medicine that puts all the stress on social relations, competition for resources, and vested class and professional interests—almost wholly neglecting the cognitive goals of the social activity being studied. Although indubitably important and often exciting, it seems to me that much of this sociologically oriented work both is missing out on one of the most fascinating parts of the history of science (the science itself) and with its radical methodological programs, is putting the cart before the horse. It is worth remembering that the "old-fashioned" idea—that the differentiating effects of an external natural world have *something* to do with the development of scientific and medical thought—has not yet been ruled out. What has happened, rather, is that all comfortable and simplistic assumptions about science's relationship to the natural world have become subject to inquiry and debate (for a useful introduction to some of the relevant issues, see Knorr-Cetina & Mulkay, 1983). Given this, the present study might be considered a modest, nondogmatic effort to make a contribution to current historiographical and epistemological debates by seeing what sorts of insights might come out of simply listening hard to what different voices of the time were saying on an issue; and by noting recurrent motifs, contradictions, and unexplained gaps in the record that point toward various problems ripe for more broadly based contextual inquiries. On a number of occasions in the course of my analysis, I attempt to make some inroads into these contextual problems myself; in other instances, I did not feel that the data I had available permitted me to do more than erect a few suggestive signposts that might stimulate other researchers.

Historians are creatures of their time no less than the people they study, and it is hardly coincidental that the present study should have been conceived and written during a period in late twentieth-century history that has seen an enormous upsurge of interest in the functional duality and asymmetry (or "laterality") of the human brain. In relative ignorance of earlier traditions, the double brain is being made to carry the burden of a wide variety of social, moral, and philosophical concerns. I cannot help but think that gaining some historical perspective on this phenomenon could offer immediate, practical benefits to scientists involved in charting future courses of research. A scientist who knows from history that the field is prone to certain easy generalizations, philosophical pitfalls, and influences from extrascientific quarters might be able to use this knowledge to bring into focus certain contemporary issues and problems that otherwise would be more difficult to see.

To be interested in the history of neurology and psychology because one has a lively interest in the sciences of mind and brain in one's own era is quite different, however, from thinking that one may use history as a vehicle to hunt for the present in an earlier age (cf. Butterfield 1931). Once having been drawn to the written records and other documents of the past, the historian must make every human effort to discipline his or her culturally colored subjectivity and take the historical evidence on its own terms. As a matter of scholarly conscience, then, this study by and large avoids drawing parallels with twentieth-century perspectives on the double brain or otherwise making reference to the contemporary research scene in the course of discussing the older literature. Only in the final chapter, which struggles with the problem of historical continuity and discontinuity between past and present traditions, does the study briefly make direct contact with contemporary views.

CHAPTER ONE

The Pre-1860 Legacy

> *The reason which persuades me . . . is that I reflect that the other parts of our brain are all of them double . . .*
> *—René Descartes, "Passions of the Soul," 1649*

1-1. The Search for the Seat of the Soul

THE DOUBLE BRAIN began its history, not so much as a scientific or a medical problem but as a theological one. The earliest physiologists, searching for the "seat of the soul," had naturally tended to look for centrally located, unitary organs in the body that could be supposed to correspond to the indivisible unity of the ruling conscious self. As the view gradually gained ground, in the seventeenth and eighteenth centuries, that the workings of the soul were intimately related to the workings of the nervous system, the latter's almost perfect bilateral symmetry had naturally caused some consternation (cf. McDougall 1911, 286; Wilks 1872, 161). It is well known, for example, that the philosopher René Descartes (1596–1650) believed that the brain's pineal gland served as the site of the soul's interaction with the body, though not the site of its physical locale, the soul being nonextended and immaterial. Precisely *why* Descartes had decided to honor an unassuming organ with such an important office is probably less well understood. In his 1649 "Passions of the Soul," Descartes explained

> The reason which persuades me . . . is that I reflect that the other parts of our brain are all of them double, just as we have two eyes, two hands, two ears, and finally all the organs of our outside senses are double; and inasmuch as we have but one solitary and simple thought of one particular thing at one and the same moment, it must necessarily be the case that there must somewhere be a place where the two images which come to us by the two eyes, [and] where the two other impressions which proceed from a single object by means of the double organs of the other senses, can unite before arriving at the soul, in order that they may not represent to it two objects instead of one. And . . . there is no other place in the body where they can be thus united unless they are so in this gland. (Translated in Vesey 1964, 47)

Descartes' second reason for choosing the pineal gland as the seat of the soul was his belief that it was not found in the brains of animals.

Since only man was endowed with an immortal soul, it seemed logical to assume that the site of this soul's operations would be a uniquely human endowment as well. In fact, the pineal gland not only exists in a well-developed form in many animals but actually was first discovered in animal brains. Descartes' critics were quick to point out this fact. In 1699, the Danish theologian and physician Nicholaus Steno also criticized Descartes' choice of organ on the grounds that the pineal gland lacked the complex mechanisms that the Cartesian doctrine of psycho-physical interaction seemed to require (Jaynes 1970). This double-pronged deprecation of the French philosopher's pioneering localization work may have contributed to a tendency among a number of later Cartesian writers to deny the pineal gland any special association with the soul and to turn instead to the handful of other more or less unitary structures in the brain—the corpus callosum, the pons varolii, the septum lucidum, the central ventricle—in their no less earnest search for some central and unitary meeting ground between mind and matter. There was a tradition in Western neurological thinking, going back to the cell doctrine of the early Church Fathers, that there had to be *some* place in the brain where all sensory messages from the outside world could come together and coalesce into a coherent unity, the so-called *sensorium commune*. Otherwise, the seamless unity of the soul's consciousness was inexplicable. The search for some *sensorium commune* was to remain one of the chief themes of eighteenth-century neurology. Just because they were not prepared to accept Descartes' solution does not mean that later generations of neurologists failed to appreciate the seriousness of the problem to which he had called attention.

Unfortunately, little by little, it became clear that the various unitary organs favored by these men had little or no immediate connection with consciousness. Early in the nineteenth century, the phrenologists, led by the Austrian anatomist Franz Joseph Gall (1758–1826), were among the first to take the growing body of evidence as they found it and map out the human soul boldly upon the convolutions of the cerebral hemispheres. Phrenology operated on the basis of three fundamental principles: (1) the brain (above all, the cortex) is the organ of the mind; (2) the brain is a composite of parts, each of which serves a distinct, task-specific "faculty";[1] and (3) the size of the different

[1] That is to say, instead of the abstract faculties of "intelligence," "memory," or "imagination" favored by classical faculty psychologists, one had faculties of "tone," "language," "number," "locality," etc. The first sort of faculty is indiscriminate; the same "memory" faculty deals equally with verbal, spatial, and musical remembering. In contrast, the second sort is individuated by the *type* of information processed. Modern philosopher Jerry Fodor believes that Gall's ability to make such a conceptual shift

parts of the brain, as assessed chiefly through examination of the cranium, is an index of the relative strength of the different faculties being served.

Young has stressed the way in which Gall sought to confirm his theory through a systematic gathering of corroborative cases; to his mind, Gall was making use of a naturalistic method not unlike that of Charles Darwin (Young 1970, 48). Something is slightly misleading about this comparison, however, because Darwin did not first conceive of natural selection, then sail to the Galapagos to confirm his idea. In contrast, the fundamental guiding principle of Gall's system, that the cortex is a plurality of organs serving different faculties, was developed at the end of the eighteenth century, *before* he had dissected a single brain or analyzed the bumps on a single skull. While the details of his localization scheme may have come later through, as Young argues, "naturalistic" means, his original conception of mental faculties and how they related to the brain should be seen, first and foremost, not as a scientific (or even pseudo-scientific) theory but as a philosophical solution to a problem in epistemology. The French sensationalists—Condillac, Bonnet, Cabanis—had argued that sensation alone was sufficient to explain knowledge. If they were right, however, it was difficult to see why the mind was not a mere "heap of impressions," as Hume had put it. What was the means by which knowledge was organized? Gall's reply was that there must exist an innate material base for the organization of sensations in the mind, just as there was a material base for the process of sensation itself (Bentley 1916).

Gall had been impressed by Charles Bonnet's emphasis upon the functional relationship between organic and psychic activity in the brain and later recalled Bonnet's remark that anyone who thoroughly understood the structure of the brain would be able to read all the thoughts passing through it "as in a book." Bonnet, though, had not gone so far as to localize definite psychic predispositions in discrete parts of the brain but rather had envisioned the brain's presumed different organs as vehicles that the immortal soul employed at will. This version of Cartesian interactive dualism was unsatisfactory for Gall, and his scheme the overruling soul—the immaterial homunculus expertly manipulating bits of cortex like a pianist at the keyboard—was abandoned (Lesky 1970). In its place was a brain-dwelling mind that resembled, as the German philosopher F. A. Lange would later put it, "a parliament of little men together, of whom, as also happens

represents one of "the great historical contributions to the development of theoretical psychology" (Marshall 1984, 214–16).

in real parliaments, each possesses only one single idea which he is ceaselessly trying to assert. . . . Instead of *one* soul, phrenology gives us nearly forty" (Lange 1881, 124).

The French physiologist Jean-Pierre-Marie Flourens (1794–1867) admired Gall's anatomical work—Gall, for example, was the first to distinguish clearly between the gray and white matter of the brain—and he agreed with the phrenologists that perception and intelligence "belong exclusively to the cerebral lobes." Experimenting on pigeons and fowl, Flourens had found that when he removed the whole of the cerebrum, the birds continued to live but apparently lost all sense of perception, instinct, intelligence, and volition. They moved away if touched or otherwise irritated, flew if thrown in the air, but when left alone, they remained motionless in a state of stupor resembling deep sleep (Lewes 1877, 469).

Nevertheless, the theological implications of Gall's "parliament of little men" were far from lost on Flourens, and he was determined to refute the materialistic heresies he saw in Gall's approach to the mind-body relationship.[2] In his critique of phrenology (Flourens 1846), Gall and his followers were declared guilty of undermining the unity of the soul, human immortality, free will, and the very existence of God. There should be no question as to why Flourens would choose to dedicate such a treatise to Descartes (cf. on this point Engelhardt 1976).

Flourens conducted his refutation of Gall on both rational and empirical grounds. On the one hand, he pointed to the testimony of "inner sense": this assured him of the indivisible unity and moral freedom of his soul, while the conceptual distinctness of mind from body led logically to a belief in its literal distinctness within each individual. Flourens did not stop there, however, but bolstered his case with what seemed at the time to be an impressive body of experimental evidence as well. Again relying mostly on birds (with the odd rodent and rabbit), he performed a series of studies that involved

[2] It is true, as Riese and Hoff point out (1950, 62), that there is no *logical* connection between cerebral localization and materialism. The former only asserts a topical connection between mind and brain and is actually silent on the nature of their relationship. Gall never ceased to insist that his doctrines were in no sense intended to challenge the teachings of the Catholic Church on the nature of the soul, which did not keep the latter from placing his works on the Catholic *Index* of forbidden books and refusing him a Christian burial. Notwithstanding his protestations, however, there is some evidence that Gall in fact did accept the link between localization and materialism and privately hoped that phrenology would contribute to the debunking of idealist philosophies (see Shapin 1979a, 174).

slicing systematically through the brain and noting resulting deficits. And, he concluded,

> 1. One can excise, either in front, in back, on top, or on the side, a fairly extended portion of the cerebral lobes, without their functions being lost. A rather limited portion of these lobes suffices therefore for the exercise of their functions;
> 2. In proportion as this excision takes place, all the functions become weaker and gradually fade; and past certain limits, they are wholly extinguished. The cerebral lobes cooperate therefore as a total unity in the full and entire exercise of their functions. (Flourens 1824, cited in Hécaen & Lanteri-Laura 1977, 63)

As Young has pointed out (1970, 60–61), methodology helps shape results, and Flourens' experiments did not support localization theory, at least in part because he chose to cut indiscriminately through the brain, disregarding structural variations in the cortex that could conceivably be correlated with functional differences. His overreadiness to generalize results obtained from pigeons into a transspecies theory of brain function is also fair target for criticism. Nevertheless, Flourens' influence on orthodox physiology was to be profound: partly because the crude experimental techniques available in the early nineteenth century meant that a limited amount of counterevidence was available; and, doubtless, partly as well because a unitary conception of mind and brain was as theologically congenial to most of Flourens' colleagues as to Flourens himself (cf. Tizard 1959). The popularization of phrenology under such men as Spurzheim, its transformation into a social crusade with left-wing, radical tendencies, its growing association with other suspect ideas such as mesmerism, also all indirectly worked to enhance the stolid, respectable appeal of Flourens' rival views on brain functioning, at least within the conservative, ingrown circles of academic physiology (cf. Cooter 1984).

Through Flourens, then, the Cartesian soul found a new seat within the cerebral hemispheres, even as it was for the moment rescued from the threat of materialization and disintegration at the hands of Gall. No longer cramped inside the tiny pineal gland, it could now be envisioned "sitting enthroned upon the cerebrum," as the English physiologist Marshall Hall would put it in 1841, "receiving the ambassadors, as it were, from without, along the sentient nerves, deliberating and willing, and sending forth its emissaries and plenipotentiaries, which convey its sovereign mandates, along the voluntary nerves, to muscles subdued to volition" (cited in Riese 1949, 123).

Yet, even if, by the close of the 1820s or so, Flourens had won the battle for antilocalization within official physiology, the outcome of the war remained undetermined. As will be seen in the next chapter, Gall's candle would be kept burning into the 1860s by Jean-Baptiste Bouillaud, when the whole question of cerebral localization would be again thrown wide open. At this time, a new generation of neurologists would partition itself on one or the other side of the battlelines drawn early in the nineteenth century, and underneath the rumble of scientific rhetoric, many of the same uneasy questions about the destiny of the soul and man's place in the scheme of things again would be very much a concern.

1-2. Brain-Duality and the "Laws of Symmetry"

In his crusade against phrenology, Flourens had not focused particular attention on Gall's views on the double brain, doubtless because the immediate dangers raised by the concept of localizing pieces of the soul in different parts of the brain had swamped all other considerations. Nevertheless, there is no question that the role of the double brain within the phrenological system raised a distinct set of theological problems—problems to which Flourens could hardly have been insensitive. From the beginning, Gall had taught that each of the mental faculties existed in perfect symmetrical duplicate, with each pair localized in corresponding regions of the two hemispheres, so that in the end each half of the brain could serve as a complete and independent organ of the mind.[3] For someone like Flourens, such a doctrine could be construed to mean that each hemisphere at least was potentially capable of generating a "soul" of its own, capable of independent will and consciousness. Gall, it must be said, had been careful to put the matter in more neutral, biological language:

> We have two optic nerves and two nerves of hearing, just as we have two eyes and two ears; and the brain is in like manner double, and all its integrant parts are in pairs. Now, just as when one of

[3] A provocative alternative to the classic phrenological conception of the double brain was published in 1849 in *Buchanan's Journal of Man*. According to its anonymous advocate (probably Joseph Buchanan): "Neurology demonstrates that every cerebral organ is balanced by an organ of the opposite function, so as to complete the symmetry of the human character, and give us an unlimited freedom of action in every direction. These opposite tendencies, the recognition of which is essential to any just conception of the human mind, are located in opposite regions of the brain; and as the exercise of any function or faculty naturally tends to over-rule or suspend its opposite, so do the organs of opposite functions, by their position and mechanical relations in the brain, counteract each other and exercise a mutual restraint" ("Anatomy of the Brain" 1849, 297).

> the optic nerves, or one of the eyes is destroyed, we continue to see with the other eye; so when one of the hemispheres of the brain, or one of the brains, has become incapable of exercising its functions, the other hemisphere, or the other brain, may continue to perform without obstruction . . . ; in other words, the functions may be disturbed or suspended on one side, and remain perfect on the other. (Gall 1822, cited in "Dr. Wigan on the Duality of Mind" 1845)[4]

Although theologically pernicious, it is important to realize that the phrenological view (seen in Figure 1) of the double brain as physically and functionally symmetrical also had very strong heuristic appeal, as even the Cartesians had to admit. If the cortex were ever to stand a chance of becoming an object of scientific investigation, it was essential to show that its convolutions were logically and lawfully constructed according to one specieswide blueprint. Now, prior to the period in which the phrenologists began to make their influence felt, it had been widely believed that the convolutions between the two sides of the human brain were organized according to no particular pattern and varied so greatly from individual to individual as to defy all attempts at classification. The French anatomist Vicq d'Azyr had even gone so far as to make these alleged convolutional irregularities a sign of the superiority of the human species, perhaps because in some way the uniqueness and freedom of each individual soul was thereby affirmed. "In the apes," d'Azyr had declared in 1805, "as in all the quadrupeds in general, the cerebral convolutions are few in number, symmetrical on the two sides, and similar in individuals of the same genus. In man, on the contrary, they are neither symmetrical on the two sides, nor similar in [any] two subjects" (cited in Hécaen & Lanteri-Laura 1977, 42). In a similar way, one reads in John Gordon's 1815 anatomy text, written before the new phrenological ideas had been carried to Edinburgh by Spurzheim, how the convolutions in the human brain were "seldom precisely alike, either in shape or size, in any two correspond-

[4] In 1854, an Italian physician, Filipps Lusanna, published a letter in the *Gazetta Medica Italiana*, entitled "Della duplicita indipendente degli emisferi cerebrali." This described several ablation studies, not unlike those of Flourens, that Lusanna believed offered *experimental* confirmation of Gall and Spurzheim's teachings on the functional duality of the brain. Hens surgically deprived of half their brains continued to behave in a typical henlike fashion, without "any lesion of the psychic functions." On the other hand, removing *both* cerebral lobes in a pigeon caused the bird to sink into an apparently mindless stupor, even though the faculties of movement and sensation were preserved. Superficial damage to the two sides of the brain in an owl had an effect intermediate between that observed in the hen and the pigeon, with more instincts becoming lost or disordered as the depth of the ablations was increased (Lusanna 1854).

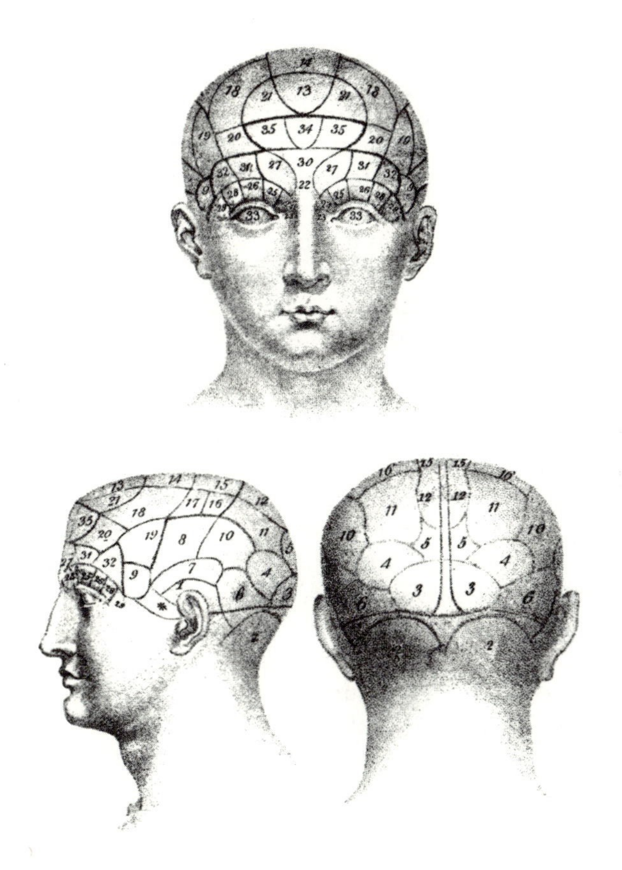

Figure 1. The phrenological model of the brain showing duplication of the mental faculties in symmetrical regions of the two hemispheres. (Source: Spurzheim 1833, frontispiece. Reprinted with the permission of the Bodleian Library, Oxford University.)

ing points of the opposite Hemispheres. In different Brains," Gordon confessed, "I do not know that any two corresponding points, in either Hemisphere, have ever been observed" (cited in Shapin 1979b, 153).

Certainly, it would be foolish to affirm that each and every early proponent of the emerging symmetry principle in cerebral anatomy was directly indebted to Gall. The manner in which new heuristic concepts arise and achieve acceptance within a scientific community is imprecisely understood, but it seems pretty certain that very few pivotal ideas are ever wholly original with one individual. What is more interesting is how a change in belief at this time was apparently able to bring about a change in what the cerebral anatomists actually *saw*. Presumably, there was no great difference between the brains that John Gordon studied and those examined by Sir Charles Bell. Yet, in his 1811 "Idea of a New Anatomy of the Brain," the latter made a point of registering his admiration for the way in which "whatever we observe on one side [of the brain] has a corresponding part on the other; and an exact resemblance and symmetry is preserved in all the lateral divisions of the brain" (Bell 1811, 118).

Now, for Gall, the two halves of the brain existed in symmetrical duplicate in order to provide a buffer against injury. Whether such was the original intention, this idea would prove extremely useful to his followers. Increasingly, phrenologists would be forced to defend themselves against critics armed with medical records that showed how virtually every individual part of the brain mapped out by Gall in its turn had been damaged, without injury to the faculty that was meant to correspond to it. One possible retort to this sort of charge was that there had been a failure to take into account the *duality* of the brain. As Alexander Combe would write in 1824, the critics in question, "by what rules of logic I know not, appear to think injury of one organ sufficient to destroy the function of both." He concluded triumphantly

> *It will be seen, from an attentive perusal of the cases quoted* [by critics], *that not a single instance is to be found, in which . . . destruction of both organs has occurred, while the alleged manifestations* [of the faculty in question] *existed.* In almost all cases, the injury or disease is expressly said to be on one side only; and where it is on both, the parts affected are different. (Combe 1824, 191–92; cf. on this same point Bouillaud 1825, 263–64; Holland 1840, 183; Davey 1844, 377; Buchanan in "Duality and decussation" 1850, 513–14; and the critical views of Lange 1881, 116–17)

One potential problem with this line of strategy, however, was that there were at least a few cases on record in which damage to only half of the brain had unmistakably led to loss or disturbance of mental functioning. When confronted with such cases, the phrenologists suddenly seemed to look at matters quite differently. How was one to explain, for example, one case (Hood 1825–26) of a left-brain damaged patient who had lost his capacity to speak, even though the language faculty in his right hemisphere was presumably sound? Well, the phrenologists replied, obviously any unilateral damage that upset the perfect simultaneous functioning of the two hemispheres would have the effect of disordering the faculties involved. In Spurzheim's words,

> From all the observations which have been made on animated nature, it may be inferred as an universal law that whenever the Creator has bestowed two organs on an animated being, the healthy condition of both is indispensable to the production of their full effect in the economy of that being. (cited without acknowledgement in Hood, 1825–26)

It must be realised that although he taught that the two hemispheres were functionally symmetrical, Gall himself had not thought it necessary to assume that they functioned in *synchrony*. On the contrary, he had suggested that each of the two brain halves was used singly, with one coming to take up the slack after the other had worn out; and he held that the paired sensory organs of the body functioned in a similar fashion. This idea failed to win many converts, partly because it seemed so obvious that, normally, we see and hear with our two eyes and ears equally. In like manner, it was reasoned, the twin halves of the brain must also normally work in tandem.

The writings of the young French physician François Xavier Bichat (1771–1802) were particularly influential in promoting this idea of a necessary connection between symmetry and simultaneity. Bichat was a vitalist whose metaphysical orientation profoundly influenced French physiology up to the time of Claude Bernard. He differed, however, from many other vitalists in not simply defining life as a principle, which explained nothing about it, but by describing it as an ensemble of functions that collectively offered a resistance to death. Those functions, he believed, could be grouped under two principle headings: (1) functions serving "organic life," or that part of existence centred entirely within the organism's internal world and unaffected by habit or experience; and (2) functions serving "animal life" or the "life of relations," that part of existence concerned with relating the organism to the external world and much altered by experience and habit. "To the former," declared Bichat, "belong the passions; to the latter the understanding" (Hall 1969, 122–24).[5]

In advanced species such as man, the organs that served organic life included the organs of digestion, respiration, generation, and circulation, with the heart being central. The organs of animal life included the arms and legs, the sensory organs, and above all, the cerebral hemispheres. The most fundamental physical difference between these two groups of organs, Bichat taught, was that, where the former were characterized by great irregularity of structure, the latter were all found in symmetrical pairs and functioned in unity, presumably with certain necessary qualifications in the case of the limbs. The aesthetic attractiveness of this notion—the centuries-old identification of symmetry and physical beauty—may have played a certain role in the development of Bichat's thinking here but, overtly at least, he offered a more practical rationale for his views. Symmetrical functioning between paired organs was necessary, he taught, because the individual had to be able to relate to the external world equally with both sides of his body. Given this, it became clear that asymmetrical functioning between the organs of animal life necessarily led to maladaptive behavior and must be considered a sign of pathology. As proof, Bichat challenged his readers to imagine a situation in which one of the two hemispheres of the brain—a double organ that was "remarkable" for its symmetry in health—had somehow managed to become larger and more powerful than its twin, with a resulting greater capacity to react to stimuli from the environment:

> I say that in such a case perception will be confused, because the brain is to the soul what the senses are to the brain. . . . Now, if lack of harmony in the exterior sensory system upsets the brain's perception, why should not the soul perceive confusedly, when the

[5] In making this psychological distinction, Bichat is not speaking of passion in the sense generally understood today, or even as it was often understood in his own time. He seems, rather, to be more or less following Descartes' conception of a "passion of the soul" as a "perception" or conscious state in which the soul is passively affected, just as in sense perception, but with the important difference that the thing perceived is attributed by the soul to itself, rather than to some external physical body. It is noteworthy that this is potentially a much more physiological approach to the emotions than that adopted by, say, the Hobbesians of Bichat's time, for whom all "passions" were motivational states of appetite or aversion, or by the faculty psychologists, for whom passion or feeling was an ultimate faculty of the mind: the faculty of being affected positively or negatively by objects cognized (Alston 1967, 480).

two hemispheres, unequal in strength, fail to blend the double impression that they receive into a single one. (Bichat 1805, 22)

It is worth noting that Bichat has managed here to evade the theological problems raised by the spectre of "two souls" in two brain halves by simply taking for granted that the mind operated *through* the double brain but was ontologically distinct from it. This is why he could declare in perfect serenity that, in the event of illness or injury to the brain, it was better for both brain halves to suffer damage than for only one to succumb; "judgment will be weaker, but it will be more exact." Bichat even went so far as to affirm that a blow on one side of the head often had the effect of restoring the wits of someone whose brain had been unilaterally disordered from an earlier injury on the other side of his head (Bichat 1805, 23). We indeed are far from the much more brain-dependent view of mind that allowed Andrew Combe to affirm that the effects of unilateral brain damage were negligible because, the brain being double, the faculties embedded within it existed in duplicate as well.

1-3. Madness and the Double Brain

As hemisphere symmetry and simultaneous functioning increasingly came to be accepted as the twin preconditions for a healthy mind, it began equally to be argued that certain forms of insanity might result from independent, incongruous action between the two sides of the brain. One strain of this sort of reasoning has just been noted, in the writings of Bichat. However, the phrenologists, as well as certain others willing to envision the mind as somehow spatially congruent with the brain, were led to think along rather different lines.[6] People began to ask whether certain circumstances might arise in which each hemisphere could take on an independent life of its own. It seemed particularly plausible to suppose that one brain-half might fall prey to a delusion (caused by unilateral disease or injury) while the other remained healthy, leading to a literal struggle between reason and

[6] Of course, before the idea of a link between madness and disharmonious hemisphere action could begin to be entertained, it is necessary that insanity be accepted as a *disease*, caused by a structural or functional disorder of the brain. Bynum (1981, 49) pointed out that there are two diametrically opposed reasons why someone in the early nineteenth century might want to look upon madness as a disease of the brain. If mind is conflated with soul, as it was for Bichat, then such a perspective puts the mind comfortably out of reach of the ravages of decay and mutability—and mortality. On the other hand, if mind is seen as a function of the brain, totally dependent upon that organ, as it was for the phrenologists, then mental derangements might naturally be expected to be caused by some corresponding neurological malfunctioning.

unreason within a single skull. One of the earliest versions of this idea dates back to 1747, when that bête noire of French Enlightenment philosophy, Julien Offrey de la Mettrie, spoke of the philosopher Blaise Pascal, who

> always required a rampart of chairs or else someone close to him at the left, to prevent his seeing horrible abysses into which (in spite of his understanding these illusions) he sometimes feared that he might fall. . . . Great man on one side of his nature, on the other he was half mad. Madness and wisdom each had its compartment, or its lobe, the two separated by a fissure. Which was the side by which he was so strongly attached to Messieurs of Port Royal? (la Mettrie 1747, 120).[7]

In the first decades of the nineteenth century, the phrenologists would collect a fair number of cases of apparent unilateral mental disorder and hold them up as proof positive of the duplex character of the brain and its faculties. Gall had reported on a Vienese minister tortured by insults on his left side, so that he was constantly turning to look in that direction, "although he knew distinctly with his right side that those insulting sounds were the simple result of a disease of the left [*sic*] half of his brain." In Paris, Gall had also treated a young lady who confessed to him her fear of going mad on one side of her head, "because she had noticed that the course of her ideas was not the same on this side as on the other." A third patient, "a woman of great talent," similarly claimed that she perceived everything quite differently on the left side than on the right and affirmed that sometimes her power of thinking would cease on the left and an "icy torpor" would grip half her skull. "It seems to me," she complained to Gall, "that from the forehead to the back of the head my brain is divided into two distinct halves." She voiced this view, according to Gall, without "the least knowledge of the structure of the brain, nor of my physiological discoveries" (Gall, cited in Elliotson 1847, 212–13; see also Caldwell 1824, 116–17).

[7] Contrast la Mettrie's interpretation of Pascal's disorder with that of Sigmund Freud a century and a half later. Speaking of a certain variety of vivid obsessions "which are nothing but memories, unaltered images of important events," Freud adds, "As an example, I may cite Pascal's obsession: he always thought he saw an abyss on his left hand 'after he had nearly been thrown into the Seine in his coach'. Such obsessions and phobias, which might be called *traumatic*, are allied to the symptoms of hysteria" (Freud 1895b, 74). The neuropsychologist John Marshall (in personal communication) made the further suggestion that Pascal might have been suffering from a form of "hemi-neglect," an organic attentional disorder associated with right hemisphere lesions (see Chapter Nine).

But the explanatory powers of Gall's double brain did not seem to be limited to cases such as these. In the United States, the pioneering alienist Benjamin Rush (1745–1813) described several cases of somnambulism and what would today be called double personality, in which the patients had acted as if they depended upon "*two* minds." He had wondered whether the explanation for these cases might lie in "all the mind being, according to Dr. Gall, like vision a double organ, occupying the two opposite hemispheres of the brain" (Rush 1981, 670). In France, the alienist Jean Esquirol (1782–1840), who had succeeded Pinel at the Salpêtrière, spoke of a form of madness in which an individual "no longer has the faculty of directing his actions because he has lost the unity of his mind. He is the *homo duplex* of St. Paul and of Buffon; impelled to evil by one motive and restrained by the other." Such a "lesion of the will," Esquirol said, could be conceived of "as resulting from the duplicity of the brain, whose two halves, not being equally excited, do not act simultaneously" (Esquirol 1838, 363).

Somewhat later, in 1854, the French alienist Follet (reviewed in Boismont 1854) would present what was hailed as the first large-scale anatomical evidence in favor of the view that certain forms of mental and nervous disorders were indeed caused by a disruption of the normal symmetrical functioning of the two sides of the brain. Examining 300 autopsies between 1833 and 1854, Follet had found that, in epilepsy (a disease that was routinely classified at this time with various forms of insanity), there was almost always a marked disparity between the weight of the two hemispheres. Perhaps oddly, the possibility that these asymmetries may have been a long-term effect of the epilepsy, rather than its "essential cause," seems not to have been considered, either by him or by the alienists and neurologists who continued to cite his report well into the last decades of the nineteenth century.

In Edinburgh, the phrenologist Hewitt Watson agreed with his colleagues that "many cases of insanity," especially perhaps "two-fold personality," might be explained by assuming pathological dissociation or disequilibrium between the two hemispheres. He went one step further than previous writers, however, and questioned whether, even in *health*, the two sides of the brain always functioned in precise synchrony. The title of Watson's 1836 article, "What Is the Use of the Double Brain?" was adopted, he said modestly, "more fully to impress that the following suggestions are to be regarded as questions or hints for the considerations of others, and not in the light of ascertained points." The problem of the double brain, he felt, seemed "likely to remain long open to discussion" (1836, 608–609).

Nevertheless, even at this early stage of inquiry, there was reason to believe that Gall's view, that the brain was double simply to provide a safety factor against injury, was inadequate. If one assumed instead that the brain was an organ whose two halves were able to engage naturally in cooperative but *independent* activity, with the attendant risk of this capacity reaching pathological proportions, then a number of features of human mental life became potentially susceptible to physiological explanation. Watson laid particular stress on what today might be called the human capacity for intersubjectivity: for having one's ideas and feelings coexisting in thought with "the presumed wishes and ideas of others." The classic example of this sort of situation is a game of chess, in which one mentally anticipates an opponent's responses to one's own moves. It seemed to Watson that probably one's own mental activities were mediated by one half of the brain, while an empathetic representation of the thought processes of some other individual was mediated by the other half. "According to this view," Watson said, "mental communication with others, as it is commonly expressed, may be just a self-communing between the hemispheres of our own brains, accompanied by signs and sounds addressed to the senses" (1836, 610).

Watson also suggested that brain-duality might be able to account for the fact that, in dreaming, we habitually "make ourselves into two parties, one of which is (always?) *self* more or less changed" (p. 611). This slightly puzzling statement was probably a reference to dreams in which the dreamer is aware of standing outside of a narrative and observing the activities of another individual whom he also recognizes as himself. Finally, Watson saw in brain-duality a possible explanation for differences in the intensity of consciousness—between simply seeing something, say, and taking careful note of it (or, in phrenological jargon, between *perception* and *attention*). In the less intense state of perception, he proposed, there was perhaps "an active state of either of the corresponding intellectual organs," while in the more intense state of attention, there was "combined activity of the two organs directed to the same matter" (p. 611). Several years later, this last proposal would receive a certain degree of apparent "experimental" confirmation by the brash English champion of phrenology and mesmerism, John Elliotson. Elliotson told how he had found, in the course of stimulating the various cerebral organs of a mesmerized subject that

> the effect was always stronger if both organs of a faculty were touched over: and once or twice there was so little susceptibility, that no effect came till both were touched over. If, for instance, I touched over only one organ of Love of Property, she often did not attempt to steal, but would talk of nothing but acquiring property, and justified theft by all sorts of arguments. (Elliotson 1845–46a, 568–69; cf. the discussion of Braid and "phreno-hypnosis" in Chapter Six)

Soon after, in England, Sir Henry Holland (1788–1873) wrote a brief essay, "On the Brain as a Double Organ" (Holland 1840). A fashionable London doctor who later served as one of Queen Victoria's private physicians, Holland, strictly speaking, was neither an alienist nor a neurologist. He was, however, convinced of the importance of psychological factors in disease and had a lively interest in all questions relating to the functioning of mind and brain. In his 1840 paper, he ventured to say that he was "not sure that this subject of the relation of the two hemispheres of the brain has yet been followed into all the consequences which more or less directly result from it." Above all, how was one to reconcile brain duality with the subjective sense of "unity or individuality, of which consciousness is the interpreter to all?"

Like most of the phrenologists, Holland concluded that unity of mind must depend upon the two halves of the brain functioning in a symmetrical, synchronous fashion. And, also like the phrenologists, he felt that at least some forms of insanity might be "due to incongruous action of this double structure to which perfect unity of action belongs in the healthy state." He spoke, for example, of cases of derangement in which there appeared, as it were, "two minds, one tending to correct . . . the aberrations of the other" and of individuals torn between two contradictory impulses, leading to "a painfully exaggerated picture of the struggle between good and ill" (Holland 1840, 185). Nevertheless, in the end, Holland rejected the materialistic implications inherent in the phrenological view of the brain's functional duality and came down in favor of a dualism reminiscent of that of Bichat. The brain of man was double, he argued, but standing *over and above* that brain was a single, immaterial mind. The appearance in mental derangement of what superficially looked like "two minds" was simply due to "the co-existence before the mind of real and unreal objects of sense, each successively the object of belief" (p. 173).

In the later decades of the century, Holland's paper was to become internationally one of the best-known pre-1860 works on the double brain. This can probably be attributed as much to Holland's social standing and reputation in the medical world as to the intrinsic merits of his ruminations in this department, for really his article said nothing strikingly novel or profound. It is, though, at least noteworthy for its very explicit emphasis on the idea that the normal unity of action that prevailed between the two sides of the brain must be possible only because the two hemispheres were bound together by commissures.

> On the connexions afforded by the Corpus Callosum and the other commissures depend, it may be presumed, the unity and completeness of the functions of this double organisation, as well as the translation of morbid actions from one side to the other. And any breach in the integrity of the union, and of the relations thus established, may tend no less than disease in the respective parts themselves, to disturb the various actions of the brain and nervous system. (Holland 1840, 175)

It was true, Holland admitted, "that observations made, through experiments or accidental lesions of the commissures, give results quite as equivocal as those on other portions of the brain." Still, the conception of the commissures as so many fibrous bridges grafting the two brain-halves together and providing a means for communication between them was so utterly logical—just by *looking* at the brain—that it was hard at this stage to be overly troubled by the lack of empirical evidence actually supporting the idea. Yet, Holland had unwittingly put his finger on an empirical lacuna that would not go away, that would continue to vex and perplex the best minds of neurology throughout the whole of the nineteenth century and well into the twentieth (see Chapters Five and Nine).

1-4. Arthur Ladbroke Wigan and the "Duality of the Mind"

In marked contrast to Holland, Arthur Ladbroke Wigan (d. 1847), an English general practitioner from Brighton, was of the opinion that the corpus callosum was "an organ of no importance, and not necessary to the functions of the brain" (Wigan 1844b, 49). Given this, there is a touch of irony in Wigan's dedication of his thick, badly digested tome on the double brain to his London colleague—"a profound thinker and able physician, a scholar, a philosopher, and a gentleman."

A New View of Insanity: The Duality of the Mind resulted from twenty-five years of pondering by Wigan and can fairly be considered the climax of the pre-1860 period of thought on the double brain. At the same time, its author's passionate claims to have opened a new chapter in the history of neurology, to have made a physiological discovery comparable in its importance and originality to Harvey's discovery of the circulation of the blood, must be treated with a certain amount of polite skepticism. Certainly, as a reviewer in *The Phrenological Journal* tartly pointed out, Wigan's first proposition, that "each cerebrum [hemisphere] is a distinct and perfect whole, as an organ of thought," dated back to "the very foundation of phrenology" ("Dr. Wigan, etc." 1845). Others before him, as we have seen, had also concluded both "that a separate and distinct process of thinking . . . may be carried out in each cerebrum simultaneously" and "that each cerebrum is capable of a distinct and separate volition, and that these are very often opposing volitions" (Wigan 1844b, 26).

The evidence Wigan presented in favor of his belief in the duality of mind and brain also was not particularly novel in its general orientation, although none of the writers earlier mentioned can compete with him for the sheer volume and variety of his clinical reports, anatomical observations, and anecdotes—some of dubious relevance. Wigan explained how his attention was first drawn to the problem of the double brain when, at an early stage of his medical career, he witnessed a certain autopsy. When the dead man's skull was cut open, Wigan had seen, to his astonishment, that one of the hemispheres "was entirely gone," even though the patient "had conversed rationally and even written verses within a few days of his death" (1844b, 40). Was only *one* hemisphere, then, sufficient to sustain a fully functioning mind? Wigan searched back in the medical literature and found that a number of earlier authors had written on cases that suggested that it was. Gall had spoken of a clergyman who "had preached and been occupied as usual in the instruction of youth three days before his death": "on examining the right hemisphere, it was found converted into a clotted substance of dirty yellowish white" (cited Wigan 1844b, 51–52). John Abercrombie had described a man whose mind remained whole and tranquil, despite an injury that opened up the right side of his skull and ultimately caused him to lose almost the entire half of the brain on that side (Abercrombie 1831, 163–64). John Conolly had recorded the case of "a man of family and independence" whose mind remained clear and rational to within a few hours of his death. At autopsy, it was discovered that one entire hemisphere (the side is not mentioned) had been lost and in its place was "a yawning chasm" (cited by Wigan 1844b, 41).

Of these authors, Gall and the phrenological alienist Conolly, at least, would have happily agreed with Wigan's subsequent conclusion that, if mind could persist undisturbed in the absence of an entire hemisphere, then the fact that we each have two hemispheres must mean that we have two minds, that our brains are functionally dual. Indeed, Gall and Conolly's purpose in recording their cases of accidental hemispherectomy was surely in part to demonstrate that very proposition. Abercrombie, on the other hand, insisted that his cases showed "The mind holding intercourse with the external world through the medium of the brain . . . but . . . nothing more." Above all, he said, his observations warranted "nothing in any degree analogous to those partial deductions which form the basis of materialism" (Abercrombie 1831, 164–65).

When Wigan came to apply his discovery of man's "duality of mind" to an understanding of various forms of madness, he also was working within the phrenological tradition, even if—again—he pushed the argument further than it had been taken before. "When the disease or disorder of one cerebrum becomes sufficiently aggravated to defy the control of the other, the case is then one of the commonest forms of mental derangement or insanity," proclaimed Wigan's seventh premise. "In the insane," he went on, "it is almost always possible to trace the intermixture of two synchronous trains of thought, and . . . it is the irregularly alternate utterance of portions of these two trains of thought which constitutes incoherence." Examples demonstrating variations on this thesis were seemingly endless. To cite only one of the more striking of Wigan's offerings,

> I knew a very intelligent and amiable man, who had the power of placing before his eyes *himself* and often laughed heartily *at his double*, who always seemed to laugh in turn. This was long a subject of amusement and joke, but the ultimate result was lamentable. He became gradually convinced that he was haunted by himself, or (to violate grammar for the sake of clearly expressing his idea) his *self*. This other self would argue with him pertinaciously, and, to his great mortification, sometimes refute him, which, as he was very proud of his logical powers, humiliated him exceedingly. I remember very well some of the conversations he related, as if taking place between himself and his other self; and though at the time they merely furnished amusement . . . yet, if such conversations were given piecemeal by a madman, they would form exactly the sort of incoherence we notice in the insane. . . . At length, worn out by the annoyance, he deliberately resolved not to enter on another year of existence—paid all his debts—. . . waited, pistol in hand, the night of the 31st of December, and as the clock struck twelve fired [the pistol] into his mouth (Wigan 1844b, 126–27)

Recalling, perhaps, the fate of the phrenologists under Flourens, Wigan tried hard to make it appear as if his doctrine, that the mind and brain of man were equally dual, in no way implied that he was adopting a materialistic perspective or denying the reality of the Cartesian soul. His reasoning, though, rings conspicuously hollow:

> It seems to me that the use of these two words [*mind* and *soul*] as convertible terms, is a serious obstacle to the freedom of investigation of the mental phenomena, and is at the root of all the difficulties which occur in the discussion of the intellectual faculties.
>
> When I speak of *Mind*, then, I wish to be understood to signify the aggregate of the mental powers and faculties, whether exercised by one brain or two; and when I have occasion to allude to the GREAT, IMMORTAL, IMMATERIAL PRINCIPLE, connected for a time with the material world by means of our physical organization—I shall call it by its proper name, THE SOUL. (Wigan 1844b, 5–6; cf. pp. 11, 32)

Such protestations of piety become even more suspect in light of a conspicuous anticlerical bias pervading much of Wigan's unpublished writings. In arguing that the cause of insanity was to be found in pathological brain functioning, particularly in the incongruous action of the two hemispheres, Wigan was self-consciously adopting the role of the enlightened medical man combating the ignorant views of such men as Johann Christian August Heinroth (1773–1843), who had argued, in contrast, that the main cause of mental illness was *sin*. This idea, Wigan exclaimed, was a "theological monstrosity" that caused the luckless lunatic no end of guilt and suffering.

> Confused by the mixture of insanity and reason in some of these unhappy beings, instead of seeing different and contradictory states of two minds, two organs of thought, he thinks the opposition to be between the natural mind of man and the spirit of evil. It is a horrible doctrine . . . on which I cannot write or even think with calmness, when reflecting on the atrocities which I have myself witnessed at the beginning of the present century. (in Winslow 1849, 504–505)

The paradox of *Duality of Mind* lies in the fact that, even as it proclaimed reassuringly that insanity was caused by physiological malfunctioning and carried no moral taint, it argued no less strongly that every individual had a *moral duty* to cultivate sufficient power over his two brains so as never to succumb to the ravages of madness himself. This, of course, made insanity a sign that one had failed to carry out this duty properly and implied that on some level the madman indeed was personally to blame for the suffering he brought upon himself and his family. In the final analysis, a case could be made that this ethical imperative emerges as Wigan's most original contribution to the pre-1860 dialogue on the meaning of the double brain.

In order fully to understand the way in which Wigan implicated the double brain in moral responsibility, a little biographical background is necessary.[8] It appears that, during the first years of his medical career, Wigan had become increasingly distressed (according to his friend, Forbes Winslow) by what he called the "motiveless crimes of the young." Searching for an explanation, his thoughts had turned to neuroanatomy. Sir Charles Bell was at that time engaged in his investigations into the functional differences between the nerves (see section 1.5) and Wigan was filled with excitement at the new horizons opening up in science's understanding of human nature. It was even not impossible, he felt, that some perspicacious student of anatomy just might soon be in a position to "*catch nature in the fact*, and discover the connection between mind and matter" (Winslow 1849, 499).

The pleasant thought that he himself might be destined to be this student had surely occurred to him and may have been at the back of his mind when he proposed that the "motiveless crimes of the young" had their source in the peculiarly cramped state of the brain during the adolescent years. During puberty, he argued, the "organ of thought" grew so rapidly that the bony cavity of the skull was unable to keep pace. This led to a state of cerebral congestion, which found relief in reckless activity. Criminal youngsters, in other words, were victims of their physiology and could not be held responsible for their actions. Wigan did recommend that the teenagers be flogged but not to punish them. Beating, rather, was advisable because it had the *physiological* effect of rousing the faculties "from the moral torpor which lets the diseased propensity take the lead" (Winslow 1849, 501–507).

It seems to have been a short, easy step from assuming that disordered brain functioning could explain the motiveless crimes of the young to deciding that it could account for the "existence of moral evil in the world" generally. All physicians knew, Wigan declared, that "a blow on the head shall entirely alter the moral character of

[8] On Wigan's life and career in general, Basil Clarke's essay, "Arthur Wigan and the 'Duality of the Mind' " (scheduled for publication in *Psychological Medicine*) is the most comprehensive source available at the moment, and it is likely to remain so.

the individual, that slight inflammation of certain parts of its structure shall change modesty, reserve, and devotion, into blasphemy and obscenity; that a small spicula of bone from the internal surface of the skull shall transform love into hatred" (Wigan 1844b, 120–21).

The crucial turning point came, however, only with Wigan's realization that the human brain (and mind) was *dual.* The discovery of brain-duality meant that mankind, for the first time in its history, was presented with a scientifically sound means of conquering the evil within its own breast. Because disease and injury rarely struck both hemispheres of the brain with equal severity at the same time, it should be possible for someone on the verge of madness or a criminal act to override or inhibit the deranged thoughts being produced by his *diseased* hemisphere using his other, *healthy* hemisphere. It was possible, in other words, for each brain half to act as a "sentinel and security for the other," steadying its fellow in health and intervening to correct the "erroneous judgments" of its fellow when disordered.

Significantly, Wigan tells us that although he had been mulling over the problem of man's double brain for some twenty-five years, he only finally felt compelled to "put pen to paper" and produce his masterwork after reading the Reverend John Barlow's well-known 1843 tract, *Man's Power Over Himself to Prevent or Control Insanity.* That "acute and able writer," Wigan declared, had come "so near the *truth*, that to use the children's phrase, 'he burns' " (1844b, 9). Barlow had argued that insanity could be prevented or controlled because human beings possessed a "peculiar force" that was capable of taking command of the cerebral functions and, in this way, of assuming control over the greater part of the bodily functions. "We find, therefore," was his conclusion, "two forces in activity together, . . . the VITAL FORCE by virtue of which [man] . . . is an animal—and the INTELLECTUAL FORCE by virtue of which he is something more" (Barlow 1843).

Although Wigan believed Barlow's intuition that individuals could subdue their own morbid impulses was absolutely on the mark, he also saw that the reverend was unable to give his theory any scientific certitude. This was because he had no inkling, in Wigan's view, of the true anatomical basis underlying the human capacity for self-control. What Barlow really meant by "man's power over himself to prevent or control insanity"—admittedly, without knowing that this was what he really meant—was "the command of one brain over the other" (Wigan 1844b, 37).

The educational implications of Wigan's findings were far reaching. Human unity of thought and conscience, which Descartes had seen as an essential attribute of the immortal soul, had now been shown by physiology to be contingent upon the synchronous action of the two brains. Capacity for synchrony of action, however, was neither innate nor inevitable. Every human being was born a double animal, "made up of two complete and perfect halves, and [with] . . . no more central and common machinery . . . than is just sufficient to unite the two into one sentient being" (1844b, 155). The ability to work one's two cerebral hemispheres in harmony was an acquired *skill*, which some learned more readily and completely than others. The "object and effect of a well-managed education," then, was "to establish and confirm the power of concentrating the energies of both brains on the same subject at the same time; that is to make both carry on the same train of thought together" (proposition 18). "If this duty be neglected," Wigan warned darkly (p. 49), "or if the discipline be defective or erroneous, the animal grows up into the most detestable combination of intelligence and physical force that infests the earth."

Our understanding of the ethical and pedagogical aspects of Wigan's work is enhanced by realizing that this doctor lived and worked during a time in British social history when the medical profession's newly won jurisdiction of the insane was being threatened by the rise of a new, nonmedical perspective on madness (cf. Harrington 1986). That alternative view saw insanity less as a somatic complaint having its source in some sort of brain dysfunction than as a *moral* defect that could be treated through the inculcations of habits of self-control. As the argument that "very little dependence is to be placed on medicine alone for the cure of insanity" was made with growing vigor, doctors increasingly felt forced to assimilate the new therapeutic ideas into their old naturalistic paradigm of madness as disease (Scull 1981, 141–43). To that end, some turned to Gall's phrenology, a system that permitted them simultaneously to refer in "lay" terms to the psychological benefit of moral treatment and to its "medical" rationale in terms of the putative development and atrophy of different brain parts (Bynum 1981). However, Wigan's *Duality of Mind* offered physicians an alternative and a considerably more ingenious neurological framework within which the efficacy of a moral approach to the prevention and treatment of insanity could be at once accepted and explained "medically."

Wigan's book was harshly reviewed by the phrenologists, not so much because they disagreed with its essential premises but because they were offended by Wigan's extravagant claims to originality and especially by his overtly dismissive attitude towards the phrenological enterprise, when it was clear that, in fact, he owed a significant intellectual debt to phrenology. John Elliotson's tirade against Wigan in the journal, *The Zoist*, is a minor masterpiece of invective:

> The inordinated love of fame, like many weaknesses, . . . serves a most useful purpose to defective or ill trained persons, like crutches and sticks to the lame. But the high human being will not require these miserable aids to enquiry and virtue, he will pursue truth for its own glorious sake. . . . [Men like Wigan] do things continuously and shamelessly which astonish and disgust those whoare good "in spirit and in truth", and would cause Christ to exclaim again that "it shall be more tolerable for the land of Sodom and Gomorrah", &c,—strong expressions, but not too strong for detestation of inconsistency and hypocrisy. (Elliotson 1847, 224–25n)

Elliotson also spitefully pointed out that "if to do two things at once, a person requires two brains, he ought to require several brains when he does several things at once, and a countryman walking the streets of London, using his stick, talking, hearing, and staring as he proceeds, could not dispense with fewer than five" (p. 233). Following a preliminary airing in the *Lancet* (Wigan 1844a), Wigan's views were discussed at a meeting of the Westminster Medical Society on March 16, 1844 and later in the pages of the *Lancet*. There, the general mood was also skeptical, when not frankly derisive ("[Discussion on Wigan (1844a)] Westminster Medical Society" 1844; Davey 1844; Ryan 1844). An anonymous reviewer in *The Journal of Psychological Medicine* was more positive in his assessment of Wigan's book, observing that "the inquiry is a novel one, and as yet *sub judice*. . . . But at the same time, we readily agree that, 'while the brain is considered as the structure of one organ only, there is not much hope of any improvement in' this branch of 'our physiology' " ("[Review of] *New View Insanity*" 1848, 222). Forbes Winslow, although conceding that occasionally Wigan "allowed his originality to overstep the bounds of prudence," nevertheless was convinced that his old friend was a genius who had discovered "great psychological truths." "Generations may roll away ere a just appreciation will be made of the suggestions contained in his celebrated treatise on the 'Duality of the Mind' " (Winslow 1848, 497).

1-5. *Changing Categories of Reference*

The twenty-odd years between the publication of Wigan's *Duality of Mind* in the 1840s and Broca's discovery of cerebral asymmetry in the 1860s saw few new publications bearing explicitly on the double brain issue. This should *not* be taken to mean however, that the mid-century period is comparatively unimportant to our story. On the contrary, the mid-century was a time of developments in medicine and biology that collectively would constitute a far more significant legacy to later theorists on the double brain than a dozen recycled Wigans could have done.

This chapter began in a theological mode, with an examination of the special problems the double brain posed for the first Cartesian physiologists, involved in the quest for the seat of the soul. The years between Descartes and Gall saw the migration of the soul from the pineal gland through various other unitary structures into the twin cerebral hemispheres. At mid-century, what had become of Descartes' soul? Its place of residence had not changed since Gall's time, but nevetheless something new was definitely happening to it. The contexts within which people referred to it were changing, its traditional ontological status was being questioned, even its old name, the "soul," increasingly was being eschewed in favor of more neutral terms. How are we to understand this?

This question forces me to pay more attention than I have so far to the relationship between debates on the meaning of the double brain and certain wider cultural and intellectual developments of the time. Frank Turner's historiography, *Between Science and Religion* (1974), analyzed the rise of "scientific naturalism" during the second half of the nineteenth century. Turner chose to focus on England, but he rightly points out that the national phenomenon with which he is concerned was just one variety of "a general cult of science that swept across Europe during the second half of the century" (Turner 1974, 13). This cult would come to be associated with such figures as Renan, Taine, and Bernard in France; Büchner and Haeckel in Germany; and Spencer, Huxley, and Tyndall in England. Broadly speaking, it was characterized by the so-called positivist conviction that the empirical methods of science represented the only legitimate means of acquiring knowledge of the universe; that within the "new nature" science was creating, unverifiable metaphysical and theological ideas, such as the "soul," had no place; and that the ultimate aim of scientific progress should be a wholesale secularization of society. As Turner puts it, "The naturalistic publicists sought to expand the influence of scientific ideas for the purpose of secularizing society rather than for the goal of advancing science internally. Secularization was their goal; science their weapon" (1974, 16).

Although the scientific naturalists differed from one another on many points of detail, what united them most nearly was a shared faith in a certain view of the natural world and man's place in it. This was summed up by James Ward near the end of the century:

> This naturalistic philosophy consists in the union of three fundamental theories: (1) the theory that nature is ultimately resolvable into a single vast mechanism; (2) the theory of evolution as the working of this mechanism; and (3) the theory of psychophysical parallelism or conscious automatism, according to which theory mental phenomena occasionally accompany but never determine the movements and interactions of the material world. (cited in Turner 1974, 15)

If Turner is right, then one can begin to realize that the mid-century years from Wigan to Broca must have been a troubling time for Descartes' soul. Sketching in very rough outline a highly complicated confluence of events, it may be said that this was a period of transition from an era in which a medical man still felt compelled to affirm that he posed no threat to Judeo-Christian beliefs and values to one in which the categories of reference in medicine and neurology were more or less thoroughly secularized. This was a time in which the various life sciences, envious of the prestige awarded the physical sciences, had begin to look to the latter for their own orientation and, consequently, increasingly were inclined to favor the idea that there was nothing in living processes that could not be explained in terms of chemistry and physics operating under somewhat complex circumstances. On both philosophical and empirical grounds, the varieties of vitalism taught in the early decades of the century by men such as Bichat, Paul Joseph Barthez, Charles Bell, John Abernethy, and Johannes Müller began to come under attack. Hermann von Helmholtz's establishment of the conservation of energy as a universal principle in 1847 played a particularly significant role in this process by making much less tenable the view that living systems could not be comprehended within the same framework as nonliving systems. Increasingly, the scientific naturalists would argue that there was one set of laws basic to all nature, and these were the laws of theoretical mechanics (cf. Mandelbaum 1971, 290–91).

At the same time, the emergence of various naturalistic evolutionary explanations of life (Lamarckian, Spencerian, and Darwinian) were bridging the ontological gap between man and the other animals, presumed by earlier generations. Evolution strictly forbade any view of human personality as "a mysterious attribute, dropped from the skies, without antecedents in nature" (Ribot 1891, 25). If, as Henry Maudsley affirmed in 1859, man was "a part of nature, and like everything in material existence, produced from particles of matter by the same forces and in obedience to the same laws," then there was scant hope that the Cartesian soul, free and immaterial, could long continue to find a home within medicine and neurology.

While all this was going on, neurology proper was witnessing the rise of a revolutionary new view of nervous functioning that seemed as if it might ultimately account for even complex purposive behavior in mechanistic terms. The work of Charles Bell (1744–1842) in 1811 and François Magendie (1783–1855) in 1822 had demonstrated that there were functional differences between the posterior and the anterior nerves of the spinal cord: the former being particularly concerned with sensation, the latter with movement. With this discovery, a way was offered for interpreting the psychological facts of sensation and motor response in physiological terms. Although the idea of reflex action went back conventionally to Descartes, in the 1830s Johannes Müller and Marshall Hall were able to reinterpret it and put it on a neurological basis, in light of this new knowledge about the functional division of the nerves. It seemed clear to them that connected sensory-motor nerves must act as the material basis for the phenomenon of automatic but purposive action. Although they themselves did not suggest that the sensory-motor reflex arc could account for anything other than simple "mindless" behavior, their work at once began to raise questions as to how far the new physiological model of reflex action might be envisaged equally as the underlying cause of complex conscious behavior (Smith 1973, 83–84).

William Carpenter, a prominent light on the mid-century Victorian medical scene, was one of the first to propose that the principle of reflex action might be able to account for certain cerebral processes, whether accompanied by consciousness or not (with the latter called *unconscious cerebration*). Shrinking back from the full implications of this idea, however, he had hastened to qualify it. The cerebral hemispheres were obviously not *only* ganglia acting reflexively; they were also the twin organs through which the human soul exerted its will on the world (Jacyna 1981, 113). Carpenter's contemporary, Thomas Laycock (1812–1876), who would later have an important influence on the thought of Hughlings Jackson, was not so sure. In an 1845 article, provocatively entitled "On the Reflex Action of the Brain," he put forward his view that

> the brain, although the organ consciousness, was subject to the laws of reflex action, and that in this respect it did not differ from the other ganglia in the nervous system. I was led to this opinion by

> the general principle that the ganglia within the cranium being a continuum of the spinal cord, must necessarily be regulated as to their reaction on external agencies by laws identical with those governing the functions of the spinal ganglia and their analogues in lower animals. (Laycock 1845, 298)

To a considerable extent, then, it is true to say that nineteenth century biology was dazzled by the paradigm of the physical sciences and consequently increasingly reductionist in its orientation. It is important, however, not to oversimplify the situation. As Roger Smith (1973) has pointed out, in many cases, "mind" at this time was being reduced to a physical world that natural philosophy had earlier come to conceive in partly mentalistic terms anyway. Because of psychology's importance within natural philosophy, one is confronted with a paradoxical situation in which even a committed "New Nature" scientist could continue to utilize certain psychological categories of understanding in his work—purposiveness, self-activation, teleology—with no clear sense of inconsistency.

Nowhere is this phenomemon more striking than in the realm of evolutionary theory. Ideas about biological evolution in the mid-century were strongly teleological in outlook, providing a focus for all that survived of Enlightenment hopes for the future moral and intellectual progress of human society. It was quite commmon for writers who professed to have a strictly secular view of the natural world, casually to speak as if there were a force at work in nature evolving ever higher forms of human existence; bringing mankind out of ignorance, selfishness, and savage ways, into knowledge, altruism, and culture. Herbert Spencer, for example, declared in 1850:

> Progress . . . is not an accident, but a necessity. Instead of civilization being artificial it is a part of nature; all of a piece with the development of an embryo or the unfolding of a flower. The modifications mankind have undergone, and are still undergoing, result from a law underlying the whole organic creation. . . . As surely as a tree becomes bulky when it stands alone, and slender if one of a group; as surely as a blacksmith's arm grows large, and the skin of a laborer's hand thick; . . . as surely as a passion grows by indulgence and diminishes when restrained; as surely as a disregarded conscience becomes inert, and one that is obeyed active; as surely as there is any meaning in such terms as habit, custom, practice;—so surely must the human faculties be moulded into complete fitness for the social state; so surely must evil and immorality disappear;

so surely must man become perfect. (cited in Mandelbaum 1971, 234)

Turner affirms that there was no place within the new secular philosophy for moral and existential concerns about "man considered as a problem to himself and within himself" (1974, 29). This may indeed have been what the secular naturalists publicly affirmed, but we begin to see here that in practice matters were rather more complex. It would be better, perhaps, to say that, having formally rejected the great majority of traditional moral and existential categories of reference, the naturalists' science came to serve as a new framework for exploring the "nonscientific" question of man considered as a problem to himself and within himself. The double brain—still, on some ultimate personal level, the seat of every man's soul—was to be far from immune to such subtle unions of estoeric scientific knowledge-claims with broader social and ethical concerns. In the chapters that follow, one of my major tasks will be to show the way in which this was so.

CHAPTER TWO

Language Localization and the Problem of Asymmetry

2-1. The Intellectual Context

THE ATTEMPT to find an intellectual context for Paul Broca's epochal localization of articulate speech in the brain takes me back to Gall and his followers in the early decades of the nineteenth century. As is well known, Gall had placed the so-called faculty of "verbal memory" in the suborbital region of the two frontal lobes. As a schoolboy, he had observed that those of his classmates who consistently outshone him in tasks involving verbal memorization—though, as he jealously stressed, he was their intellectual superior in other respects—tended to have bulging "cow's eyes." It occurred to him that bulging eyes must be a sign of a good verbal memory (Young 1970, 13, 136). Although it is doubtful that this insight led him to make any immediate inferences about his schoolmates' brains, it was enough to arouse his interest in the idea that any number of physical features might be used reliably to identify various mental strengths and weaknesses.

Later, within the context of his mature phrenological system, Gall was able to explain the link between bulging eyes and good verbal memory by affirming that unusually large frontal convolutions could affect the development of the orbital cavities. By this time, he had localized not just language but *all* the intellectual faculties in various parts of the anterior region of the brain, arguing that a man was intelligent in proportion to the development of the frontal area of his brain (Soury 1899, 514). He and a few of his followers pointed to some clinico-anatomical evidence supporting a frontal localization of language: cases where injury to the brain's frontal area had resulted in speech loss without any paralysis of the vocal organs. By and large, though, the phrenologists were wary of relying too heavily on proofs offered by pathology. Gall emphasized the unreliability of the data, but one may be forgiven for suspecting that part of his distrust stemmed from the failure of the greater part of these data to support the details of the phrenological system. In any event, most phrenologists seem to have been happy to sidestep the whole contentious issue of the meaning and validity of various clinico-anatomical findings and to concentrate instead on comparative anatomy, combined with the soon-to-be infamous correlations between behavior and external skull contours.

One admirer of Gall's views who made himself an exception, however, was the eminent French physician Jean-Baptiste Bouillaud (1796–1881). A student of François Magendie, who inspired him to undertake physiological research, Bouillaud actually did his most important work in cardiology (Rolleston 1930–31). Nevertheless, he is also important in the history of cerebral localization as the intellectual link between Gall in the 1820s and Paul Broca in the 1860s. Having become convinced, as a young man, of the correctness and importance of Gall's vision of the brain as a plurality of organs, Bouillaud rapidly conceived a powerful sense of mission to use the proof of both clinic and laboratory to convert colleagues, most of whom were followers of Flourens, to his point of view. Above all, Bouillaud was certain that Gall had been right to localize the faculty of language in the frontal area of the brain.

> All that my own observations have indicated to me up to now, relative to the localization of the intellectual cerebral organs, or to the determination of the seat of these organs, is that the anterior lobules of the brain are the organs for the formation and memory of words or the principle signs representing our ideas. (Bouillaud 1825, 284)

During the long winter of phrenology's official disrepute, Bouillaud, as the French aphasiologist Moutier would later put it, was "the soul of the doctrine of localization. He managed during a period of almost fifty years, to affirm with an incredible vigour and steadfastness, that 'the seat of the organ of articulate language' is situated in the anterior lobes of the brain" (Moutier 1908, 13). From 1825 through 1860, Bouillaud defended Gall in debates before the French Académie de Médecine, performed experiments,[1] and collected well over one hundred clinical cases that he felt supported his views. Over the years, as he ripened into an imposing grand old man of medicine, he became increasingly strident, increasingly intolerant toward those who opposed him. According to a contemporary, he was "found in an eternal

[1] In the late 1820s, for example, Bouillaud described an experiment in which the anterior part of a dog's brain was pierced with a gimlet at a spot roughly corresponding to what would later be called Broca's area. The dog survived the operation, but Bouillaud claimed that it was much less intelligent than before and, most significantly, had entirely lost the power of *barking*, though it could still cry out in pain (Bateman, 1869–70, 386),

state of discontent and excitement" and believed that "posterity [was] waiting for him" (cited in Schiller 1979, 171).

Not only did Bouillaud keep Gall's language localization doctrine in the consciousness of his colleagues, he also extended and refined it. For Gall, language had been strictly an intellectual faculty. For Bouillaud, however, it was clear that spoken language consisted of two distinct components: an intellectual component, concerned with the creation of words to represent ideas ("internal language"); and a mechanical, executive component, concerned with coordinating the special movements involved in articulation ("external language"). Although he assumed that both components of language were situated in the frontal lobes (perhaps, he speculated, in the gray and white matter, respectively), his clinical investigations convinced him that it was possible to suffer loss of only the articulatory aspect of speech, with intelligence and verbal comprehension remaining undisturbed (Bouillaud 1825; Soury 1899, 578).

In spite of Bouillaud's best efforts for posterity, however, opinion on the language localization issue remained highly polarized. Perhaps, it would be useful at this juncture to try to bring into focus some of the less obvious reasons why this might have been so.

One must realize, to begin, that the great majority of Bouillaud's colleagues found nothing unacceptable or even particularly novel about the general idea that the mind could be conceived as an aggregate of "faculties." The concept of the faculty as an abstract or immaterial attribute of the human mind or soul was deeply entrenched in French psychology at this time. The leading contemporary speaker for the faculty theory probably was Adolphe Garnier, whose treatise on the human faculties was praised widely as a more or less definitive compendium of the range of behavior available to the soul. Garnier taught that a foresightful Providence had distributed the different faculties unequally among the population and that such natural inequities, which made some people suited to be kings and others suited to be peasants, were what ultimately made society possible. All this seemed to make sense to many people (Zeldin 1981, 10–11). It is true that later, in his 1870 *De l'intelligence*, Hippolyte Taine would launch a withering attack on the French faculty approach to the mind, ridiculing "those little spiritual beings hidden under phenomena as under garments" and "that idea of the infinite which comes from reason, the faculty of the infinite" (Ribot 1877, 376). In the 1840s and 1850s, while Bouillaud was championing the phrenological cause, however, such harsh words were still years away.

All this might incline one to suppose that the French would have been quite sympathetic to phrenological doctrine, with its own decided emphasis on the faculties of the mind; but this was not the case. Instead, one essentially finds a situation in which a faculty approach to the *mind* (or soul) managed to flourish in an intellectual environment strongly inclined to favor a unitary approach to the *brain*. This coexistence of seemingly incompatible approaches will seem less strange once it is realized that faculty psychology in France was harnessed throughout the first half of the nineteenth century to a school of philosophy known as eclecticism. Reflecting the religious and political climate that had emerged with the resurrection of the monarchy under Louis Philippe, eclecticism was for a while (at the injunction of Victor Cousin) taught in the schools as the official philosophy of France. Its basic premise was that everything worth saying in philosophy had already been said, the age of systems was past; all that was necessary now was to take from already existing systems what was true and form a *perennis philosophia* out of the bits and pieces. How was one to decide, though, which bits were true? The eclectics replied that the criterion to be used in sorting the intellectual wheat from the chaff was spiritualism (or idealism), a system that itself apparently was exempt from critical scrutiny. Invoking the spirit of Descartes to give a further patriotic character to their philosophy, idealistically oriented eclectics thus argued that the foundation of their venture was psychology. Introspection or reflection on its own nature was the only sure way for the soul to discover the truth about all things: the laws of the human mind; the nature of right and wrong, beauty and truth; the nature and attributes of God; and so on (Ribot 1877, 367).

It should now be easier to understand how a mid-nineteenth-century French neurologist might accept a faculty model of the mind (or soul) but rise up in arms against any attempt to localize any of these faculties in a physical structure like the brain. At stake for such a neurologist was not only the scientific truth or usefulness of a particular approach to brain functioning (i.e., the unitarian model associated with Flourens) but the continuing validity of the time-honored beliefs in religious dualism and human immortality. It was seen in the last chapter how Flourens was fighting Gall for Descartes' soul, back in the first decades of the century. By the 1850s, the religious dimension of the brain localization controversy increasingly sustained a social dimension as well. It is important to understand that, for Frenchmen living in the middle of the nineteenth century, the quarrel between science and religion had become far more than simply a struggle between abstract opposing views of man and nature. It also was inextricably tangled with a long string of political grievances against the church and monarchy. For many discontented, left-leaning intellectuals, science, and the positivist philosophy with which it had become so closely associated, was widely perceived as a means of undermining the social and political status quo (Zeldin 1981, 263–64). If the old philosophy could be shown to be rooted in error and superstition, if a system of ethics based upon the transcendent was impossible because everything in the universe was natural, then the church and other traditional authorities were stripped of authority—and must give way to a new rational order, ruled by a new "priesthood of science" (Jacyna 1981, 129). For this reason, debates on language localization not only raised various religious bogeys but could, in a strange way, serve as a litmus test of the participants' politics and ethics. The modern French neurologists Hécaen and Lanteri-Laura have summed up the situation neatly:

> In the nineteenth century, the cleavage is clear: the localizers are in favor of regicide, hostile towards the [political] order, adversaries of the death penalty, advocates of the lowering of the electoral census, deniers of the immortality of the soul and of freedom, despisers of papal responsibility, and of the restoration of the Jesuits, atheists, even republicans; the antilocalizers ["unitaires"] are legitimists, they admire Bonald, J. de Maistre and the premier Lamennais,[2] they take delight in the coronation of Charles X and in the death penalty for blasphemers, and so on. (Hécaen & Lanteri-Laura 1977, 54)

It is finally worth realizing that, under Bouillaud, the localization debate had taken on an added poignancy because the faculty under discussion was not just any faculty but *language*, that uniquely human capacity which Descartes had considered to be the most important reason for denying souls to animals; "the true difference between man and beast" (Young 1967, 122). Although not many people still were willing to affirm that speech was a divine gift bestowed upon mankind at the dawn of history, nevertheless, a feeling lingered on that this most noble human capacity must be protected against materialistic reduction. The physiologist François Magendie, for example, spoke in 1836 of certain colleagues who, "sustained and inspired by religious creeds," regarded speech and other "sublime features of the human character" as aspects of the soul, derived from "the Divine essence, of which immortality is one of the attributes" (cited in Young 1970,

[2] Bonald, J. de Maistre, and the premier Lammennais are dryly described by one French historian as "theocrats . . . for whom politics was the handmaid of religion and kings were the servants of the Pope, in whom was embodied the divine will on earth" (Cobban 1965, 78).

82). Similarly, the philosopher Maine de Biran, witnessing with some trepidation the new discoveries on nervous functioning being made in his time by such men as Magendie and Bell, agreed that physiology might be able to study the role of nerves and the brain in those elementary, reflex aspects of behavior that human beings shared with the lower animals. He stoutly denied, however, that there was any similar connection between the nervous system and such uniquely human forms of thought and behavior as language and abstract reasoning. Here, he said, one must recognize an absolute hiatus between spirit and body, which the natural sciences would never be able to bridge (Mandelbaum 1971, 285).

When Paul Broca (1824–1880) formally entered the language localization debate in 1861, he was a seasoned and respected scientist of thirty-seven. Born of Huguenot stock in the Gironde region of France, he had entered the University of Paris at age seventeen to study medicine and spent the first ten years of his career doing research in pathology. Since founding the Paris Société d'Anthropologie in 1859, his restless professional life had circled around physical anthropology and neuroanatomy, his interest in the latter having been raised through his contact with the distinguished neuroanatomist Pierre Gratiolet (1815–1865).

As the driving force behind the Société d'Anthropologie, there seems little doubt that Broca must have found Bouillaud's doctrine of cerebral localization both philosophically and politically attractive. Since its inception, the Société had become notorious as a focus for left-wing, anticlerical activity in French science.[3] Its members were united in their active distaste for anything that smacked of clerical mysticism and theological obscurantism and in their strong desire to bring French philosophy down to earth by firmly grounding it in a material base (Hammond 1980, 122). If the brain sciences could establish a theory that involved breaking up the soul and localizing bits of it in different parts of the cortex, this obviously would go some way towards the goal of "materializing" French philosophy and discrediting right-wing religious authorities. Broca was almost surely aware of this when he declared in 1861 that

[3] Broca's Huguenot background may not be wholly irrelevant to his active role in orienting the Société in various implicitly anticlerical directions. Theodore Zeldin has remarked that Protestants, whether actually active believers or not, tended to play a leading role in the Second Empire anticlerical movement. Having seen their faith and traditions largely stamped out by the joint persecution of the church and the monarchy, their hostility was two-pronged and ran deep (Zeldin 1981, 270).

> the doctrine of cerebral localization was the natural consequence of the philosophical movement of the eighteenth century, for the days were over when one could say without hesitation, in the name of metaphysics, that the soul being single, the brain, in spite of anatomy, must be single also. Everything concerning the connection of mind with matter had been called into question, and in the midst of the uncertainties that surrounded the solution to this great problem, anatomy and physiology, until then reduced to silence, finally had to raise its voice. (Broca 1861a, 191)

The positivist orientation in nineteenth-century European science was closely tied up with a certain methodological bias towards the belief that the best or most scientific way to understand a phenomenon was to break it down to its essential building blocks. In the life sciences, this bias had recently been vindicated in the successes of Rudolf Virchow's cellular pathology. It found expression of a different sort in the elementary "idea-particles" employed by associationist psychologists. For Broca, part of the a priori attraction of the doctrine of cerebral localization was that it suggested that problems of higher brain function might be susceptible to this same analytic—i.e., truly "scientific"—approach (cf. Riese 1949, 118; Tizard 1959). In his view, the work of such men as François Magendie and Brown-Séquard had already demonstrated that, on the spinal and subcortical levels, the central nervous system was indeed a plurality of physiologically distinctive parts, variously concerned with sensory and motor functioning. On the cerebral level, where one was dealing with the problem of localizing aspects of *mind*, naturally everything became much more complicated. Still, he declared in 1861, the anatomical diversity of the cerebral hemispheres, whose lobes and distinctive fissures had in recent years been carefully mapped by Gratiolet and his colleague François Leuret, pointed strongly to the conclusion that localization of some sort should be possible, since one of the "general laws" of physiology was that structurally distinct organs have distinct functions. He admitted, though, that "nothing so far permits a determination of the precise relationship between these diverse organs and these diverse functions" (Broca 1861a, 195–96, 309, 319).

Finally, it must be pointed out that Broca was biased powerfully to agree with Bouillaud that, if a faculty of language were to be localized *anywhere* in the cortex, it would be in the frontal region. Comparative animal anatomy, evidence from pathology, and comparisons of the brains of different races with their "known" intellectual capacities, had all colluded to convince him that, even though it had not yet been possible to locate specific cerebral faculties, a broad distinction might be made between the functions of the frontal lobes and those of the posterior lobes. On this point, it seemed, Gall had been proved right.

> The most noble cerebral faculties, those that constitute understanding properly speaking, such as judgment, reflection, the faculties of comparison, and abstraction, have their seat in the frontal convolutions, whereas the temporal, parietal, and occipital lobe convolutions are appropriate for the feelings, penchants, and passions. (Broca 1861c, 338)

In Broca's view, Gratiolet had provided some of the most powerful new evidence in favor of this old phrenological idea. In his comparative anatomical work with Leuret, Gratiolet had been especially interested in trying to discern some relationship between the intellectual capacities of different species and human races and the differing structures of their nervous systems. The correlation between brain development and suture closings in the skull had led him to believe that he had found a direct, accurate way to measure differential growth of various brain areas. With the Negro races, he thought he had detected earlier closure in the frontal (intellectual) regions, with later closure in the occipital regions, presumably permitting continuing development only of those inferior functions located in the back of the skull. He came to designate this group *race occipitale*. The sutures of Mongoloid groups seemed to close last in the parietal regions; they were consequently classified under the heading *race pariétale*. Only white, European man was found to have late closure in the frontal area. Therefore, Gratiolet extrapolated, only European man (as a member of the *race frontale*) had the capacity to develop intellectually over a long period of childhood development (Harvey 1983, 125).

2-2. *Tan and the Localization of "Articulate Language"*

Broca's formal entry into the language localization debate was precipitated by a paper Gratiolet presented before the Société d'Anthropologie on April 4, 1861, dealing with the relationship between brain volume and intelligence among the various human races. During the discussion period, the question was raised of cerebral localization of specific mental faculties. Gratiolet professed himself opposed to the idea. Ernest Aubertin, son-in-law of Bouillaud and a loyal disciple, argued that if the localization of just *one* faculty could be demonstrated, it would decide the matter in principle. He then declared that Bouillaud already had shown that the faculty of *language* resided in the anterior lobes of the brain. He spoke of a particular patient in the "hospice des incurables" who had lost the ability to speak; and then, he told the assembly that he would renounce Bouillaud's doctrine of language localization if the autopsy of this particular patient failed to reveal anterior lobe lesions (Carrier 1867, 18–19)!

Broca, already highly sympathetic in principle to localization theory, although perhaps not yet fully decided about the value of Bouillaud's more specialized research, resolved to accept Aubertin's challenge. The patient in question, M. Leborgne (known as "Tan," because that was the only word he could say), was transferred on April 11 to the Bicêtre, where he was put under Broca's care. He died just six days later, and the results of the autopsy seemed to confirm Aubertin and Bouillaud's predictions. Broca admitted that the brain actually had suffered damage, both to the anterior lobes and the parieto-temporal region. Nevertheless, for reasons having to do with the apparent relative age of the different lesions (see Marie 1906c), he tentatively linked the loss of speech, which he called *aphemia*,[4] with damage either to the second or the third frontal convolution.

A preliminary report to the Société d'Anthropologie (Broca 1861b) was followed by a detailed presentation before the Société Anatomique a few months later (Broca 1861c). There, Broca paid tribute to Bouillaud for his recognition of the importance of the phrenological doctrine of language localization and for his lonely defense of that doctrine over so many years. "This view . . . would doubtless have disappeared with the rest of the [phrenological] system, if M. Bouillaud had not saved it from shipwreck by subjecting it to important modifications,

[4] Broca explained that he had chosen the word *aphemia* (from the Greek word *phēmi*, meaning "I speak") only after much deliberation. He was originally attracted to the word *aphrasia* (a combination of the negative *a* with the verb *phrazō*, meaning "to speak clearly"). This seemed appropriate to him, since most of his patients were not wholly speechless, but rather had lost the ability to combine their words into true sentences. He ultimately decided against *aphrasia*, however, and opted for *aphemia* instead because he wanted a name that had retained its pure, ancient Greek significance. (The verb *phrazō* had spawned the word *phrasis*, which had then found its way into Latin and most modern Romance languages.)

In 1864, one of Broca's colleagues, Armand Trousseau, launched an attack on Broca's nomenclature. A Greek doctor in his service, M. Crysaphis, had informed him that, etymologically, Broca's word meant "bad reputation" or "infamy." The philologist M. Littré had further told Trousseau that *aphemia* was a worthless word because, when one wished to form a substantive from a Greek verb, one should never take the present indicative form. Littré suggested that the word *aphasia* (*phasis* meaning "speech") be used instead. Trousseau took his advice and managed to get the new term almost universally adopted. Broca's defense of his original choice (in which, among other things, he pointed out that *aphemia* only meant "infamy" in *modern* Greek) was eloquent but ineffective (Ryalls 1984).

and surrounding it with a retinue of evidence taken above all from pathology" (p. 330). At the same time, Broca pointed out that the phrenologists had localized language at the most forward point of the anterior lobes, whereas with his own patient, a considerably more posterior localization was indicated. His own findings, then, were incompatible with Gall's "system of bumps." However, they would be perfectly compatible with some sort of "system of localization by convolutions" (p. 367).

Six months after the death of "Tan," Broca had the opportunity to study a second case of speech loss; that of the elderly M. Lélong, who came to the Bicêtre on October 27 and died on November 8. Reporting again before the Société Anatomique, he professed to feel considerably more confident about his original localization proposals, for the lesion in this second case had been in precisely the same point as that in the first, with the third frontal convolution much more powerfully damaged than the second. He admitted that

> two facts are not much when it is a matter of resolving one of the most obscure and controversial questions of cerebral physiology; nevertheless I cannot keep from thinking, pending fuller results, that the integrity of the third frontal convolution (and perhaps the second) appears indispensable to the exercise of the faculty of articulate language. (Broca 1861d, 406)

These were the modest circumstances by which Broca became "the most distinguished partisan of this doctrine [concerning the frontal localization of language], to which he was to give the strongest impetus," as one of his students, Carrier, would put it in 1867. Soon after the Lélong case, Broca felt himself justified in localizing articulate language even more precisely, in the posterior portion of the third frontal convolution. His work galvanized the medical community, both in France and abroad, and the next several years saw him both continuing to gather evidence in support of his doctrine (e.g., Broca 1863) and trying to account for a few troubling cases reported by others (e.g., Armand Trousseau and Jean-Martin Charcot), which at least at first glance seemed to contradict his views.

What exactly did Broca think he had localized? From the very beginning, he had stressed that the faculty with which he was concerned was simply that of "*articulate language*, which one must be very careful to avoid confounding with the *general language faculty*" (1861c, 331). Somewhat like Bouillaud, he believed that there were two sides to language: a generalized receptive and generative side, which was closely allied with intelligence, and a more specialized mechanical or executive side, which seemed to be independent of intelligence. Because it seemed that aphemics could still understand language and because many could still write, it was clear to Broca that they had suffered no damage to their general language faculty. What they had lost, rather, was "*the faculty of coordinating the movements appropriate for articulate language*" (1861c, 333).

Although independent of intelligence, articulate language was nonetheless a *mental* faculty. At one point in his 1861 report on "Tan," Broca raised the possibility of aphemia being a disease of locomotion, a *motor* disorder, only to dismiss the idea as highly improbable. (1861c, 335). Aphemics, rather, were to be understood as victims of a peculiar form of memory disorder. Their lips and tongues were not paralyzed, they could move them to perform any number of other tasks, but they were unable to remember how to articulate words with them. Broca's argument for an "articulatory center" in the cortex, in short, entailed no belief that parts of the cortex might serve motor functioning. On the contrary, it is clear that he accepted the reigning doctrine of his time, which held that the sensory-motor constitution of the central nervous system extended only as far up as the subcortex and that the cerebral hemispheres were concerned strictly with mental activity ("it is beyond all question," he declared in 1861, "that the cerebral lobes, properly speaking, are alone set apart for thought" [1861a, 309]). Broca explained the fact that "Tan" was aphemic *and* suffering from right hemiplegia—the latter, of course, a blatant motor disability—by supposing that the anterior lobe lesion responsible for the aphemia in this case had extended down into the corpus striatium, "the motor organ most near to the anterior lobes" (1861c, 347). He felt it likely that the site of the *general* faculty of language ultimately would prove to be somewhere in the anterior lobes too; he refrained, however, from speculating as to its exact location.

2-3. *The Académie de Médecine Debates*

Events took a new turn in 1863 when a physician from Montpellier, Gustav Dax, brought to the attention of the Académie de Médecine in Paris a memoir that his deceased father, Marc Dax, had read before his medical colleagues in 1836. In more than forty cases involving speech loss without paralysis of the articulatory organs, Marc Dax had noticed left-brain damage, but he had been unable to find a single case that involved damage to the right side of the brain alone. He thus concluded that, while damage to the left hemisphere does not inevitably alter what he called *verbal memory*, when such an alteration occurs, "the cause should be sought there even if both hemispheres are affected together" (Dax 1836, 243). Gustav Dax was convinced that his father's ideas were important, and he asked the Académie to evaluate them.

The Académie agreed, and appointed a commission—composed of Bouillaud, M. Béclard, and M. Lélut—to undertake the task. Of these three men, Louis-Francisque Lélut (1804–1877) had long been a particularly violent opponent of the idea of cerebral localization. He was the author of various polemical tracts opposing phrenology (*Qu'est que la phrénologie* in 1836 and *Rejet de l'organologie phrénologie de Gall et des ses successeurs* in 1843), and he had always been outspoken in his criticism of Bouillaud. Lélut's report on Dax was read before the Académie by Béclard in December 1864. The author, who could not trouble himself to attend in person, emphasized that he was presenting his own opinions only and declared his colleagues free to contradict him later if they wished. This said, he proceeded to inveigh in general terms against the doctrine of cerebral localization and summarily dismissed Dax's work as so much phrenological hogwash (Lélut 1864–65). Bouillaud, not surprisingly, was incensed by this treatment of the subject, but since the hour was late, he declared that he would postpone his response until a later session. He finally was given an opportunity to take the floor on April 4, 1865. The passionate debate on the "faculty of articulate language" that followed in the wake of his presentation ("Discussion sur la faculté du language articulé" 1864–65) would last for more than three months and can be considered, as Riese (1947, 326) argued, the climax of the early history of aphasia. It is a climax, however, in which Broca himself would ironically play no direct role, not being a member of the Académie and indeed being away from Paris and ill while the greater part of the debates were in progress.

Bouillaud began by attacking Lélut for merely ridiculing the doctrine of language localization instead of truly weighing its pros and cons, "One might say, at the risk of slipping into parody, that our learned author was voting the *mort sans phrase* of the doctrine." Bouillaud then launched into a defense of cerebral localization in general and language localization in particular, with all the zeal of an evangelist at a revival meeting. Lélut's opinion notwithstanding, Gall's view of the brain as a plurality of organs was *not* impossible and ridiculous; on the contrary, in the light of current knowledge in physiology, it was the only view that was truly feasible. Bouillaud compared Gall with such giants of science as Kepler and Newton (p. 588) and drew attention to Broca's recent "dazzling and fruitful conversion" to the phrenological doctrine of language localization (p. 584). Continuing in this religious vein—and with a nice little play on words—he affirmed that Broca had now made himself a veritable "St. Paul" (p. 726) of Gall's truths; truths Bouillaud was convinced were destined to revolutionize medical thought.

It is worth noting that Bouillaud spoke rather little about Marc Dax—even though Dax's work, *not* Gall's or Broca's, was meant to be the main focus of discussion. It is not hard to guess the reason for Bouillaud's negligence in this respect: Dax had not said a word about an anterior localization of language but simply had hoped to demonstrate a special role for the left hemisphere in "verbal memory." Although Bouillaud did not completely ignore the unilateral issue, as will be seen, there nevertheless seems little doubt from the Académie proceedings that he essentially used the Dax paper for his own purposes: seizing on it as an excuse to reopen before the Académie the question of language localization, and then doing his best to shift the main focus of debate away from the left hemisphere to the anterior lobes.

In the long debates that followed Bouillaud's speech, the skeptical and somewhat arrogant voice of Armand Trousseau (1801–1867) emerged as a dominant presence. A medical professor at the Hotel-Dieu, Trousseau's three-volume *Clinique Médicale de l'Hôtel-Dieu en Paris* had become a standard reference work in clinical medicine and would remain so for some fifty years. Earlier he had challenged Broca's localization work on clinical grounds, reporting on a speechless patient whose brain at autopsy was found to contain numerous lesions but whose inferior frontal convolutions were apparently undamaged. By the time of the debates, he had gathered information on 135 cases of speech loss, while Broca had a mere 32 (Ryalls 1984). Clearly, he was a man to be reckoned with, and he knew it.

Trousseau had a number of objections to pose to the company. In the first place, he said, Broca's word for speech loss, *aphemia*, was inappropriate, since in Greek it meant "infamy" (cf. note 3). He proposed to replace it with a word of his own, *aphasia* (a proposal that, of course, would prevail). Trousseau also pointed out that simple localization of a lesion was not enough; one must also consider the *nature* of the lesion localized. Finally and, in his view, most importantly, he argued that Broca was very wrong to speak of aphasia as a disorder simply involving loss of speech and distinct from general intelligence. When speech was lost, he said, learning and memory necessarily were disturbed as well; and he compared the aphasic brain to the sea, upon which "the prow of the ship can leave no trace" ("Discussion sur la faculté, etc." 1864–65, 648, 670, 674–75).

Parchappe asked the question What is speech? and felt that this must be examined before one could attempt to localize anything. Velpeau spoke of a certain garrulous barber who was later found to be missing both of his frontal lobes, and he inquired about the implications of this case for Bouillaud and Broca's theory. Cerise stressed that there is a difference between *symptoms* and *functions*, pointing out that one is not justified in assuming that, simply because damage to a part of the brain causes speech loss, this same brain part therefore is the site of speech.

In spite of these various and sometimes significant criticisms, the bias within the majority of the assembly was clearly tilted in favor of the localizers. By the end of the debates, even Trousseau had been forced to concede that it was at least possible that the anterior lobes did play a special role, albeit not necessarily an exclusive one, in the production of speech. The localizers departed from the Académie buoyed up with a fresh sense of optimism. As one of them, M. Bonnafont, lyrically put it, there was a certain elated sense among these men "that the modest skiff that transports [the doctrine], after having long traversed stormy seas and further battled with hurricanes, its sails distended at last by the wind of progress and new facts, will finish by successfully reaching the port" ("Discussion sur la faculté, etc." 1864–65, 867). Such, indeed, was more or less how matters would unfold: the establishment of localization as a fundamental principle in brain functioning—helped along, not only by Broca's work but by various other developments in clinical and experimental neurology (see Chapter Three, section 3-1)—ranks as one of the more compelling success stories of nineteenth century physiology.

For the historian, this effective triumph of Gall's old doctrine raises a difficult question: Why do certain ideas in the history of science and medicine suddenly become "ripe" for general acceptance, after possibly years of neglect or repudiation? Let us remember that Bouillaud had been arguing for a frontal localization of language for some thirty-five years and had collected more than one hundred cases to back up his views, yet he had never managed to convince more than a modest portion of his colleagues. Broca's evidence was considerably less plentiful than Bouillaud's and his claims, for a localization of language in the foot of the third frontal convolution, if anything, were even more radical. Why, then, was his work apparently able to break the stalemate over the localization question and turn the tide of medical opinion increasingly in its favor?

This question admits no cursory, cut-and-dried reply. Sociologists in the field of scientific knowledge point out how a number of alternative theories generally can be constructed in science that more or less account for a pool of presumed-to-be-relevant observation and data. Because of this, any theory that in the end achieves wide acceptance is said to be "under-determined" by cognitive factors alone. This fact opens the door to the possibility that divergent social, metaphysical, and professional interests play a hidden role in the theoretical preferences and interpretative jockeyings of science.

There can be no denying that the triumph of cerebral localization theory in the 1860s under Broca suggests itself as a rich source of grist for the sociologist's mill, although a proper attempt to grind through the evidence will not be undertaken here. Certain political and philosophical factors (mentioned in section 2-1) seem to have predisposed Broca to look more favorably upon the doctrine of localization than was strictly warranted by the evidence alone. Careful analysis might well reveal that many of Broca's supporters were more or less unconsciously influenced by some of these same political and philosophical factors. One clear signpost pointing toward a link between the acceptance of localization and certain wider social issues can be found in Pierre Marie's (admittedly biased) description of the early French response to the localization question (Marie 1906c). Writing at the turn of the new century, Marie recalled the cult of personality that surrounded Bouillaud in the 1860s and the army, "*the crowd* . . . with its divinatory instinct and its profound ignorance," that chose to follow him. He argued that, in those days, Broca's proposal that speech had its seat in a tiny portion of one of the convolutions of the brain seemed so extraordinary, so innovative, that one longed to believe it true. The medical students of the day seized upon the doctrine in part because, by its very novelty and distastefulness to the older generation, it seemed to represent scientific progress. They also seized upon it, however, according to Marie, because it had become a symbol of free thought and liberal politics: "for a while, among the students, faith in localization was made part of the Republican credo" (p. 570).

2-4. Broca and the Question of Human Uniqueness

In spite of the general momentum towards an acceptance of Broca's work, it should not be thought that, after 1865, his critics were wholly silenced. In English, the alienist Henry Maudsley felt that the localizers in France were glossing over the complexity of speech and language as aspects of mind. Denouncing Broca's speech faculty as a "wonderful metaphysical entity," he acidly opined that its localization in the third frontal convolution had not seen its like in psychology "since Descartes located the soul in the pineal gland" (Maudsley 1868, 690–92). Charlton Bastian in England also (initially) was skeptical. That there should exist a special convolution for the faculty of articulate language was, he believed, an "a priori improbability," and he spoke further of "the shifting and unsatisfactory nature of the evidence" (Bastian 1869, 209). But the most important British dissenter from the French faculty approach to speech disorders was John Hughlings Jackson. Very early on, he began to have doubts about the reality of Broca's speech faculty and to question Broca's interpretation of his data. He then went on to develop a complicated alternative approach to speech loss and higher nervous functioning, based on a view of the brain as a sensory-motor apparatus operating according to associationist laws. His views, and the various factors contributing to their development, receive extensive consideration in Chapter Seven of this study.

In France, one of Broca's colleagues at the Société d'Anthropologie, M. Gaussin, questioned Broca's belief that the third frontal convolution was concerned only with articulate language. Gaussin was far from convinced that speech was a primary faculty. He suggested that it was, instead, merely one form of a more global "faculty of expression": other forms being human writing and pantomine, the "natural language" of deaf-mutes, and the "rudimentary cries" and expressive facial contortions of animals. Was it not likely, Gaussin asked, that the third frontal convolution actually served, not just articulate speech, but *all* modes of the "faculty of expression"?

The chief interest of Gaussin's objections, for our purposes at least, lies in the response they provoked from Broca during the discussion session following. Broca said that he fully agreed with Gaussin about the need to speak of a general "faculty of expression" common to animals and men. Nevertheless, he was still inclined to think—and not on clinical grounds alone—that articulate language was not merely an aspect of that other, general faculty but really was a thing apart. Articulate language was unique because it alone had no analogue in the animal world; it alone was a strictly human acquisition. "I know of nothing else that animals lack and that we have." Consequently, "to know how the intelligence of animals resembles that of man and how it differs from it; there is something that the question of aphemia will be able to help [us] discover" ("Discussion sur l'aphémie" 1865, 413–14).

Something new had entered the debate. Broca had suddenly reminded the company that the faculty which he had managed to localize was not just *any* faculty, but the one by which human beings were most clearly identified as human, by which they were set apart and above the dumb beasts of the planet. Of course, in contrast to at least some of his critics, Broca would not have accepted for a minute the theological view that the uniqueness of language somehow precluded it from scientific investigation or was evidence of man's immortal soul. Nevertheless, no less than his critics, Broca seems to have believed that language continued to give a certain degree of substance to mankind's old claims to pride of place in the natural world.

Broca, however, went one step further. By focusing attention on the neurological conditions underlying the functioning of this unique faculty, he implied that the science of aphasiology stood in a particularly favored position to cast light on certain fundamental questions about the "essence" of human nature. Humans were unique, he seems to have been suggesting, in no small part because they possessed a certain remarkable faculty located in the posterior half of their third frontal convolution. In light of this, one is inclined to think that Maudsley was right. There is indeed something about Broca's localization of articulate speech in the third frontal convolution reminiscent of Descartes' localization of the soul in the pineal gland.

It fell to a member of Broca's Société d'Anthropologie, M. Pruner-Bey, to bring into sharp focus the implications of Broca's seemingly casual remarks to Gaussin. Pruner-Bey asserted that any man who had the good fortune to discover the cerebral organ responsible for articulate language

> enriches science on a double score; he establishes the anatomical basis for the most imposing difference between man and animal, and he brings unshakeable support to the doctrine of cerebral localization. I will go even further: such a discovery would raise the corner of the veil that envelops the organ of human intelligence in its totality; for it is evident that, in the creation and use of human language, all the intellectual faculties, viz. perception, feeling, volition, have to give assistance. Now, does there exist in the encephalon a focus, an anatomical center, or at least something that offers the appearance of such, which one might suppose to be the reunion point of all these faculties? You know, Gentlemen, the singular hypotheses expressed on this subject. (Pruner-Bey 1865, 558)

2-5. *The Discovery of Asymmetry*

The motif of human uniqueness and man's relationship to the animals would ultimately swell to a climax in Broca's solution to the vexing problem of cerebral asymmetry. This story occupies a peculiar position in the early history of language localization. Initially, the fact that lesions causing speech disorders tended to be found almost exclusively on the left side of the brain was simply seen as an unexpected, and thoroughly unwelcome, complication to the effort being made to localize speech in the frontal lobes. On theoretical grounds (as outlined in Chapter One), the idea that the two halves of the brain might function differently was completely unacceptable; indeed, it was generally regarded as absurd. Nevertheless, by the end of the 1860s, the asymmetry problem had not only forced a transformation in the way neurologists regarded higher mental functioning in the human brain, it also had altered how they sought to measure the value of different classes of humanity in the natural scheme of things.

As was noted in section 2-3 of this chapter, the association between speech disorders and damage to the left cerebral hemisphere actually seems to have been noticed first by that obscure country doctor, Marc Dax, in 1836. Dax's work was brought to the attention of the Académie de Médecine in 1863 by his son, Gustav Dax. In that same year, Broca also first began to struggle with the problem of asymmetry. This coincidence of dates has led some historians (e.g., Souques 1928; Joynt & Benton 1964) to inquire into the possibility that Broca may have known of Dax's work, notwithstanding his own frequent protests to the contrary. The truth of the matter, however, is unresolved.

In any event, the published record shows that on April 2, 1863, Broca reported on eight new cases supporting a localization of speech in the third frontal convolution. "This figure," he declared, "seems to me sufficient grounds for strong presumptions." One remarkable thing about these cases, he then went on to say, was that in all the lesion had been on the *left* side. Significantly, it does not seem that eight cases suggesting a left-sided localization of speech, which he was in no hurry to demonstrate, were sufficient for drawing any "strong presumptions." "I do not dare draw from this any conclusion," Broca said, "and I await new facts" (1863, 202).

But the peculiarity of the coincidence continued to trouble him. "I hope," he asserted at this time, "that others more fortunate than I will finally find an example of aphemia caused by a lesion of the *right* hemisphere. Up until now, it is always the *left* third frontal convolution that has been affected." The stakes in the quest for an aphemic with a right-brain lesion were high, for "if it were necessary to admit that the two symmetrical halves of the encephalon have different attributes, that would be a *veritable subversion* of our knowledge of cerebral physiology" (in Broca 1877, 509, italics added).

Several months went by with no such cases being unearthed. On the other hand, in July 1863, a physician in the Paris hospital service,

Jules Parrot, reported on a patient who had suffered complete destruction of the right third frontal convolution *without* any obvious damage to speech or intelligence ("Atrophie, etc." 1863). While he stressed that this case did not absolutely disprove the possibility that the right hemisphere played a role in speech, Broca nevertheless realized that matters were not looking too promising. "If further observations continue to prove on the one hand that certain lesions of the left hemisphere are accompanied by aphemia, and that the same lesions do not produce aphemia when they are situated to the right, it will be necessary indeed to acknowledge," he said reluctantly, "that the faculty of articulate language is localized in the left hemisphere" ("Atrophie, etc." 1863, 380–81).

In making this admission, Broca was not simply calling into question the continuing empirical viability of Bichat's "laws of symmetry," which would have been bad enough. The degree of resistance that initially would greet the idea of a unilateral localization of language leads the historian to suspect that—at least for some of the participants in question—other, more deeply rooted aesthetic and philosophical beliefs were also being threatened. If it were established that the brain was functionally lopsided, this would make a mockery of the classical equation between symmetry, on the one hand, and notions of health and human physical perfection, on the other. It would force the consideration of the absurd, if not scientifically blasphemous, idea that nature would create two (apparently) identical organs that functioned differently or would allow an organ to come into existence that was incapable of discharging the functions for which it was intended (cf. Robertson 1867, 520). It might even undermine all recent efforts to bring logic and lawfulness to the study of the cortex, raising the spectre of retrograde movement toward the implicitly theological view of the cerebral cortex as an organ beyond all scientific classification (see Chapter One). In other words, asymmetry could be seen as reopening the door to just the sort of religious obscurantism Broca and his fellow anthropologists were dedicated to abolishing.

M. Laborde, responding to M. Parrot's case, further reminded the Société Anatomique that belief in the functional identity of the two hemispheres had always figured prominently in the theoretical writings of phrenologists and other advocates of cerebral localization (see Chapter One). "It is thus that one sought, indeed, to explain the persistence of a part or the totality of a function, in spite of the alteration of the organ or the part of the organ that was reputed to preside over this function, by supposing that similar, still healthy portions replaced those that were diseased" ("Atrophie, etc." 1863, 386).

In addition to all its other undesirable implications, then, it turns out that Broca's failure to establish the bilaterality of his language faculty threatened to discredit one of cerebral localization's main strategies of defense against its critics.

Of course, it was not absolutely necessary to see the idea of unilateral localization as a threat to the phrenological principles of localization itself. Lélut, for example, reporting on Dax's paper in the winter of 1864, had not hesitated to denounce the idea of a left-sided locus for speech as phrenological obscurantism:

> If such a fact were true, the brain, that mysterious organ, would be much more mysterious still. Each of its two hemispheres, each portion even of each of its hemispheres, could potentially become the seat of different functions. Nothing forbids matters being [rien ne s'oppose à ce qu'il en soit] the same for the other double organs of the body, and one might thus come to prove, always guided by observation, that there is only one eye, the left for example, which sees, [while] the right was used for some wholly different thing. But as is well known, and speaking seriously, matters are the same for the hemispheres as for the two eyes; they serve the same functions; the left is neither more nor less injured than the right in speech disorders. . . . (Lélut 1864–65, 174)

There was at least one potential response to all these various arguments against the possibility of asymmetry (from both sides of the localization debate) and, during the debates that followed in the wake of Lélut's report, Bouillaud was the first to rise to Dax's defense. On April 4, he declared that an asymmetrical localization of language actually was not so absurd as M. Lélut and many of the others seemed to think, for it was not without precedent in human physiology. Were there not certain acts for which we normally, even exclusively, employed our right hand in preference to our left? Would it be absolutely impossible, then, that for certain mental functions such as speech, we similarly favored our *left* hemisphere? ("Discussion sur la faculté" 1864–65, 543). Riese has said that Bouillaud's words here mark the "birthday of the doctrine of left cerebral dominance" (1947, 331), in that the link between right-handedness and left-brain speech was made for the first time. In fact, nothing in Bouillaud's remarks suggests he was advocating a causal link between handedness and unilateral language localization; he seems, rather, to have been simply drawing an analogy.

Even more damning to Bouillaud's reputation as the "founder" of the doctrine of left cerebral dominance is that, although he spoke up

for Dax at the debates, he was not at all decided on the unilateral theory and, in his heart of hearts, tended to think it highly improbable. Soury notes that, after 1865, Bouillaud would "insist more than before, and ever increasingly on the possible double origin of speech loss" (Soury 1899, 592). Why though would Bouillaud have made a plea before the Académie for a doctrine in which he did not truly believe? One answer is that he sensed that to do so would be politically expedient. We may recall our earlier argument that Bouillaud was in some sense exploiting Dax's work for his own purposes, using it as an excuse to promote his and Broca's views on frontal lobe language localization. If he were suddenly now to concede that Lélut, the sworn enemy of all localization doctrines, might have been partially justified in his ridicule of Dax, he risked giving the antilocalizers a chance to seize the offensive and swing sentiment back in their direction. Consequently, it probably seemed safer to oppose Lélut on all counts across the board.

On April 24, Trousseau, perhaps rather surprisingly in light of his general hostility towards the localizers' cause, spoke out in support of Dax and Broca. The statistics, he said, clearly *did* demonstrate an overwhelming, though not absolute, tendency for language disorders to result from left-brain lesions only. Moreover, at least, he did not find it particularly remarkable that one side of the brain should be prone to certain disorders to the exclusion of the other side. "I know perfectly well that the right and the left side of the body are subject to different diseases, and that man has in the past been described in terms of a right man and a left man [on a décrit autrefois dans l'homme un homme droit et un homme gauche]."[5] Unilateral hysteria and intercostal neuralgia, for example, both tended to strike more frequently on the left than the right side of the body (cf. Chapter Three). The left-sidedness of aphasia, Trousseau suggested, was simply a phenomenon of the same order. The only difference was that, in hysteria and neuralgia, there was as yet no known reason why one side of the body should be more vulnerable than the other. In the case of aphasia, however, an explanation for asymmetry might conceivably be found in some special anatomical disposition possessed by the left hemi-

[5] Although, unfortunately, I have been unable to look at the document first-hand, Trousseau is very likely referring here to a work written in 1780 by Meinard Simon du Puy, *De affectionibus morbosis homini dextri e sinistri*. There the case had been strongly made that man is a double being, consisting of a left- and a right-sided individuality. In his landmark 1865 paper, Broca would also refer to this work, denying that his views were in any way comparable to du Puy's extreme position.

sphere, perhaps the result, he speculated, of differential blood-flow to the two sides of the brain. ("Discussion sur la faculté" 1864–65, 664).

On May 9, Paul Briquet, the physician whose celebrated 1859 *Traité clinique et thérapeutique de l'hystérie* effectively brought hysteria into the realm of French neurology, took the floor to launch a full-scale frontal assault on the asymmetry theory. He began by calling Trousseau to task for failing to base his ideas on Bichat's anatomical work, which, he felt, had demonstrated unequivocably that "all the organs of the life of relations"—the cerebral hemispheres being, of course, paramount among them—"were double, because their functions had to make connection with the exterior world through both sides of the body." The idea that paired organs might function differently was too ridiculous even to warrant consideration. In his indignant words,

> Is it possible that the right eye would see only blue, black, and red, whereas the left eye would be constructed for seeing only green, yellow, and blue? Could the right ear hear in music only *do, re, mi, fa*? and the left ear *sol, la, ti*?—could the right nostril be built to smell pleasant odors, and the left for the unpleasant?—and similarly, for the tongue, which would taste sweet flavors only on the right side, and acids on the left? ("Discussion sur la faculté" 1864–65, 714)

Now it was true, as Trousseau had observed, that neuralgia and hysteria tended to manifest their symptoms more frequently on the left side of the body. It was wrong, however, to see in this fact any evidence for a view of the nervous system as functionally asymmetrical. "M. Trousseau is still working out of the science of thirty years ago, an era . . . when no one had the slightest idea what hysteria was" (p. 715). In the first place, so-called intercostal neuralgia was not a true spinal-based neuralgia at all but simply an hysterical "hyperaesthesia of the muscles." Hysteria was a perversion of the passions, and since the muscles served as agents of emotional expression, they were always powerfully affected in this disorder. Why, though, the preference for the muscles on the *left* side? The explanation was obvious. "Owing to the fact of education and habits, the muscles of the trunk and limbs on the left side are weaker than those on the right side; consequently . . . on account of their lesser strength, the muscles fatigue much more than those on the opposing side, and for them fatigue leads to pain. . . . The hypothesis of M. Broca," Brique concluded sternly, "can in no way be granted" (pp. 717–18).

On June 6, the alienist Jules Baillarger took the floor to consider the asymmetry question. His quiet, pragmatic manner was a welcome respite from the melodrama and lyricism favored by some of the earlier speakers. The statistics in favor of a left unilateral localization of language *were* compelling, Baillarger said. Accepting, then, the *fact* of asymmetry, what did it mean? Was it really so contrary to common sense and physiological law as some were saying—akin to saying that we see with only one of our two eyes or hear with only one of our two ears? Or was it possible that the consequence of accepting the unilateral theory was "less severe that it appears at first glance"?

> It is with the right hand that one writes, and if that hand comes to be paralyzed, the faculty of writing is lost; long, continuous efforts are needed to succeed in making use of the left hand. In the presence of this fact that lesions [causing] aphasia exist fifteen out of sixteen times on the left side, couldn't one, Gentlemen, imagine something analogous for speech as that which exists for writing? ("Discussion sur la faculté" 1864–65, 851)

Baillarger then referred to Trousseau's attempt to find a structural basis for hemisphere functional inequality in differential blood flow to the two sides of the brain. He felt that there might be something in that idea, but he himself had had another thought. Gratiolet, he recalled, believed that there were developmental differences between the two hemispheres, with the left frontal area growing in advance of the right (see Gratiolet & Leuret 1839–57).[6] Perhaps this developmental difference could explain both why people were right-handed (according to him, Broca was already an advocate of this idea) and *also* why lesions causing aphasia were found fifteen times out of sixteen on the left side of the brain (pp. 851–52). Much more clearly than Bouillaud had done, Baillarger thus proposed a link between speech asymmetry and handedness by seeking an explanation for both in alleged developmental differences between the two hemispheres.

[6] Gratiolet had actually gone further than this, claiming that the right occipital area grows in advance of the left, making in the end for a kind of balance between the two sides of the brain. As will be seen in Chapter Three, this idea would come to figure importantly in perceptions of the relative functions of the two brain-halves. In Gratiolet's own words, cited from the second volume of his work with Leuret, of which he was the sole author: "It has seemed to me, following a series of conscientiously made observations, that the two hemispheres do not develop in an absolutely symmetrical manner. Thus, the development of the frontal folds appears to form [se faire] more quickly on the left than on the right, whereas the inverse takes place for the folds of the occipital-sphenoidal lobe. At least, in all the cases that I observed, I saw the parallel scissure that distinguishes the marginal inferior fold, take shape on the right before appearing on the left" (Leuret & Gratiolet, 1839–57, II:241–42).

2-6. *The Doctrine of "la Gaucherie Cérébrale"*

No one at the debate chose to respond to Baillarger's remarks, and he did not broach the subject again. Baillarger's proposal for a link between speech asymmetry and differential frontal lobe development is not encountered again until it appears quite suddenly as the centerpiece of Broca's paper, "Du siège de la faculté du langage articulé," presented before the Société d'Anthropologie on June 15, 1865. Although Broca acknowledged Baillarger's formal priority in proposing a link between language asymmetry and Gratiolet's findings, he declared that he actually had considered the same idea quite a bit earlier, in the course of a casual discussion with his students back in November 1864. There is no real reason to doubt the truth of this. We may recall Baillarger's own testimony that Broca was already inclined, apparently even *before* he had become involved in the language localization controversy, to see in Gratiolet's work a solution to the puzzle of handedness. If, as is not impossible, the analogy between speech asymmetry and handedness had occurred to him, it is only reasonable that he would again turn to Gratiolet's work in search of an explanation.

One must be very clear, however, exactly what sort of solution to the problem of speech asymmetry Broca would find in Gratiolet's work. The doctrine of what today is called left cerebral dominance, and what he called "la gaucherie cérébrale," was born a considerably different child than generally is realized. Broca's famous 1865 declaration that "the majority of men are naturally left-brained, and . . . exceptionally some among them, those people we call left-handers, are on the contrary right-brained" was *not* intended to represent a radical "subversion" of Bichat's "laws of symmetry." Broca was forced to account for the clinical data, but for many or all of the reasons proposed in section 2-5, he proved unwilling in the end to undermine such a fundamental tenet of French physiology.

He argued, then, that Bichat had been essentially right: there were no *innate* functional differences between our two cerebral hemispheres. The asymmetry of language localization was just an artifact of the developmental differences between the two frontal lobes, which Gratiolet had been the first to notice. The two lobes began life with identical intellectual potential, but in simple terms of physical growth, the left was the more precocious side. In childhood, then, when we were forced to master the complex intellectual and motor skills that characterized civilized, human life—articulate speech being perhaps the most complex of all—we tended to rely upon our slightly more mature frontal lobe, and thus we *educated it to the exclusion of the other*. In this way, we became right-handed and left-brained.

> In sum, the two halves of the encephalon, being perfectly identical from an anatomical point of view, cannot have different characteristics: but the more precocious development of the left hemisphere predisposes us, during our first groping activities [tatonnements], to execute with that side of the brain, the most complicated physical and intellectual acts; among these it being certainly necessary to include the expression of ideas by means of language, and more particularly, by articulate language. . . . [B]ut this specialization of functions does not imply the existence of a functional disparity between the two halves of the encephalon. (Broca 1865a, 393)

Early in the twentieth century, the French aphasiologist Moutier would make the telling observation that Broca's theory linking right-handedness and left-brain speech, with left-handers speaking from their right brains, in no way could be seen as a conclusion derived from the irresistible force of statistical evidence; in those early years, no statistics were available. After all, before 1865, no one had seen any reason to record the right-handedness or left-handedness of a patient suffering from speech loss (Moutier 1908, 24). To save the "laws of symmetry," Broca therefore can be said to have constructed a theory in advance of facts, gambling that later evidence would bear him out. To a certain extent, it would; and almost a century would pass before neurologists would begin to argue that the model linking preferred hand to hemisphere specialization for speech was too simplistic, particularly in the case of left-handers.[7]

[7] At a later point in his 1865 address, Broca had qualified his original statement so far as it applied to left-handers. In these people, he said, the connection between speech localization and manual dexterity was not necessarily hard and fast. "It seems to me in no way necessary that the motor part and the intellectual part of each hemisphere must be bound up, the one with the other, in regard to the precocity of their respective development in the two hemispheres" (1865a, 386). Few people at the time took much notice of this cautionary note; and, as a modern-day commentator, P. Eling (1984), pointed out, most modern historians writing on the early history of aphasia have equally ignored it. It is true, as Eling says, that Broca's disclaimer of a clear relationship between speech-asymmetry and handedness seems perspicacious in light of modern findings. One wonders, however, if during the last three decades of the nineteenth century, it is likely to have seemed less wise than out-dated. The distinction Broca drew in 1865 between "motor" and "intellectual" parts of the brain was derived from the reigning doctrine of his day, that the cerebral convolutions were the terra incognita of the mind and not analyzable in terms of sensory-motor functioning. With Fritsch and Hitzig's 1870 work on the electrical excitability of the cortex (see Chapter Three) and the rise of the distinction between "motor" and "sensory" aphasia, belief in the special status of the

Not only was the theory premature, but its dependence on Gratiolet's data concerning differential frontal lobe growth also involved Broca in a flagrant contradiction, which seems to have gone unrecognized by contemporaries. We have seen that Broca was an adherent of Gratiolet's tripartite division of different human groups into *occipital, parietal*, and *frontal* races. Gratiolet had argued that European man's superior capacity to develop himself intellectually was related to the lengthy development of his frontal lobes in childhood, as indicated by the late suture closings in the frontal part of his cranium. Broca not only accepted Gratiolet's reasoning here, but it had influenced him in his feeling that a prime intellectual faculty like language "must" have its seat in the frontal part of the brain. It seems pretty clear, though, that if *slower* frontal lobe development can be causally linked to intellectual superiority, then the alleged *faster* development of the frontal area on the left side of the brain can hardly explain the left hemisphere's superiority in language and skilled manual tasks.

There was a corollary to Broca's "gaucherie cérébrale" theory, and it would muddy the conceptual waters still further. This was the idea of "suppléances cérébrales" (cerebral substitution), which Broca originally developed to account for the rather embarrassing fact that, in cases of aphasia, there often seemed to be little direct correlation between the extent of a lesion and the severity of the resulting speech defect. In one case, a small lesion might completely annihilate speech, while a lesion ten times larger might merely cause articulation difficulties. If the consequences of damage to the third frontal convolution could vary so widely from individual to individual, how could Broca claim to have identified *the* site of articulate language?

Broca's reply was that, under certain conditions, not yet clearly understood, the two hemispheres could substitute for each other—a phenomenon that, somehow, did not undermine specific left-sided localization. He now rather awkwardly speculated that the right hemisphere probably played some sort of reinforcing role in the production of speech even in health, though the *left* hemisphere was still the principal site of this function. In cases of injury or disease to the left side of the brain, the right side was consequently able to step in and shoulder at least some of the duties normally directed by its twin (Moutier 1908, 24–25). Soon, the principle of substitution was regularly being invoked to explain not only the lack of correlation between lesion and speech defect but the phenomenon of *recovery*, in which

convolutions was no longer tenable, and there must have seemed little a priori reason to doubt the existence of a firm connection between handedness and speech asymmetry.

individuals made speechless through destruction of Broca's area, rapidly regained a considerable portion of their capacity to speak (for an early discussion of substitution, see Parant 1875).

It is important to notice the extent to which the introduction of this principle of "substitution" made Broca's third frontal convolution resistant to criticism on clinical grounds. It was now possible for a patient injured in that region to suffer little speech defect or to recover more or less completely from any defect originally experienced, all without calling into question the fundamental correctness of Broca's original localization work. Once one further had taken into account the possibility of microscopic alterations to the third frontal convolution that normal inspection at autopsy would not reveal *and* had allowed for the possibility of left-handedness that had gone unrecognized, then all one's bases pretty much had been covered. "From now on," as Moutier (1908, 25) put it, "the observers or the brains were going to be in the wrong when observations would not confirm the reigning doctrine; but the third frontal [convolution] could no longer be caught out [être en défaut]."

This, then, was the situation in France in 1865. It turns out, however, that the French were not the only ones troubled by asymmetry at this time and trying to account for it. In 1866, an explanation for the unilaterality of language localization, an explanation strikingly similar to Broca's, was put forward "independently and almost contemporaneously" by a British physician, W. Moxon. It is worth noting that both Hughlings Jackson and William Broadbent were accustomed to speak of the "unilateral education" theory as a *joint* brain-child of Broca and Moxon (Jackson 1869b, 345; Broadbent 1872), and Jackson always stressed the importance of Moxon's work in the formation of his own thought. Since, in spite of these endorsements by his countrymen, the name of Moxon seems to be virtually unknown in the early history of language localization, it may be worthwhile pausing a moment over his 1866 paper.

Moxon began, as had Broca, by declaring that the idea "that language . . . has its organ on one side only of the brain" was completely unacceptable. "For no animal, vertebrate or invertebrate, is an exception to the law, that all 'organs of relation'—all organs whose function is to communicate with or move in relation to the world outside the creature's body—are bilaterally symmetrical." Indeed, the law of bilateral symmetry was "perhaps the most general truth in all the science of animal construction." The idea, then, that "the highest expression of communicating organs, the organs of human speech," should be asymmetrically localized was so "contrary to the course of nature" that one should "expect some secondary explanation to be found which shall re-establish harmony between the ground-plan of construction of the human organs of speech and that of animal communicating organs generally" (Moxon 1866, 482).

The "secondary explanation" that Moxon then proposed can be summarized as follows: Just as we have a right and left hand, so do we have a right and left *tongue*, each side being controlled by a different brain-half. The only important difference between our two hands and our two tongues is that, in the case of the former, there is a "union of the tongues of the two sides down the middle where they come in contact." Two hands, two tongues—but only *one* "attention" or "will": "We have one attention and two tongues, just as we have one attention and two hands." The unity of attention has certain implications for the education of our bilateral organs. Because it is impossible for attention to focus simultaneously on two separate limbs, it tends to educate one to the exclusion of the other (education being "the mark left in the brain by former acts of attention"). We generally are not handicapped by such lopsided training, however, since "the accumulation of skill or memory on the one side will suffice for the two in consensual motion . . . so that in regard of simultaneous motions . . . there would be no reason for educating both sides." To understand speech asymmetry then, "we have only to perceive that the right tongue tends to guide the left" to realize that "all the store of recollections of movement-associations which constitute the power of speech will be localised in the left brain, which corresponds to the right tongue. . . . Yet, in the right brain there will be all the organs, which if educated would become the seats of speech-power; so that the ground-plan symmetry of the organs of speech is preserved" (Moxon 1866, 483–85).

One major conceptual weakness of Moxon's theory is that, unlike Broca's, it is unable to offer any reason why, given two symmetrical brain-halves, the great majority of children tend to delegate responsibility for the education of their bilateral organs to the *left* hemisphere, rather than indifferently to either side. Maudsley had a different objection to Moxon's proposal. He argued that the two "nerve-nuclei" that control the symmetrical halves of the tongue must, because they have always acted in concert, "have such a close commissural connexion as to act like a single nucleus. What justification, then, have we for refusing the same power to one nucleus which we assign to the other over the bilaterally symmetrical action of the instruments of speech?" (Maudsley 1868, 722).

Moxon's vision of such mental entities as "attention" and "will" interacting with brain-halves also raises certain sticky philosophical issues. The conclusion seems unavoidable that we are meant to see attention as a reified entity distinct from the brain and its functions, something which somehow "chooses" to educate one of its hemispheres over the other. Certainly, attention cannot dwell in the hemispheres themselves, because it would then have to exist in bilateral symmetry, one "attention" in each hemisphere. At least Broca's version of the unilateral education theory, whatever else we may think of it, manages reasonably well to avoid having to take an explicit stand on the hoary question of mind/brain relations, since it proposes an initial physiological cause for unilateral learning: unequal development of the frontal lobes. It is then possible to assume that, once begun, the process of lopsided education is self-perpetuating. There is no necessary implication (albeit a possible one, depending on one's reading of his paper) that the "self" in childhood somehow chooses the left hemisphere in preference to the less developed right one.

Reflecting on the contrasts between Broca in France and Moxon in England, one wonders how far they might be understood as a reflection of broader differences in French and British styles of physiological speculation. Daston (1978) has argued that the response in Britain to various problems in psycho-physiology was shaped by social pressures to harmonize psychology with ethical imperatives. The Victorian preoccupation with the moral role of human volition, in particular, made British psycho-physiologists markedly more reluctant than their continental counterparts to adopt even an implicitly deterministic approach to issues that touched on the human spirit.

2-7. Asymmetry and "Perfectibility": Why Some People Are More Unique Than Others

In the paper by M. Gaussin, "Sur la faculté d'expression" (discussed in section 2-4), a certain amount of attention inevitably was directed to the problem of asymmetry and Broca's explanation for it. Gaussin began by expressing his "genuine satisfaction" with Broca's realization that the physical symmetry of the body necessarily required an (innate) functional symmetry between all its duplex organs. He reminded his audience how symmetry was widely recognized to be a sign of excellence within nature. Everyone knew, he felt, that it "is more and more accentuated in proportion as one ascends the taxonomic ladder [échelle des animaux] or that of the functions." The notion, then, that human intelligence could be served by a duplex organ *lacking* in such symmetry, when clearly intelligence was the most exalted of all animal functions (he rated it a full notch higher than morality), was an idea he found frankly repugnant (Gaussin 1865, 399–400).

Nevertheless, Gaussin wondered whether Broca's "gauchers du cerveau" theory was really plausible. The fact that the third frontal convolution of the right side of the brain was no less developed or healthy-looking than that on the left suggested that the former *must* be used in some capacity. To affirm otherwise was to ignore "that fact of observation that organs develop through exercise and atrophy through idleness. . . . It therefore seems difficult to me, if we have in our brain two symmetrically placed organs for the function of articulate language, to admit that one of these organs could slumber over long years without displaying a difference relative to the other" (Gaussin 1865, 400). Gaussin then proposed that one might be able to reconcile the clinical data with the apparently healthy condition of the right frontal convolutions, by assuming that both sides of the brain in fact normally did cooperate in speech. There was, however, perhaps a slight difference in *strength* between them, the left being a bit more powerful. In cases of left unilateral damage, then, the right hemisphere, abruptly deprived of the support of its sturdier twin, would be rendered at least temporarily impotent. In cases of right unilateral damage, on the other hand, the strong left hemisphere would still be capable of carrying out speech acts on its own.

This perception of the brain atrophying in response to disuse like a neglected muscle is a notion that can be traced back to at least the phrenologists, but it will not be discussed further here. What is significant, from our point of view, is not so much what Gaussin himself said, as the response his remarks prompted from Broca. The latter began by reaffirming that the localization of speech bore on the ancient problem of what separated man from the beasts—but then went on to suggest a direct link between this first assertion and the asymmetry issue. *If*, Broca said, articulate language were an innate faculty—part of the biological heritage common to animals and men, like M. Gaussin's "faculty of expression"—then there would be no reason to expect it to be asymmetrically localized either on the right or the left side of the brain. The asymmetrical localization of language was a direct consequence of the fact that this faculty was *not* innate but rather was part of man's cultural heritage, acquired by each individual human being only after long training. The fact of handedness had shown that the manual skills of civilized existence tended to be developed unilaterally. Now the aphasiological data had shown that a similar rule applied to the development of at least some intellectual skills, such as speech: "just as in our external life, we carry out 80 percent of our activities with the help of our right arm, so similarly the left hemisphere ends up, by force of habit [en suite d'une sorte d'habitude prise], exclusively undertaking the function of language" ("Discussion sur l'aphémie" 1865, 412).

Man, in short, was born no better than the other animals. What set man apart was his highly advanced capacity to become something better than he was born, to civilize or "perfect" himself to a higher level than any other animal. Born an animal, man became a human being by acquiring the skills of civilized existence. And those skills, as handedness and speech asymmetry showed, had their seat in just *one half* of the brain.

This meant that both cerebral and manual asymmetry were in some sense a tangible reflection of the effects of education and civilization upon the individual and the species, Lamarckian principles being believed to operate. Thus, even while he maintained the general truth of Bichat's "laws of symmetry" so far as mundane physiological functioning was concerned, Broca nevertheless now suddenly could make the startling claim that, "asymmetry of the folds or secondary convolutions constitutes in my eyes a characteristic of superiority." He was quite certain that cerebral asymmetry did not decrease but actually increased in primate brains as one ascended the taxonomic ladder. Naturally, it was most pronounced of all in the human species.[8] Indeed, he proclaimed, "lack of symmetry constitutes unquestionably one of the *principal traits of the human brain*" (1869, 392–93, italics added). As he summed matters up in 1877,

> Man is, of all the animals, the one whose brain in the normal state is the most asymmetrical. He is also the one who possesses the most acquired faculties. Among these faculties—which experience and education developed in his ancestors and of which heredity hands him the instrument but which he does not succeed in exercising until after a long and difficult individual education—the faculty of artic-

[8] Prior to Broca, one finds a few figures who said or implied that asymmetry (manual and cerebral) was a sign of human superiority. An 1855 pamphlet by a French writer, Tony Moilin, *Quelques considérations sur l'homme droit et sur l'homme gauche*, had gone so far as to declare that "Man is the most asymmetrical of the animals, just as he is the most perfect of them" (cited in Bérillon 1884, 10). In the United States, an 1850 article by the mesmeric explorer of the human mind and brain, Joseph Buchanan, also stressed that "the correspondence of the hemispheres is never very exact," but the author clearly implied that these cerebral irregularities were far from bad. Differences between the hemispheres were said to be most pronounced in man, rather less visible in quadrupeds, still less noticeable in birds, and finally, imperceptible on the level of reptiles ("Duality and Decussation" 1850, 518).

ulate language holds pride of place. *It is this that distinguishes us the most clearly from the animals.* (Broca 1877, 528, italics added)

It is important to note exactly what Broca, consciously or not, has done here. In distinguishing man from the animals, he has pressed data from clinical neurology (speech loss coinciding with left-sided brain damage) into the service of certain liberal republican ideals concerning the possibility of progress through enlightenment. The flexible matrix of neo-Lamarckian concepts, which allowed for the inheritance and gradual strengthening of acquired traits across generations, provided the framework that made such an act seem both scientifically plausible and ideologically neutral, while actually giving a strong ethical and social orientation to the whole asymmetry issue.

This is seen most strikingly in the final elaboration Broca gave to his intepretation of the speech asymmetry data. He had long held to the view that certain races of man were "perfectible," by which he meant capable of progressing to ever higher levels of civilization, while others were much less so, stubbornly resisting even the most patient efforts to educate them (Broca 1860; Harvey 1983, 135). As a liberal, he regretted that nature had fashioned such a world, but a scientist, after all, could not let egalitarian sentiments cloud his judgment and subvert the facts (cf. Gould 1981, 84). Consequently, even as he and his followers now made asymmetry a chief distinguishing mark of the human brain, they realized, given this fundamental inequality between different human groups, that asymmetry could not be equally pronounced in all human races and societies.[9] It must be not only a distinguishing mark of humanity in *general* but a means of sorting out the "superior" vintage groups of humanity, those capable of civilization, from the substandard. "I was able to assure myself," Broca told the Société d'Anthropologie in 1869, "that it is greater in the brains of whites than in those of negroes" (1869, 393).

Such a proclamation was hardly likely to have been viewed as contentious by most of his fellow anthropologists. Belief in the biological inequality of the races was one of the most deeply imbedded dogmas of nineteenth-century European consciousness. In France, as elsewhere, this dogma conveniently served a variety of political and social ends. For example, Joy Harvey has remarked (1983, 133) that, at any point during the second half of the nineteenth century, it would have been possible to delineate the scope of the French colonial Empire

[9] D. Bender has pointed out that, in French, the word *race* can be applied either to biological races or to social groups, and he notes that "in the minds of these nineteenth century anthropologists the distinction was not at all clear" (1965, 142).

simply by marking the addresses of the corresponding national members of the Société d'Anthropologie upon a map of the world. These individuals must have found a certain comfort in this new evidence, clearly etched on the brain and verifiable by anyone, that the nonwhite races were incompetent to govern themselves. And even Société members unconvinced of the wisdom of further colonial expansion would have seen little reason to question Broca's findings. After all, it was not necessary to be a wholehearted advocate of colonial expansion to be convinced of the biological inequality of the races, and to concede certain inevitable social consequences to flow therefrom.

It is important to realize that Broca's belief in the incorrigible "nonperfectibility" of certain human races coexisted with a generous conviction in a limited capacity for "perfectibility" among certain nonhuman species. "Whereas the Australian resists all attempted efforts to civilize him," he remarked blandly at one point, "the rabbit trapped in a snare consents readily to domestication" (Broca 1866a, 69). Such a transspecies expansion of the perfectibility concept not only had the effect of implicitly stressing the evolutionary continuity between man and beast; it also bore significantly on the asymmetry question. As Broca explained,

> If one is right to consider asymmetry of the convolutions to be a characteristic of increasing improvement ["perfectionnement"], it is not only because it is very pronounced in man; it is also because one finds it again, to a lesser degree certainly, but still very evident, in the great anthropoid apes. I will add that certain species of lower rank, but improved by man, have much more asymmetrical brains than congeneric species that live in a wild state. Thus, whereas the brain of the fox (*canis vulpes*) is almost symmetrical, that of the dog, which is constructed on the same pattern, is much less so. . . . This result would be inexplicable if one did not attribute it to the influence of man, who has developed through education, according to his needs and tastes, the cerebral faculties of the dog, his most faithful servant. . . . (Broca 1877, 527–28)

It did not take long for the argument linking asymmetry and "perfectibility" to obtain a secure foothold in the international medical literature. In his classic work, *The Brain as the Organ of the Mind* (1880), one finds the English neurologist H. Charlton Bastian affirming that cerebral asymmetry is only a little more marked in the nonwhite human races than in the higher apes but reaches its maximum form in the highest or most civilized (white) races (p. 400). The French physician Gaëtan Delaunay felt that "in a general manner, asymmetry is less pronounced in the female than in the male, something which is moreover explicable by this consideration: that the female, from an evolutionary viewpoint, is situated between the almost symmetrical child and the [adult] male whose asymmetry is now known to us" (1874, 67).[10] This idea continued to be echoed right through the end of the century. The authoritative British *Dictionary of Psychological Medicine* still saw fit to inform its readers as late as 1892 that "female heads are much more symmetrical than male heads," while going on to note cheerfully that "even-headedness among idiots and imbeciles is certainly a mark of low type of head; animals are notably even-headed" (Clapham 1892, 579; cf. Pierret 1895, 271).

The French anthropologist Gustav le Bon meanwhile had concluded in the early 1880s that ambidexterity (an indirect index of deficient cerebral asymmetry) was more frequent in women, savages, and small children than in adult European men. Le Bon was not a man, moreover, to shrink from drawing social conclusions from his data. To his mind, the inferior brains of women (as deduced from a variety of evidence) showed clearly that they were not capable of benefiting from higher education and should remain in the home. So far as colonialism and the political management of the different races were concerned, he felt that "the level of government proper to a group [i.e., republican vs. totalitarian] could . . . be determined from the brain or skull" (cited in Harvey 1983, 148).

Inspired by le Bon's findings on ambidexterity, the Italian criminologists A. Marro and C. Lombroso, in 1883, undertook an inves-

[10] Authors routinely compared the degree of asymmetry supposedly found in the brains of various intellectually "inferior" groups (women, nonwhite races, etc.) with that allegedly found in the brains of (small, uneducated) children. They drew their philosophical justification for this from that great nineteenth-century concept of "recapitulation," encapsulated in the biological saw "ontogeny recapitulates phylogeny." Briefly, this idea held that the individual, in the course of development, passed through a series of stages that effectively "recreated" the evolutionary history of its species as a whole; as Gould (1981, 114) put it, the organism "climb[ed] its own family tree" (for a detailed scholarly treatment of this theme in biology, see Gould 1977). Now, if women and adult members of nonwhite races stood at a lower level of evolution than white men, they might be expected to resemble the *children* of white men, who were, after all, (temporary) living representatives of a past stage in the evolution of their kind. And, if it could be established that there was a parallel between the amount of asymmetry found in the brains of both children and the "inferior" groups, then it would have been pretty clear to nineteenth-century aphasiologists that Broca had been right to find a solution to the asymmetry problem in the evolution of human civilization. Indeed, it would not have been unreasonable to suppose that insight into the rise of cerebral asymmetry in the evolution of the human species as a whole might come from studying the ill-formed brains of the lesser human groups (cf. Delaunay 1878–79).

tigation of the incidence of ambidexterity and left-handedness (soon, the two were often conflated) among madmen and criminals. Differences in frequency compared with the normal population were negligible for the former but highly significant for the latter (Marro & Lombroso 1883). This discovery would later be incorporated by Lombroso into his influential theory of the "born criminal," which held that certain individuals were biologically incapable of learning to conduct themselves according to the ethical codes of civilized life. Born criminals tended towards left-handedness and ambidexterity because, as atavistic reversions to the type of mankind that existed before the invention of law, property, and the family, they lacked the asymmetrical nervous system that identified a man of morality and civilization (Lombroso 1903; Kurella 1911; cf. Lombroso 1884).[11]

In 1883, one of Jean-Martin Charcot's more distinguished students, Charles Féré, wrote a short report in which he criticized the current vogue for linking asymmetries and anomalies of the brain to criminality, intellectual capacity, etc. "There is no necessary connection," he affirmed, "between the moral and intellectual state and the gross morphology of the brain" (Féré 1883, 65). There is no evidence that this admonition made any significant impression upon the colleagues to whom it was so pointedly addressed.

[11] There is a sense in which Lombroso's views on the born criminal's deficient asymmetrical development might seem to contradict his teachings elsewhere that the presence of gross physical asymmetries, including asymmetries of the brain, could actually be used to *identify* individuals possessed of a twisted or depraved character. The Lombroso school was not unaware of this tension in its work, and at least one of Lombroso's colleagues, L. Lattes (1907), took it upon himself to explain why the presence of certain sorts of pronounced asymmetries in the criminal brain, obviously, must not be considered a sign of the criminal's high level of evolution, as Broca's teachings might seem to suggest. Lattes basically argued, then, that in contrast to the modest, left-leaning cerebral asymmetry associated with the development of civilized man, asymmetry in the criminal brain was almost always atavistic or pathological in character. In general, such asymmetry was exaggerated to abnormal proportions, consisted mostly of peculiar localized disparities between congruent brain regions and generally tended to manifest itself in the "wrong" direction. It seems, in short, that criminals could be identified by their lack of normal, healthy asymmetry; *or* they could be identified by their pathologically exaggerated and twisted asymmetry. Such bet hedging naturally had its uses.

CHAPTER THREE

Left-Right Polarities of Mind and Brain

3-1. From Asymmetry to Left-Brain Superiority

AS THE CENTURY progressed, the original interpretation of asymmetry taught by the Broca school underwent a subtle but significant mutation. Because Broca had insisted that the left hemisphere was not *congenitally* endowed with any functional capacities denied to the right, his strict followers tried to stress, in the words of the Parisian alienist Benjamin Ball,

> Indeed, the important point . . . is not at all the preponderance of the left hemisphere over the right hemisphere; it is the superiority of one of the two halves of the organ [over the other]. In general, man chooses the left brain, in a few exceptional cases, he gives preference to the right side; but the thing that is necessary to establish, above all, is that man is not at all naturally ambidextrous like the animals; he is essentially unilateral. (Ball 1884, 35)

Many people, however, failed to heed this injunction that the *left*-sidedness of cerebral asymmetry was really beside the point, that the fact of asymmetry pure and simple was what really mattered. This is understandable. It must have been all too easy to lose sight of the details of Broca's argument and to focus simply on the remarkable fact that left-brain damage produces speech loss while right-brain damage does not.

New developments in the history of cerebral localization doubtless also played an important role in encouraging a tendency to see the left-sidedness of cerebral asymmetry as an essential factor in the equation; to perceive the left side of the brain as frankly superior to the right. In 1870, the German physicians Gustav Fritsch and Eduard Hitzig applied galvanic currents to the cortex of laboratory dogs and demonstrated the existence of "motor centers" in a part of the brain traditionally held to be inexcitable (Fritsch & Hitzig 1870). The superiority of their electrical method of localizing over the old use of experimental lesions attracted almost as much attention as their actual findings. Other neurologists, notably David Ferrier in England, hailed the Fritsch and Hitzig method as a technological breakthrough and adopted it to continue the search for both motor and sensory centers in the brain (Ferrier 1876). Various animal studies, pursued through the 1870s and 1880s, soon made it clear that all these centers were *bilaterally* represented. At the same time, new clinical studies were increasingly reinforcing the idea that most or all of the higher "intellectual" functions, presumably associated with human beings alone, were housed exclusively in the left hemisphere.

This well-documented phase in the history of aphasia, which will occupy us only briefly here, opens with the 1874 monograph by the young German neurologist Carl Wernicke, *Der Aphasische Symptomencomplex. Eine Psychologische Studie auf Anatomischer Basis.* Wernicke started off from the premise that the frontal area of the cortex (anterior to Rolando's fissure) served motor function and memories of movement, while the posterior area (the temporal and occipital lobes) was involved with sensory processes and memories of sensory impressions. This view was justified, in his view, both by the recent localizations of motor functions in the frontal region by men like Fritsch and Hitzig and by the morphology of the temporo-occipital area of the brain, which closely resembled that of known sensory areas—experimental attempts to localize various sensory centers in the brain had not yet yielded any clear results. Recognition of the sensory-motor division of the cortex, Wernicke declared, opened the door to a new, more comprehensive understanding of the clinical picture of aphasia. The pure motor aphasia discovered by Broca and associated with damage to the frontal convolutions had been thoroughly reviewed in the literature, and was no longer disputed. Wernicke believed, however, that there existed a second "pure" form of aphasia, sensory aphasia, associated with damage to a second, anatomically distinct language center localized in the sensory (temporal) part of the left hemisphere. In contrast to Broca's motor aphasia, in which a patient could generally understand what was said to him but was more or less speechless, sensory aphasia was characterized by "a comparatively large fund of words in association with essentially complete loss of comprehension of the spoken word" (1874, 124). Noting his belief that "the pure sensory form [of aphasia] . . . has not been represented by a single typical case in the literature," Wernicke went on to present two detailed examples of the syndrome that had come to his attention.

But Wernicke hoped to do more than simply place a new syndrome on the clinico-anatomical map. His 1874 monograph also represented an attempt to offer a comprehensive, predictive model of aphasia, capable of incorporating all the various, confusing clinical pictures in the literature, including writing and reading disorders. This model, moreover, differed from the efforts of certain earlier authors in that it was firmly grounded in an anatomical base, consistent with Meynert's conception of the brain as a system of cells and fiber tracts operating according to reflex laws. The old phrenological attempt to localize complex mental attributes, Wernicke argued, had been misguided. What was actually localizable were much simpler perceptual and motor functions. These were stored as sensory and motor "traces" or "impressions" (nascent memories), in specialized areas of the brain located adjacent to corresponding elementary zones. That is to say, auditory memories for words were supposed to have their seat close to the primary auditory center, memories used in reading were supposed to be located near the primary visual center, the part of the brain damaged in Broca's aphasia was thought to be close to the centers for the mouth and tongue, etc. These primitive memories, according to Wernicke, served as the basic units that, interacting via the fiber tracts according to laws of "association," gave rise (somehow) to the rich complexity of mind. In his words,

> There is a significant difference between the invention of various theoretic centers (coordination center, concept center)—with complete neglect of their anatomic substrates, because the unknown functions of the brain up to the present have not completely warranted anatomically-based conclusions—and an attempt, based on an exhaustive study of brain anatomy and the commonly-recognized laws of experimental psychology, to translate such anatomical findings into psychological data, seeking in this way to formulate a theory by use of the same kind of material.
>
> The theory of aphasia, here proposed, permits a consolidation of the diversified picture of the disorder. The diverse variety of symptoms which formerly had presented new riddles to each investigator does not now seem so striking, but may be predicted according to the laws of symptom-combination.
>
> *However, one feature is common to all. The underlying basis of the disorder lies in a disruption of the psychic reflex-arc necessary for the normal speech process.* (Wernicke 1874, 142–43)

The theoretical orientation outlined in Wernicke's "psychological study on an anatomical basis" was to prove enormously attractive, especially in the German-speaking countries, which would soon take the lead in neurological research. Not only did it draw strength from the studies of reflex action that were then at their apogee, but it also profited by being closely modeled after the conceptions of a familiar and remarkably popular psychology, associationism. This held that the human psyche was a type of immaterial "mind-space" where various images interacted, or associated, with one another. These images were themselves simply secondary elaborations of sensations coming from the outside world. The arrival of any given sensation into the mind-space awoke the memory of other sensations according to fixed "laws" of similarity, contiguity, complementarity, etc. The association thus created resulted in the birth of an "idea"—how is not really clear—that in turn, was associated with other ideas giving rise to proper sequential thought. Associationism, therefore, was really an elaborate variation of sensualism, allowing one to posit sensation as the catalyst that precipitated and maintained a sort of chain reaction of mental events (Jeannerod 1985, 72). All of this was now "neurologized" in Wernicke's hands.

Convinced that their young colleague was on the right track, clinicians responded with alacrity to Wernicke's clarion call for the creation of a comprehensive model of all the different syndromes of aphasia. In the space of virtually a single decade, sometimes dryly refered to as the "splendid seventies," all sorts of cerebral centers for different mental functions (reading, writing, calculation, abstract conceptualization) were happily identified and incorporated in one or another variation on the Wernicke model. The discrete clinical pictures associated with lesions of these alleged centers (agnosia, agraphia, alexia, acalculia) became firmly ensconced in the medical literature. One important, if perhaps unintended, effect of all these developments, again especially in the German-speaking countries, was to encourage a rather single-minded fascination with the continuously unfolding talents of the let side of the brain alone.

Hugo Liepmann's description of "apraxia" in Germany at the end of the century perhaps represents a crowning piece of evidence for the growing perception of the left hemisphere as the superior side of the brain. The important story of apraxia, and its role in the rise of Liepmann's belief in what would come to be known as the "callosal syndrome," receives more comprehensive treatment in Chapter Five. At this point, however, I will simply say that, from a variety of evidence, Liepmann had been led to conclude that the left hemisphere, acting via the right hand, predominated in voluntary, purposeful movements, although not quite to the same degree as in language. More important, he later came to believe that the activity of the right hemisphere was under the controlling influence of the left (Liepmann 1905, 1907; Liepmann & Maas 1907). It is hard to avoid the conclusion that the human will was here on some level being localized in the left side of the brain. One is not surprised to learn, then, that Liepmann believed left cerebral predominance to be the chief source of man's superiority to animals. In his words,

> Among the animals, the left and right hemispheres are still equally endowed [gleichwertig]; this has recently been established even for talking animals (parrots).
>
> The chief superiority of men over the animals is tied up with the superposition in us of unidentified mechanisms, as far as their material constitution is concerned, in the left hemisphere. We see now also that the left hemisphere's extra capacity is not limited to language. (Liepmann 1905, 46)

In spite of the Germans' growing domination of clinical neurology during the last decades of the century, one finds in France a considerable tendency to give short shrift to German perspectives in aphasiology and an enduring loyalty to the original lines of orientation laid down by Broca. This phenomenon can probably be understood as part of the heightened sense of defiant nationalism and anti-German sentiment that so marked French consciousness in the decades immediately following the Franco-Prussian War. In France, then, it seems as if a growing tendency to celebrate the left side of the brain as inherently superior to the right was significantly fostered, not so much by the clinical developments across the Rhine, as by the strong Lamarckian element in French evolutionary thought of the time. This rapidly tended to blur the distinction between left hemisphere superiority acquired by individual cultivation and left-brain superiority as part of man's unique native endowment (cf. Broca 1872). At the dawn of civilization, it might not have mattered which hemisphere was "chosen" for language and skills requiring manual dexterity. Once the choice had been made in favor of the left, though, the cumulative effects of left-sided education across countless generations would have transformed that hemisphere into the evolutionarily more advanced side of the brain, the repository of humanity's most civilized and uniquely human faculties. As the well-known Boston physician George Gould would put it in 1907,

> It becomes plain that, in the right-handed, intellectual life and progress are by means of the mechanisms of the left cerebral hemisphere. There is no intellect as we understand it except through speech, vocal and written, and the instruments of this function exist only in the left brain of the right-handed, and in the right brain of the left-handed. Mentality of the dextral therefore lives preponderatingly in and through the left half-brain. The fact strongly emphasizes

> and capitally illustrates the great biologic law that all progress consists in differentiation. *In the evolution of civilization each bit of cortical brain substance* [on the left side] *is being told off to a certain peculiar office.* That large parts [on the right side] are still without particular and discoverable duties argues plainly for the great progress and differentiation of function in the future of humanity's advance. (Gould 1907, 599, italics added)

3-2. "No Grin Without a Cat": The Structural Basis of Left-Brain Preeminence

Regardless of whether it found its ultimate origins in nature, nurture, or some combination of the two, it was not long before the left hemisphere's functional superiority over the right had become an unquestioned tenet of human cerebral physiology. In France, logically enough, it generally was felt that the old view of the two sides of the brain as structurally identical was no longer tenable in light of their dramatic functional differences. Thus, the quest was undertaken for the structural conditions underlying left cerebral predominance.[1]

In 1865, Broca had called attention to Gratiolet's report on developmental differences between the frontal (and occipital) lobes of the two hemispheres (for the Gratiolet text in question, see Chapter Two, note 7). He was doubtless quite pleased when, a year later, he learned of the work of a German osteologist, Hans Carl Barkow, which he felt importantly confirmed the Gratiolet data. In his *Bermerkungen zür Pathologischen Osteologie*, Barkow, looking at the phenomenon of plagiocephaly (*plagio* meaning "slanted" or "oblique"), had found in his samples of human skulls a tendency for the frontal region to be slightly more pronounced on the left side, and the occipital region slightly more pronounced on the right side (Barkow 1864). In the wake of Broca's endorsement, Barkow's work was widely seized upon as pointing strongly to the conclusion that the frontal convolutions in man's brain were more numerous on the left side, and the occipital convolutions more numerous on the right. From the way in which later authors (including Hughlings Jackson) cite Barkow, it seems pretty certain that many had not seen the original work and were unaware that it was concerned with skull configurations and not with direct brain measurements.

[1] The word *predominance* is used here in an effort to be faithful to the terminology of the time. The currently popular term, *dominance*, seems to have entered the lexicon of neurology rather late, although no one seems to know who actually was the first person to use it (cf. Oppenheimer 1976, 11).

In his 1866 report, Broca praised the exactness of Barkow's results, obtained "avec une patience toute germanique," mentioning that he had had the opportunity to confirm them himself on two series of twenty brains, male and female respectively (Broca 1866b, 196).[2] Three years later, the case for left frontal/right posterior convolutional differences between the two hemispheres was given an explicitly *functional* cast, following a report to the Société d'Anthropologie claiming that the left frontal lobe tended to be more abundant in gray matter than the right and the right occipital lobe more abundant in gray matter than the left, gray matter being seen as the "stuff" of intellect. In a complementary fashion, the right frontal and left occipital lobes both allegedly possessed relatively more white matter than their homologues (Roques 1869).

It is important to realize that other authorities—for example, the physiologist Alexander Ecker (Ecker 1868) and the neuroanatomist Carl Vogt—flatly denied the Gratiolet-inspired theory of unequal development between the frontal and posterior lobes of the two hemispheres. Léonce Manouvrier's (1883) detailed analysis of plagiocephalic skulls also contradicted any conclusions one might have been tempted to draw from Barkow's work, finding no consistent tendency for asymmetry to be oriented in one direction or the other (see also the negative results of le Bon 1878a). Perhaps, partly in reaction to such unhelpful studies, an 1879 report measuring lateral differences in the brain's composition quietly dropped the frontal/occipital distinction and contented itself with the claim that, in general, the left hemisphere from a very early age possessed more gray matter and the right more white matter (Parrot 1879).

Later, during the post-Wernicke years when attention was no longer so exclusively riveted on the role of the frontal lobes in speech, Jules Bernard Luys would contend that actually structural asymmetries between the two hemispheres tended to be most pronounced in the "spheno-temporal regions" (Luys 1879). This proposal, which went almost unnoticed by contemporaries, would receive a measure of support in 1890 from a Viennese anatomist, O. Eberstaller, who claimed to have found that the left sylvian fissure in man tends to be slightly longer than the right. Research by D. J. Cunningham, published two years later in Ireland, agreed with that of Eberstaller (Cunningham 1892). Cunningham, however, denied that these differences in fissure

[2] This study of Broca's was reported in the "Correspondence" section of the *Bulletins de la Société d'Anthropologie* but does not seem to have been formally published.

length had any implications for understanding the structural conditions underlying speech-asymmetry and handedness. This is because he had found the same asymmetries in the brains of various higher apes; and he did not believe that apes displayed any evidence of right-handedness (Cunningham 1902, 293). The idea that hemisphere functional asymmetry might in some way be explicable in terms of temporal lobe asymmetries then went underground for some eighty years, to be resurrected in the 1960s with the work of Norman Geschwind and his colleagues in the United States (Geschwind & Levitsky 1968).

Since, to many people, more numerous convolutions on the left side implied a heftier piece of brain matter, studies investigating possible differences in weight between the two sides of the brain complemented and overlapped with the convolutions studies. An extensive investigation by Robert Boyd in England, which predated by several years Broca's asymmetrical localization of language, claimed to have found that "almost invariably the weight of the left cerebral hemisphere exceeded that of the right by at least an eighth of an ounce" (Boyd 1861). In contrast, John Thurman was only one of a number of writers who found a tendency for the right hemisphere to outweigh the left (Thurman 1866). This might have been deemed a serious setback for left-brain superiority, had it not been for the fact that Broca's theory only required that the left *frontal* lobe be functionally dominant over the right. Over a period of more than ten years, Broca studied 440 brains and concluded in his final 1875 report (which actually only looked at a small sample of his data) that, while the overall weight difference between the two hemispheres was negligible (and indeed in his table one finds a slight bias in favor of the right), the left frontal lobe outweighed the right by an average of 4 grams (Broca 1875). In an early reference to these researches, Broca had suggested that the overall lack of difference in the weight of the two hemispheres was probably due to "a sort of compensation between the weight of the two frontal lobes and of the two occipital lobes" (Broca 1866b, 196).

Towards the end of the century, one writer struck upon another way of getting around the fact that weight comparisons of the two brain-halves had failed unequivocably to favor the left. He simply turned the standard argument on its head. Researchers had agreed that the left hemisphere's higher ratio of gray to white matter was what made it superior to the right hemisphere. What they had failed to realize, though, was that gray matter, being less dense, weighed *less* than white matter. Consequently, the lighter weight of the left hemisphere actually was proof positive of its functional superiority (van

Biervliet 1899, 296). The supporting data changed; the foregone assumptions remained unaltered (cf. Gould 1981, 73–112).

For some nineteenth century neurologists, it was not enough simply to affirm that certain developmental or compositional differences between the two sides of the brain favored the unilateral education of the left. What, some asked, was the immediate physiological *cause* of these differences? For a while, one finds a considerable amount of attention being paid to a theory, first formally proposed by Armand de Fleury of Bordeaux (de Fleury 1865) and later by William Ogle in England (Ogle 1871), that the left hemisphere was structurally superior to the right because it was nourished by a more vigorous blood supply. This in turn was possibly caused by differences in the mode of origin of the two carotid artiers or by differences in their respective circumferences. De Fleury's blood-supply theory made no attempt to accommodate itself with the French argument, being made during this same time, for unequal but *compensatory* anatomical differences between the two sides of the brain. No one seemed to mind. It was enough, perhaps, that a bit of biochemistry had been introduced into the debate. Enamoured as they were with neuroanatomy, French neurologists also were fascinated by the arterial system of the brain and the "nutritional" processes of the blood. The former bias perhaps had its roots in the successes of Charcot's "clinico-anatomic" method (see Chapter Six); the latter may have been partly inspired by the equally impressive achievements of the Claude Bernard school of physiology.

In any event, it turns out that Broca himself was rather lukewarm about de Fleury's blood-supply theory. In an 1877 review of de Fleury's work, he opined that "the mode of origin of the two carotids exerts a certain influence on the distribution of work between the two hemispheres"—but "not a decisive influence." Reasonably enough, he pointed out that if asymmetry of the carotid arteries were the sole cause of cerebral functional asymmetry, then it followed that all left-handers must suffer from an inversion of the normal vascular structure; and "it is quite certain that, in the overwhelming majority of cases, left-handers are free from this rare anomaly, and that, so far as the origin of the aortic vessels is concerned, they do not differ appreciably from right-handers" (Broca 1877, 526). The blood-supply theory became even less creditable as it was increasingly realized that the two cerebral arteries were connected by the anterior communicating artery, meaning that the blood contributed by each was pooled, equalizing both supply and pressure to the two hemispheres. Efforts to demonstrate appreciable and consistent differences in the size of the carotid arteries were also unsuccessful (Harris 1980, 22).

3-3. The Left Hemisphere and the "Other Side of the Brain"

The first two sections of this chapter should have made clear that earlier historians had been quite correct to affirm that the nineteenth century perceived the left hemisphere as the superior side of the brain, "predominating" in most or all higher mental and moral activities. What I intend now to challenge, however, is the common corollary argument that, with one or two honorable exceptions, scientists and medical men interested in human cerebral functioning paid little or no attention to the right side of the brain during the immediate post-Broca years. Arthur Benton, for example, declared categorically that neurologists at this time believed that the right hemisphere had "no distinctive functions" but was "merely a weaker version of the left" (Benton 1972, 8).

Certainly, as Benton stresses, there is no denying that the nineteenth century by and large overlooked the possibility of particular right-hemisphere involvement in certain visuo-spatial tasks, which are today (in the late twentieth century) generally seen as one of its chief functional strengths. In light of the early association pointed out between the right hemisphere and occipital lobe functioning, this omission may seem all the more curious. John Hughlings Jackson would argue, beginning as early as 1864, for the existence of special "visuo-perceptual" capacities in the right hemisphere, but (as will be seen in Chapter Seven) his claims were not based specifically on a view of the occipital area as "visual cortex." Munk did not localize his "visual center" in the occipital lobes until 1877, correcting David Ferrier's earlier localization in the supra-marginal and angular gyrus. In 1895, T. Dunn, an American physician from Philadelphia, wrote a paper in which he suggested that there might exist "a center (which may, for convenience, be named the geographical center) on the right side of the brain for the record of optical images of locality, analogous to the region of Broca for that of speech on the left side in right-handed persons" (Dunn 1895, 54; reprinted in French in Hécaen, 1978). Like the work of Jackson, however, this work was not based on a consideration of presumed or actual anatomical differences between the two occipital lobes but was derived from clinical observation.

Anterior versus Posterior Activity. Nevertheless, the fact that the possibility of special right-brain visual functions was mostly overlooked does not mean that the right side of the brain was ignored. As was discussed in Chapter Two, while the anterior lobes were identified with human intelligence and reason, the posterior (occipital) lobes were equally seen as the site of sensibility, emotion, and the animal instincts. Quite a few writers went still further and branded the back of the brain a refuge for all of mankind's uncivilized, darker tendencies: criminal impulsiveness, animal savagery, unbridled lust. Even if Broca had not already cast suspicion on the right hemisphere, then, by implying that it was permitted to remain in an uneducated, half-savage state, the reputation of that side of the brain would doubtless have suffered from its associations with occipital lobe functioning. Thus, it began to be argued that the right hemisphere played a predominant role in passive sensibility, emotion, activities serving trophic, instinctual life, sleep, unconscious thought processes, criminality, and madness—in this sense, neatly complementing the assertive, civilized, intellectual activities of the left hemisphere.

Motor versus Sensory Functioning. In fairness, the old phrenological notions about posterior versus frontal lobe functioning were not the sole source of such a lateralized view of brain function. It became known, beginning in the 1870s, that cortical sensory centers were located predominantly in the posterior regions of the brain and motor centers in the frontal regions. A certain amount of clinical and experimental evidence was reported, which seemed to suggest that these centers were differentially represented in the two hemispheres. The Austrian physiologist Sigmund Exner, who later was Sigmund Freud's university instructor, used clinical data to make inferences about motor and tactile representation in the human brain. He came to the conclusion that motor representation was both more intensive and more extensive in the left hemisphere, while sensory representation was more intensive and extensive in the right hemisphere (Exner 1881, 64). In France, Armand de Fleury had similarly concluded that, all other things being equal,

> in right hemiplegia, which depends on an apoplectic ictus [stroke] in the left cerebral lobe, it is mobility and muscularity [myotilité] which are above all affected: swallowing, respiration, the ocular motor nerves, the motor fibers of the tongue, etc. are damaged, but tactile, internal [profonde], and sensory sensibility generally remain intact.
>
> In contrast, if the apoplectic hemiplegia has struck the limbs on the left side, one finds . . . that the functions of sensibility . . . touch, vision, sense of smell, hearing are more or less anaesthetized on the left side only. (de Fleury 1872, 840)

Intelligence versus Passion. Not only did the clinic give some credence to ideas about the right hemisphere's greater "sensibility," it also seemed to support a view of that side of the brain as more "emotional" than its rational homologue. In 1881, Jules Bernard Luys wrote a paper arguing for an "emotion center" in the right hemisphere. A neuroanatomist who in his younger years had done important work on the thalamus, Luys also worked as a clinician at the Salpêtrière, was appointed chief physician at the Charité hospital in 1886, and later became director of the prestigious Maison de Santé at Ivry (founded by Esquirol in 1828). He began to be struck by the fact that there seemed to be definite and consistent personality differences between those of his patients suffering from right-hemiplegia (caused by a left-sided cortical lesion) and those suffering from left-hemiplegia (caused by a right-sided lesion).

> Whereas ordinary hemiplegics, *right hemiplegics*, are more or less apathetic, more or less silent, passive and stricken with hebetude;—left, emotional hemiplegics, are more or less afflicted with an abnormal impressionability. They respond, when one questions them, in limping tones broken up with a type of sobbing. . . . In other circumstances . . . they are boisterous and loquacious, their face is congested, their eyes sparkle, they are perpetually in motion. . . . [I]t is not unusual to see some of them in the grips of a veritable fit of excitation, maniacal, having false conceptions, delusions of persecution, and even attempting suicide. (Luys 1881a, 380)

Luys proposed that the hyperemotionality and manic tendencies of his left-hemiplegics might be explained on the hypothesis that some normal "inhibiting" center for emotion in the right hemisphere had been destroyed by a lesion supplementary to that responsible for the left-hemiplegia. This lesion he tentatively localized in the right temporal lobe. At Ste. Anne's hospital, a colleague and friend, Benjamin Ball, claimed in the wake of Luys' 1881 report to have observed in his own patients the same sort of contrasts Luys had seen: left hemiplegia associated with affective abnormalities, and right hemiplegia complicated by language or intellectual disturbances (Klippel 1898, 56).

The growing case for a special right hemisphere role in sensibility and emotion was given an important boost by repeated observations that hysterical disorders had a marked tendency to manifest their symptoms on the *left* side of the body (see Chapter Six for more on the role of hysteria in shaping ideas about the double brain). Paul Briquet (1796–1881), a hardheaded physician at the Charité hospital who had come to the study of hysteria with some initial distaste, recorded in his authoritative 1859 *Traité clinique et thérapeutique de l'hystérie* that, in 430 cases, hysterical hemianaesthesia and hemiplegia had been observed three times more frequently on the left side than the right. Briquet defined *hysteria* as "a neurosis [physiological malfunctioning] of the portion of the encephalon assigned to receive emotional impressions and sensations" (cited in Morel 1983, 596). He, however, declined to see any special physiological significance in the asymmetry of its symptomatology. It is not hard to understand why, when we recall his stalwart defense of Bichat's "laws of symmetry" at the 1865 Académie de Médecine debates on language localization (Chapter Two).

In the mid-1870s, when the Charcot school turned to Briquet's *Traité* as the chief point of departure for its own view of hysteria, Briquet's data on left-sided hysteria were noted, but his remarks about their total lack of physiological significance seems to have been rarely, if ever, recalled. Following Briquet's lead, Charcot and his followers developed a concept of hysteria as a disorder brought about by "functional" lesions of the central nervous system: a neurological disturbance interfering with the patient's sensory-motor functioning but causing no discernible structural damage to the brain. In Paul Richer's 1881 *Études cliniques sur la grande hystérie ou hystéro-épilepsie*, one of the most important works to come out of the Charcot school during these years, hysteria's reputed predilection for the left side of the body was mentioned a number of times; Richer even went so far as to call this predilection "Charcot's rule," although he warned that there were several exceptions. How he got away with this appellation is not clear, since the journal sources show that Charcot recently had become rather skeptical of the "rule" Richer tried to attribute to him. In an 1878 paper, he had remarked that, after believing in the asymmetry of hysteria's symptomology for a long time, he now thought that it was found almost as frequently on the right side as the left (Charcot 1878b, 1075).

Still plenty of others both in and out of Charcot's immediate circle were prepared to defend Briquet's original figures and to interpret them in explicitly neurological terms. Indeed, in 1874, the prominent physiologist Charles Edouard Brown-Séquard had stated the case in no uncertain terms: "the right side of the brain serves chiefly the emotional manifestations, hysterical manifestations included. . . . We have collected," he continued, "cases of paralysis in one-half of the body, caused by hysteria, and this proportion has been found; . . . in

121 of these cases, there was disease of the brain on the right side 97 times and disease on the left 24 times" (Brown-Séquard 1874a, 10, 14).

Even earlier, in 1871, Armand de Fleury, the man who originated the blood supply explanation for functional asymmetry in speech (Chapter Two), not only accepted the idea that hysteria favored the left side of the body, but proposed an explanation for it. In contrast to Brown-Séquard, de Fleury argued that the right hemisphere was not actually the more emotional side of the brain. Its apparent greater affectivity was simply an artifact of its being generally weaker than the left hemisphere, presumably because of its less plentiful blood supply:

> Each cerebral hemisphere consequently builds up in perception and eliminates by transformation into action, the [sense] impressions impelled toward the encephalon, and, of the two organs, the one that is less skilled in *discharging* [through motor activity] is manifestly the weaker in *dynamic* strength.
>
> This *is the fate of the right hemisphere* [ce role *est celui de l'hémisphère droit*], which is not really, more than its congener, an organ of *affectivity*, but which, less powerful in its effective reaction, is more naturally vulnerable than its homologue to neurotic [i.e., "functional"] afflictions of the sensibility. (cited in Bérillon, 1884, 138–39, italics original)

Right Hysteria versus Left Hysteria. At the very end of the century and in an intellectual environment that had seen considerable change since the triumphant days of Charcot, Pierre Janet was still familiar with the statistics indicating a prevalence of left over right hysteria; and he seems to have seen no reason to question at least the empirical validity of this "rule." At the same time, he did not ignore the fact that at least one third of all cases of unilateral hysteria were *right*-sided. In his 1898 *Névroses et idées fixes*, he even stated his belief that hysterical mutism and aphasia tended to occur in conjunction with right-sided anaesthesias and/or paralyses but rarely with left. It seemed to him that this association, which he pointed out had been remarked upon by others, had relevance for any attempt to determine the association between hysterical ailments and organic ones and bore upon the hypothesis concerning the anatomical localization of "functional" lesions. In 1899, then, he undertook a systematic study of the case records of 388 hysterical patients in an effort to see if there were any

Table 1. Percentage of patients suffering various symptoms, by type of hysteria

Symptoms	Mixed	Left	Right
Attacks of sleep, somnambulism, fugues	57.9	72.2	48.0
Disturbances in the movement of the limbs	39.1	37.8	71.5
Disturbance of the nutritional functions	23.9	33.1	25.4
Aphonia [loss of voice], aphasia, language difficulties	9.4	2.7	22.5
Coughs, hiccups, respiratory disorders	11.5	1.3	16.7
Intellectual disturbances	32.6	30.4	31.3

SOURCE: Data from Janet 1899, Tables 2 and 3.

consistent relationships between certain types of hysterical symptoms ("accidents") and the side of the body afflicted with permanent anaesthesias and paralyses, the so-called stigmata (Janet 1899).[3]

Janet began by dividing the patients into three groups: (1) patients exhibiting mixed, protean, or bilateral stigmata, of which there were 138; (2) patients exhibiting left-sided unilateral stigmata, of which there were 148; and (3) patients exhibiting right-sided unilateral stigmata, of which there were 102. Janet noted that, although there *were* more cases of left than right hysteria in his sample, he had been surprised to find that the ratio was only something like 3:2, instead of the 3:1 found by Briquet. He made no comment about the possible significance of this.

Division having been made, Janet subjected the symptoms associated with each group to statistical analysis. His results, adapted from Tables 2 and 3 in his article, are given in Table 1.

As Janet had predicted, there was a significant tendency for language-related disorders to occur in conjunction with right-sided hysteria. The statistics, however, also revealed a few other interesting trends: somnambulism, fugue, and epileptiform attack were more frequent with left hysteria than with right (lending credence to the right hemisphere's image as inward-looking and hypersensitive); difficulties with mobility occurred far more frequently in conjunction with right hysteria than with left (in seeming conformity with the idea that the left hemisphere was more intensively involved in motor functioning than the right). The statistics offered no support to the view of the left hemisphere as more or less exclusively concerned with intellectual

[3] The name F. Raymond, Janet's supervisor, appears as coauthor of this study, but such use of joint signatures was only a formality (see Ellenberger 1970, 341).

functioning, a fact Janet also pointed out. Nor did "nutritional" functions appear to be differentially lateralized, in spite of certain opinions of Brown-Séquard, which will be examined shortly. In the end, Janet himself was most struck by the finding that respiratory disorders were *ten* times more frequently associated with right hysteria than with left. In light of this, he ventured to suggest that there might exist a certain rapport between the more voluntary levels of respiration and articulate speech, making for a more pronounced cortical representation of the former on the left side of the brain (p. 854).

Yet, at this time, Janet already was beginning to have doubts about the validity of the strict neurological approach to hysteria in general. While he continued to accept Charcot's view that hysteria required a "suitable soil" (i.e., a weak or abnormal nervous constitution, which could be inherited), he was also increasingly inclined to favor the use of dynamic psychological models in explanations of this disorder. Abandoning talk of "functional" lesions, he began to focus instead on seeking the origins of hysterical symptoms in traumatic events in the patient's life-story. In 1901, Joseph Babinski, formerly a loyal member of the Charcot school, went further towards discrediting the Charcot school's neurological approach to hysteria with a catalytic paper read before the Paris Société de Neurologie, "Définition de l'hystérie." This paper declared that there was no relationship between disorders caused by organic lesions of the encephalon, and so-called hysteria phenomena (Babinski 1901; see also Chapter Nine of this study).

Yet, even with the decline of the Charcot school and the rise of alternative, psychological approaches to the neuroses, a number of authors continued to adhere to the view that hysteria favors the left side of the body. In 1905, one writer, Arthur Whiting, would put the left/right ratio as high as 4:1, while confessing that he had "no valid explanation" for such a dramatic asymmetry of symptom representation (Whiting 1905, 506). In 1908, Ernest Jones apparently was sufficiently annoyed by the continuing adherence in the literature to Briquet's old "rule" of hysteria that he undertook an investigation of his own. Reviewing 277 cases of hysterical hemiplegia, he found *no* evidence that left-sided symptoms predominated over right-sided ones; in fact, his own findings actually indicated a slight bias in favor of the right side (Jones 1908). Having just discovered Freud's psychoanalysis, with its quite different, psychodynamic view of hysteria, Jones may have hoped that publication of his statistics would manage to set this particular neurological bogey to rest once and for all. The bogey did indeed rest, more or less, for a while. In the 1970s, however, an alleged tendency for conversion disorders (hysteria) to favor the left side of the body would begin again to attract notice and would begin again to see explanations in terms of functional differences between the two hemispheres (Stern 1977; Galin, Diamond, & Braff 1977).

"Life of Relations" versus "the Organic Life." In Brown-Séquard's view, the right side of the brain not only served emotional manifestations, including hysterical ones, but also was concerned with "the needs of the nutrition of the body in various parts" (Brown-Séquard 1874a, 10). On conceptual grounds, this was an attractive idea. Recognition of a deep biological link between passion and the visceral, organic processes of the body dated back in French physiology at least to the work of Bichat (see Chapter One). And indeed, Bichat's influence upon Brown-Séquard's thinking is clear. Seeking a means of broadly characterizing the fundamental distinction between the two brains, left and right respectively, Brown-Séquard had felt justified in resurrecting Bichat's old distinction between "the life of relations" on the one hand and "the organic life" on the other (1874a, 10). He summed up the evidence for such a functional left/right formula in two reports presented before the Académie des Sciences (Brown-Séquard 1870, 1871). There he claimed that various "troubles des nutrition"—bedsores, edema, involuntary evacuation of feces and urine—were more frequently found to accompany right-sided lesions than left, at a ratio of 2:1. Pulmonary congestion, conjugate deviation of the eyes, and convulsions also were more common with damage to the right hemisphere, and he generally felt that lesions from that side of the brain were more life-threatening than lesions from the left. On this last claim, he was contradicted by the work of Armand de Fleury, whose own comparative studies had led him to conclude that the prognosis in cases of right-brain disease was generally *less* grave than in cases of left-brain disease (de Fleury 1872).

Although obviously not unchallenged, Brown-Séquard's argument for a special right-brain role in nutrition and other "organic," life-maintaining functions was fairly influential at the time and received a modest amount of apparent corroboration from outside sources. According to Edgar Bérillon, Charcot had also concluded, from his studies of bedsores developing in conjunction with hemiplegia of cerebral origin, that the "nutritional functions of the skin" were more frequently disordered by lesions of the right hemisphere than of the left (Bérillon 1884, 100). In England, William Ogle cited the work of a Mr. Callender, who had found, like Brown-Séquard, that "convulsions are a common accompaniment of disease of the right hemisphere, occurring in 39 out of 61 cases, . . . but rarely produced by disease of the left hemisphere, having been present in only 7 cases out of 48" (Ogle 1871, 293). Also seemingly consistent with Brown-Séquard's findings and views was the curious claim made by the Bonn ophthalmologist J. L. Budge that the "cerebral centre for the movements of the stomach" was on the right side of the brain: irritation of this side causing the stomach to move, while irritation of the corresponding parts on the left side of the brain produced no effects whatsoever (cited in Brunton 1874, 418).

Male versus Female. The rise of the idea that the two hemispheres of the brain could be compared and contrasted in sexual terms took its point of origin from nineteenth century views on the biological status of women. As noted in Chapter Two (section 2-6), a white European woman was typically regarded in scientific circles as intellectually more or less on par with a man from one of the "inferior" nonwhite races; that is, somewhere just above the higher apes, but definitely below the white European man. Among Broca's group, Gustav le Bon had declared that the brains of many Parisian women resembled more closely those of gorillas than those of adult white men. He admitted that there did exist a few intelligent women in the world, some of whom were even superior to the average man, but he argued that they were monstrosities, like a gorilla with two heads. Consequently, they need not be taken into consideration (le Bon 1879, 60–61).

In Chapter Two, Gratiolet's classification of human groups into *races frontales, races pariétales*, and *races occipitales* was described. Members of the occipital race were perceived more or less the same way as the posterior lobes themselves: emotional, unstable, dominated by instinct (especially sexual), and deficient in intelligence and reasoning power. In David Ferrier's classic 1876 *Functions of the Brain*, we learn that European women were also deemed to be dominated by their occipital lobes. Ferrier explained,

> As the reproductive organs in women form such a preponderant element in their bodily constitution, they must correspondingly be more largely represented in the cerebral hemispheres, a fact which is in accordance with the greater emotional excitability of women, and the relatively larger development of the posterior lobes of the brain. (1876, 262)

Given what we already know (1) about the assumed link between the right hemisphere and occipital lobe functioning and (2) about perceptions of the right side of the brain generally, it is not surprising that this should become the "female" side of the brain. Quite early on, Victor Meunier had affirmed his belief that "there are between the respective modes of activity of the left brain and the right brain, differences analogous to those currently existing between the respective modes of activity of the male brain and that of the female" (cited in Delaunay 1874).[4] Armand de Fleury had also called the right hemisphere female and the left male; indeed, there seems to have been a feeling among at least some neurologists that assigning genders to the two hemispheres of the brain was one convenient way of summing up their respective characteristics. As late as 1898, one can still find it being said that "the terms 'male hemisphere' and 'feminine hemisphere' should render rather well the differences in nature between the two brains, of which one, more intellectual, is more stable, and of which the other, more excitable, is also more readily exhausted" (Klippel 1898, 56–57). One cannot help but reflect wryly that, once one has given the two hemispheres gender identities, the idea of cerebral dominance becomes a rather apt metaphorical encapsulation of the social and economic relationship between the sexes in nineteenth century Europe.

It does not seem that all writers necessarily believed the association between the left hemisphere and masculine thought, the right hemisphere and feminine thought, was anything more than a pleasant sim-

[4] A good fifteen years before most people had any reason to suppose that the two sides of the brain functioned in a complementary fashion, Joseph Buchanan in the United States also had argued for the relative masculinity and femininity of the left and right hemispheres. His is the only pre-Broca argument of this sort I found (I exclude the related but separate belief that the left and right halves of the human *body* are "feminine" and "masculine" respectively; the history of that idea in medicine—especially embryology—can, of course, be traced back to classical times). Buchanan based his vision of complementary masculine/feminine functioning between the two brain halves on certain craniological data he had gathered himself. As he explained, "When you examine the head of a right-handed man, you will find that the left hemisphere is the best developed, particularly in those portions which give muscular power and energy of character."

". . . But while the basilar developments are stronger on the left side of the head, it is not so with the moral organs. The right side of the brain and the left side of the body seem to have a gentler and more effeminate character. The right body and left brain have a stronger and more masculine character."

In the closing paragraphs of his paper, Buchanan waxed increasingly lyrical, describing exuberantly how "the two hemispheres of the cerebrum, which I have already shown to have relatively a masculine and a feminine character, sustain to each other in action the same relation which exists in a harmonious family, where a perfect marriage has a thoroughly united husband and wife" ("Duality and decussation" 1850, 522, 524, 528).

ile. There is no question, however, about the position on this point of the French physician and "comparative biologist" Gaëtan Delaunay. By his own account, Delaunay had first become interested in the question of hemisphere differences as a medical student during the ill-fated 1870–71 Franco-Prussian War. Assigned to ambulance service at Tours, he was compelled to treat a considerable number of soldiers with frostbitten feet, and it began to seem to him that the left foot of each of these men was always more severely afflicted than the right. He relayed this curious fact to Brown-Séquard, who assured him that, according to certain researches of German origin, the right side of the body did tend to be hotter than the left.

The frostbite affair was the starting point for what was to become for Delaunay a virtual obsession with the "comparative biology" of right and left. Delaunay's medical thesis, submitted in 1874, was entitled *Biologie comparée du côté droit et du côté gauche chez l'homme et chez les êtres vivants.* It opened with the following bold assertion of Broca, taken (without reference) from the latter's 1869 article: "Asymmetry is a characteristic of superiority." Although presumably this citation was intended to set the tone for the rest of the thesis, it is a bit misleading. One quickly realizes that Delaunay believed only asymmetry in favor of the *right* side of the body, and left side of the brain, was a sign of superiority. Asymmetry favoring the *left* side of the body, and right side of the brain, seems to have been as bad or worse than no asymmetry at all. In fact, lack of asymmetry and asymmetry in the "wrong" direction are conflated throughout the thesis.

Delaunay stated in his thesis that his aim originally had been to compare the different pathologies to which the two sides of the body were subject. In the course of pursuing his researches, however, he had found himself increasingly struck by the "remarkable" parallels he kept finding between the right and left side of the brain and body, and the "known" physical and mental differences between men and women. It was impossible, he felt, that these could all be mere coincidence. As he began to look further into the matter, he seemed to find reason to believe that, in fact, women lacked a good portion of that right-sided/left-brained asymmetry that Broca had taught signified a high level of evolution. His researches, for example, had convinced him that left-handedness was significantly more common among women than men. Even women who were right-handed, he said, tended to have many left-handed habits (e.g., women buttoned their coats left over right, while men did the reverse).[5]

[5] For what it is worth, recent research on handedness and hemisphere differences

In a later work (1878–79), Delaunay went on to suggest that "the primary cause" of the link between lateral asymmetries and gender differences was probably embryological. In light of research showing that every embryo was a "fusion" of his mother and father, it was not unreasonable to suppose that the "male element" always developed into the right side of the body (and apparently the left half of the brain), while the "female element" always developed into the left side of the body. Presumably, other forces were at work that caused either the female or the male "element" to predominate in the embryo, giving rise to sexual differentiation, though this problem was not addressed. Among the more remarkable pieces of evidence Delaunay cited in support of his argument were two cases apparently reported by Dr. Sibley in the United States. The first of these described a young girl with black hair on the right side of her skull like her father and red hair on the left side of her skull like her mother; the second was concerned with a mulatto child, born of a white father and a black mother, who had smooth hair with fair skin on the right side and kinky hair with dark skin on the left (Delaunay 1878–79, 72).[6]

White Superiority versus Nonwhite Inferiority. As Gould pointed out (1981, 102), "inferior" groups are interchangeable with one another within the framework of biological determinism. Thus, the next few years found Delaunay busy at the Salpêtrière in Paris, seeking to prove through a series of simple yet pointed studies that the left hemisphere predominated in individuals at a superior level of evolution and the right in individuals at an inferior level of evolution. Included in the latter category were not only women and nonwhite races, but children, the lower classes, the aged, and even certain European national groups: notably those from the south and the Germans, who had so decisively defeated Delaunay's countrymen in the war ten years earlier.

Having sorted out his superior and inferior groups in advance, Delaunay came to the conclusion that members of the superior groups tended to gravitate towards the right in walking, presumably owing to the controlling influence of their more developed left frontal lobes. Members of inferior groups, on the other hand, tended to direct them-

between the sexes seems to have found that actually left-handedness is rarer among women than men.

[6] Delaunay claimed that these remarkable cases were reported in some (unspecified) issue of the now-defunct American medical weekly *The Clinic*. However, several hours at the New York Academy of Medicine library searching through (nonindexed) issues of this journal from the 1870s failed to turn up the primary source.

selves to the left (Delaunay 1879). Similarly, superior races tended to rotate to the right ("In France, in all our national dances, we turn to the right"); while middling inferior races (the Chinese, Japanese, Turks, Mexicans) turned towards the left; and somehow the most inferior races of all (negroes, for example) managed to not turn at all (Delaunay 1883)! Superior individuals also tended to cross their right legs over their left, thus sitting predominantly upon their left buttocks; inferior individuals did the reverse. This last was determined by examining the amount of wear-and-tear on the rear of the individuals' trousers (Delaunay 1884).

Delaunay's most ingenious and best-known work from this period involved an attempt to relate the different evolutionary levels of the two hemispheres to the different sorts of dreams each tended to produce. Unilateral dreaming was solicited by having the individual sleep on his right or left side, in this way allegedly favoring blood flow to the side of the brain nearest the pillow.[7] Right hemisphere dreams were said to be illogical, absurd, and rich in sensory detail, presumably owing to the right hemisphere's more developed sensory area; they were generally concerned with remote memories and often tended to be nightmarish. Left hemisphere dreams, on the other hand, were less absurd and sometimes even intelligent, were concerned with recent happenings, and often contained a considerable amount of discourse, "which is understandable since the faculty of articulate language is seated on the left" (Delaunay 1882).

[7] The idea that asymmetries in blood-flow cause only one hemisphere to be active during sleep, the right or left depending upon how one rests one's head, seems to have been a German innovation. It was apparently inspired by the physiological musings of the alienist Schroeder van der Kolk but may have first been clearly articulated by asylum physician Max Huppert, whose writings were reviewed in the French medical literature. In 1869, Huppert declared, "Just as the blood exerts a stimulating influence on the nervous system as a whole, so too [will it stimulate] the capillary-rich cortical substance, the workshop of ideas and thought. Now, if we lie on one side, the blood will, in accordance with the law of gravity, sink to the part of the brain lying lower, flowing from the upper hemisphere into that lying below it and collecting there. It will thus have more effect on the lower hemisphere, in regard to the working of the cortical cells there, and so exaggerate their specific functioning; while in the other hemisphere, the contrary will apply, and a weakening of activity in the ganglionic cells will result. It thus follows that during sleep undertaken in the lateral position, on the whole only one hemisphere, if at all, will be active, while the other will show very little activity or be almost inactive; thus . . . only one of the hemispheres will produce dreams" (Huppert 1869, 544). I am not aware that either van der Kolk or Huppert ever explained why gravity did not cause the blood to drop down to the base of the brain every time the individual sat up, making intellectual activity possible only in the supine or upside down position.

It would be easy to turn Delaunay into an object of ridicule, for the extent to which he managed to map the social world onto the natural one seems painfully clear with the vision gained by a century of distance. It must be realized, however, that this man was no crank but was published in the highly reputable *Lancette française* and attracted considerable respectful attention, both in France and abroad (see, e.g., Urquhart 1879–80a; Descourtis 1882; Bérillon 1884; Hall & Hartwell 1884; Manacéïne 1894, 1897; Klippel 1898; Lombroso 1903). With perfect ease of conscience, he offered up his work to the altar of disinterested science—and it was accepted, because the weltanschauung it sanctified was so deeply-rooted in late nineteenth century consciousness that its ideological nature had become essentially invisible.

Hemisphere Differences and Bisexual Consciousness. While the French were speaking of male and female hemispheres, the case for a connection between lateral asymmetry and gender differentiation was being argued along quite different lines in Germany. The Berlin nose and throat specialist Wilhelm Fliess is best known for his relationship, both personal and scientific, with Sigmund Freud during the formative years of psychoanalysis. He was, however, also a prolific theorizer and systematizer in his own right. Among other things, he was convinced that human beings were fundamentally and permanently bisexual and that recognition of this fact could illuminate many otherwise puzzling aspects of human mental life. How he came to entertain this view, and the evidence he offered in support of it, has been treated well by Frank Sulloway (1979) and will not be addressed here. Sulloway has also documented a number of ways in which Fliess' theory of bisexuality had a lasting influence on Freud's thought. In the spring of 1897, for example, Fliess proposed an explanation for the motive force underlying repression in terms of the constitutional bisexuality of the individual. Freud, who had been struggling with this problem, was startled and enviously impressed by the idea's "bold simplicity." As he later told it, Fliess had suggested at this time

> that the motive force of repression in each individual is a struggle between the two sexual characters. The dominant sex of the person, that which is the more strongly developed, has repressed the mental representation of the subordinate sex into the unconscious. Therefore the nucleus of the unconscious (that is to say, the repressed) is in each human being that side of him which belongs to the opposite sex. (Freud, 1919, 200–201)

Neurosis, in short, was a state of repressed homosexuality. Although, in the end, Freud did not adopt this idea, Sulloway argues that Fliess' views on the relationship between bisexuality and neurosis directly inspired Freud's later belief that neurosis, broadly speaking, was a state of repressed or "negative" perversion (Sulloway 1979, 184). There is no doubt, at any rate, that Fliess convinced Freud that the concept of bisexuality was crucial to psychoanalysis. In 1905, Freud declared that, without taking it into account, "it would scarcely be possible to arrive at an understanding of the sexual manifestations that are actually to be observed in men and women" (Freud 1905, 220); and later, the idea of bisexuality would come to play a key role in the psychoanalytic conception of the Oedipal situation (Freud 1923).

What is particularly interesting to us about Fliess' theory of bisexuality is that, sometime in the autumn of 1897, he decided to link it to a theory of bilateral dominance. That is to say, he proposed a relationship between the bisexual proportionalities possessed by each individual and his or her degree of right-handedness or right-lateral dominance. In both men and women, Fliess argued, the right side of the body served as the anatomical locus for the dominant sex, while the left, generally subordinate side served the opposite sex.[8]

The original inspiration behind Fliess' decision to link bisexuality with bilaterality is not immediately clear. Some commentators have suggested there was no good rationale for it at all, that the linkage was strictly "arbitrary" (Schur 1972, 139), a "manufactured wretched" piece of reasoning (Mahoney 1979, 90). This is perhaps doing Fliess an injustice. Almost certainly, he was intrigued by the formal correspondence between the idea of gender "dominance," with the opposite sex persisting in repressed or unconscious form, and bilateral dominance (handedness). But, somewhat like Delaunay, he may also have found reason to believe from his research into embryology, which was relatively extensive, that the male and female "germs," which together made up the embryo were mixed together

[8] It may be worth noting that several authors (Schur 1972, 139; Mahoney 1979, 80) have stated incorrectly that Fliess believed every human being, male and female, to possess a masculine/right body-half and a feminine/left body-half. In fact, such a differentiation, by Fliess's theory, only would have held for men. *Women*, in his system, possessed a (dominant) feminine/right body-half and a (subordinate) masculine/left body-half. The error may have arisen from Fliess's rather annoying tendency, which he shared with quite a few authors in this study, to take for granted that all his readers were going to be men and, consequently, to speak companionably in general terms about a situation that was meant to refer only to the male sex.

in such a way that the "dominant" sex developed into the "dominant" side of the body and the "subordinate" sex into the subordinate side of the body.

It must now be frankly admitted that, in speaking of the relationship between bisexuality and bilaterality (handedness), Fliess never directly discussed the neurophysiology underlying bisexual consciousness. Nevertheless, the conclusion is irresistible that, within Fliess's system, the "predominant" left hemisphere (serving the right lateral half of the body) mediated the mental processes of the"predominant" sex, while the "subordinate" right hemisphere (controlling the left half of the body) mediated the unconscious or repressed mental processes of the "subordinate" opposite sex. It is very hard to believe that Fliess would have denied that the cerebral hemispheres mediated mental processes nor is there reason to think that he would have opposed himself to the standard view of his time, which correlated handedness with cerebral dominance for language. Indeed, a number of remarks in his writings referring to the tendency of left-handers to write in mirror-image (e.g., Fliess 1906, 264) suggest an implicit acknowledgement of the relationship between lateralized functioning in the body and lateralized functioning in the nervous system (cf. the subsection that follows). For all these reasons, the view will be taken here that Fliess' views on bisexuality and bilaterality imply a certain necessary perspective on right and left hemisphere differences: one that Freud, who was very well versed in the aphasiology literature (Freud 1891), would certainly have recognized, even if he did not accept it.

Once the theory of bisexuality and bilaterality had been established, Fliess became particularly concerned to examine its implications for understanding the phenomenon of left-handedness. Although left-handedness had received considerable attention in the scientific and medical literature, most observers, he declared, had overlooked its most essential feature:

> Namely, that left-handed men display much more marked female secondary sexual characteristics, and that male secondary sexual characteristics are much more marked in left-handed women, than with fully right-handed men or women. . . . Effeminate men and masculine women are [always] entirely or partly left-handed. (1906, 260, 262)

The implications are plain. Left-handedness, and presumably right-brainedness, was a reflection of incomplete sexual dominance, a sign that an individual possessed an abnormal bisexual streak running through his physical and his mental constitution. Apart from his lateral preference, such an individual allegedly betrayed himself through certain physiological peculiarities ("The best-known . . . are the female beard and the male 'women's breast' " [1906, 266]) and by behavior "obviously" inappropriate to his or her gender. First and foremost, this latter referred to homosexual behavior, but Fliess spoke also of the hypothetical case of a woman lawyer who got herself involved in politics, rode a bicycle, smoked cigarettes, and climbed mountains. Such a woman, he affirmed, was overwhelmingly likely (he put the odds at 100:1) to be dominant on the left side of her body (Fliess 1914, 67).

While left-handedness was most important as an index of incomplete sexual dominance, Fliess also argued that "If someone is a born artist, he necessarily displays a left-sided emphasis" (1906, 285). Why should there be a relationship between left-dominance and inborn artistic capacity? The reason, Fliess asserted, lay in the fact that the consciousness of the "born artist" was more broadly bisexual than that of the normal individual. Within his (or her) mind, the psychical forces of the opposite sex were not repressed in the unconscious but interacted with the forces of the dominant sex in a way that, in contrast to the bisexual consciousness of the average left-hander, allowed both forces free expression. As Fliess later would lyrically explain, this exalted bisexual consciousness was actually the source of artistic creativity. Art, like love, stemmed from the "tension between the sexes"; the artist's male and female sides "embraced each other within his soul," breeding that which made for genius immortal (1914, 73).

I do not wish to go into the question of Freud's attitude toward Fliess' theory of bisexuality and bilaterality, except to note that he was deeply ambivalent towards it and that it played a key role in his ultimate estrangement from his friend. Nevertheless, his well-known 1898 remark to Fliess about Leonardo da Vinci, and that artist's "bisexual consciousness," may perhaps be recalled here: "Leonardo," he pointed out, "of whom no love-affair is recorded, was perhaps the most famous case of left-handedness. Can you use him?" (Freud 1887–1902, 268).

Reason versus Madness. The general belief in the right hemisphere's evolutionary inferiority, in its essentially animalistic qualities, almost certainly played a crucial role in the rise of still another perception of it—as a natural breeding-ground for madness. For centuries, as Foucault has pointed out, madness had "borrowed its face from the mask of the beast." It was not, however, until the spread of the new evolutionary ideas in the mid-nineteenth century that "this presence of animality in madness would be considered as the sign—indeed, as the very essence—of disease" (Foucault 1973, 72–73). Thus, Henry Maudsley in England came to consider all mental disorder to be a type of phylogenetic regression; a loss of civilized standards of behavior accompanied by a welling up of what he called "the brute brain within the man's."

> Whence [else] come the savage snarl, the destructive disposition, the obscene language, the wild howl, the offensive habits displayed by some of the insane? Why should a human being deprived of his reason ever become so brutal in character as some do, unless he has the brute nature within him? (Maudsley 1870, 53)

If madness is defined as loss of reason, as Maudsley has here so defined it, and if to all extents and purposes only the *left* half of our brain is reasonable, then it becomes possible to envision the "brute brain within the man's" as lying on the *right* side of the skull. In 1879, Jules Bernard Luys became one of the first implicitly to argue along just those lines with his declaration that, in the insane, the natural disparity in weight between the two hemispheres was increased to pathological proportions—*and* "completely reversed." Instead of the left lobe slightly outweighing the right as was generally true (he believed) in the sane, "it is the right lobe that has become on average the heavier, and that has drawn into itself alone the vigour [sêve] and nutritive activity of the encephalon" (Luys 1879, 554). Two years later, after undertaking a systematic study of fifty-five brains of persons judged insane at death, Luys reported that the right-lobe had been found to outweigh the left, sometimes dramatically so, in 71 percent of them (Luys 1881b).

A younger colleague of Luys from Marseille, Marandon de Montyel, studied a sample of eighty-nine brains, excluding cases of general paralysis, which he felt were exceptions to the rule. He managed to push the figure up to 81 percent (Montyel 1884). He argued, however, against Luys' interpretation of these statistics as meaning that, in the mad, there was a reversal of the normal processes of nutrition in favor of the right hemisphere. More likely, Montyel thought, the left hemisphere in madness tended to *atrophy*, and this permitted the uncivilized, brutish right hemisphere to seize control of mental life. Why, though, should the more advanced side of the brain be more vulnerable to mental disorder? For the same reason, perhaps, that the man of advanced civilization was felt to be more susceptible to nervous and mental illness than the simple peasant. "The lobe that functions the most actively in the normal state," Montyel speculated, "could well be that which also raves in delirium the most actively, wears itself out and atrophies the earliest."

In an 1887 follow-up article on comparative weights of the hemispheres in madness, Montyel felt himself in a position to declare, rather grandly, that, ever since Luys' statistics were first published, in 1881, the question of a link between madness and the right side of the brain "has not ceased to be the question of the day." In Italy, Enrico Morselli from the University of Turin and Dr. Seppilli from the Asylum of Imola independently set out to test Luys' hypothesis. Though both found a strong trend towards pathological asymmetry between the two hemispheres in madness, there was less evidence for any consistent favoring of the right side. Morselli found the right lobe to be heavier in 52.6 percent of his cases. Sepilli, working with a larger sample, found a statistically insignificant trend in favor of the right and concluded that insanity does not favor one lobe over the other (Montyel 1887). Brown-Séquard had also failed to confirm Luys' hypothesis, and in England an early study by Robert Boyd, still considered relatively authoritative, had been similarly negative (Boyd 1861). On the other hand, an 1878 English investigation by Crichton-Browne claimed to have found, in 400 cases of insanity, that the right hemisphere outweighed the left "in both sexes and in all decades of life from twenty to eighty" (Crichton-Browne 1878).

Meanwhile, the idea of the right hemisphere as the "mad" or irrational side of the brain was finding apparent support from other, less strictly anatomical sources. John Turner's 1892 study of "asymmetrical conditions met with in the faces of the insane," for example, seemed consistent with the French idea that normal left-hemisphere control was weakened or destroyed in the insane. Turner reported how, in "seventy-eight cases of asymmetry in the upper part of the face . . . there was paralysis of [upper facial] . . . movements in the right half of the cerebrum in twenty-seven, and in the left half in fifty-five; and with the twenty-one instances of asymmetry in the lower part of the face the right hemisphere was weakened in four and the left in seventeen" (Turner 1892, 209, see Figure 2).

In 1875, the Glasgow alienist Alexander Robertson read a report before the British Medical Association on unilateral auditory hallucinations, believed to be of cerebral origin. Both in his own cases and in several he cited from the German literature, he noted "the frequency with which the hallucinations were referred to the left ear" was "very striking" (Robertson 1875). He had no explanation as to why this might be. In 1883, the French alienist Valentin Magnan, today mostly remembered for his application of degeneration theory to insanity,

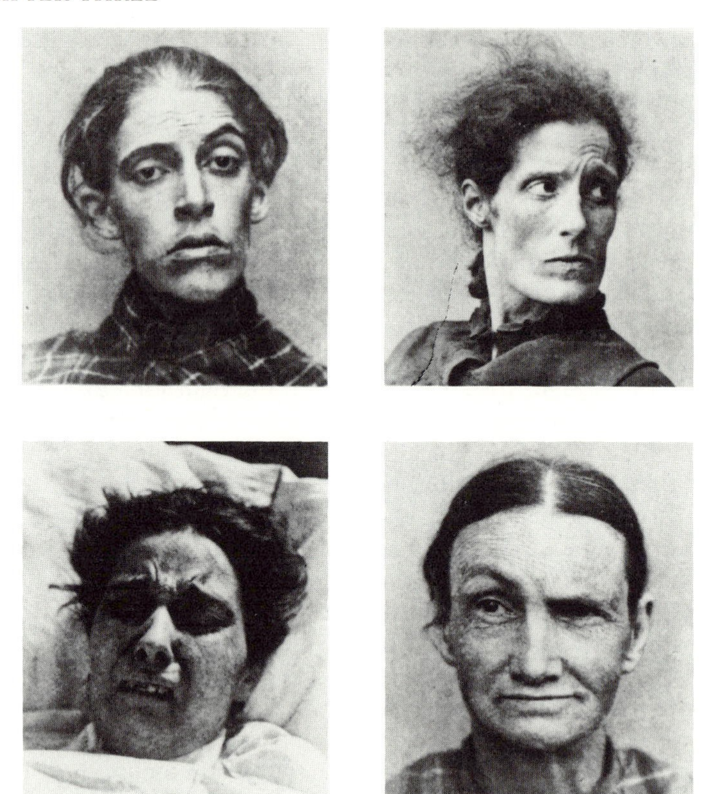

Figure 2. John Turner's photographs of "asymmetrical faces in the insane." (Source: Tuke 1892, plate II, opposite p. 948. Reprinted with the permission of the Bodleian Library, Oxford University.)

wrote a report on bilateral but contrasting auditory hallucinations in four patients suffering from "chronic delirium." He found that, in three of them, disagreeable or deprecating remarks were consistently heard in the left ear, while in the right ear, the individual would hear himself described in extravagantly flattering terms. Since "theomania" represented a more advanced stage of delirium than "demonomania," Magnan concluded that the left hemisphere (serving the right ear) was in a later stage of degeneration than the right hemisphere. He asked, somewhat along the lines of Montyel, "*Is the left hemisphere, by virtue of its preeminence, stricken first in the evolution of delirium?*" (Magnan 1883, 346; Kiernan 1883).

Broca's theory of hemisphere substitution (see Chapter Two) provided a framework for the interpretation of one last species of data that may have reinforced the "mad" right hemisphere argument. It was known that patients suffering from aphasia and right hemiplegia could often regain some speech and learn to write again using their left hands, although they were frequently unable to prevent themselves from tracing the letters mirrorwise, or backwards. In conformity with Broca's teachings, both the new speech-capacity and the mirror-writing were widely held to be a product of the right hemisphere. Mirror-writing, though, was not only associated with hemiplegia: most authorities agreed that it also sometimes manifested itself in certain types of neurotic disorder, in dementia praecox, and during post-epileptic twilight states (Savage 1892; Critchley 1928, 72; cf. Bianchi 1883).

There was also evidence that at least some aphasics spontaneously regained the capacity to speak during spells of delirium, fits of passions, or when in some other irrational state such as dreaming. Brown-Séquard in 1884, for example, cited the case of an English pastor, originally reported by Sir William Jenner in 1862, who had been rendered aphasic and agraphic with incomplete right-hemiplegia. "Afflicted with symptoms of meningoencephalitis, he was seized by a rather violent delirium, during which he spoke clearly, employing a large number of words to express his delirious ideas" (Brown-Séquard 1884, 256). If Broca was right, this temporary recovery of speech could be taken as a sign that the right hemisphere had been spurred into unaccustomed action, and that this same side of the brain was consequently also responsible for the delirium (cf. Hughlings Jackson's views on the right hemisphere and "emotional" speech, Chapter Seven).

3-4. *Concluding Thoughts: The Brain as Myth and Metaphor*

When one begins to pass in review over the various ways in which the nineteenth century came to classify and characterize the two sides of the brain, a very distinctive pattern begins to emerge. Broadly speaking, one finds that the left/right polarities of mind and brain "discovered" by nineteenth century medical men can be summed up as in Table 2.

What is one to make of such stark functional polarities of mind and brain? Anthropologist Rodney Needham (1979) has drawn attention to the fact that simple relations of opposition serve as one of the basic resources in the articulation of symbolic or mythical categories. He has pointed out that linking categories by antithetical pairs is more than just a way of sorting out experience. As a rule, linked pairs are not only perceived as opposites but as *unequal* in rank or worth. Consequently, dual oppositions systems are also a way societies enshrine certain moral and social discriminations.

Table 2. Left/right polarities of mind and brain

Left Hemisphere	Right Hemisphere
humanness	animality
frontal lobe	occipital lobe
motor activity	sensory activity
volition	instinct
intelligence	passion/emotion
life of relations	organic life
male	female
white superiority	nonwhite inferiority
consciousness	unconsciousness
reason	madness

Ever since Robert Hertz wrote his classic essay on "Death and the Right Hand" (1909), anthropologists have recognized that the two hands, identical in appearance yet differing in strength and agility, function as a potent symbolic resource in such dualistic systems. Hertz believed that the different qualities traditionally associated with right and left derive originally from the religious distinction between the sacred and the profane. The right hand, because of its superior strength and agility, tends to act as a reference point for the positive, sacred side of existence, while the left hand, somewhat weaker and more clumsy, tends to act as a symbolic focus for everything dangerous and profane.

Like the hands, the two hemispheres of the brain are at once structurally dual and functionally asymmetrical. For this reason, one may suppose that, also like the hands, they possess symbolic potency to which nineteenth-century medical men seem, all unwittingly, to have responded strongly. For this reason, I conclude that the rise of a perception of the brain as bilaterally polarized in late nineteenth-century neurology, especially French neurology, is more than just a story of changing ideas about the functions of the brain; though it is that, too. It also is a story about how the language and imagery of science and medicine may unconsciously be used by a society to express and sanction certain of its cultural "truths."

What truths have I in mind here? Well, I would start by noting how clearly this model of the brain sorted out the "good guys" and the "bad guys" of the nineteenth century bourgeois imagination. Social undesirables and social inferiors (women, madmen, criminals, recalcitrant natives)—all as much feared as scorned—were shuttled off into the inferior right lobe. There, they coexisted with all the suspect and dangerous dimensions of the human mind: those irrational, nonvoluntary, emotional processes that modern civilization could tame, perhaps, but that it was incapable of wholly suppressing, just as it was incapable of wholly controlling its more violent and unpredictable inferior members.

What is interesting is that, in their model of lateralized brain functioning, nineteenth-century neurologists seem to have recognized implicitly these limits of civilization. The walls they erected between the two sides of their polarized brain were shifting, problematic structures. Although there was a general feeling that the left side of the brain usually predominated over the right, this predominance was by no means cut-and-dried, as clearly shown, for example, in the French concept of hemisphere "substitution" (Chapter Two). A certain amount of hemisphere functional independence was inevitable, and the risk that it might grow to morbid proportions was always present (see Chapters Four and Five). In light of this, one feels compelled to raise the possibility that the polarized brain was not *only* a vehicle for symbolically rendering and passing judgement on certain perceived oppositions of mind and society. It may have been also, and equally, one of the ways in which French neurology gave voice to its uneasiness about the apparent fragility of civilized man's victory over his brutish origins and the social consequences likely to result.

The idea that human beings are an uneasy composite of conflicting opposites is a very old theme in the history of Western thought but, with the rise of evolutionary theory in the second half of the nineteenth century, it is a theme that would become increasingly "biologized." In the final paragraph of *The Descent of Man*, Darwin affirmed that the human animal still bore within him "the indelible stamp of his lowly origin." Most French biologists, reflecting in part the heightened nationalism in their country following France's defeat by the Prussians in 1870, were loyal neo-Lamarckians and tended to be highly ambivalent towards Darwin's views. Their orientation, however, was no less evolutionary for all that. And there is no question that they, like their English counterparts, were acutely aware of the "beast in man" specter raised by evolutionary theory (on the English, see Durant 1981). This awareness probably was articulated most clearly in the French concept of degeneration, the perceived sinsister flip-side of the evolutionary process. It was defined by one of its key proponents, Valentin Magnan, as

> a pathological state of the organism which, in relation to its most immediate progenitors, is constitutionally weakened in its psychophysical resistance and only realizes in part the biological conditions of the hereditary struggle for life. That weakening, which is revealed in permanent stigmata, is essentially progressive, with only intervening regeneration; when this is lacking, it leads more or less rapidly to the extinction of the species. (cited in Nye 1984, 124)

As Robert Nye (1984) has shown, ideas about degeneration came to acquire a common cultural meaning for French scientists and nonscientists alike. It is no accident, he feels, that the concept came into greatest prominence during a period of French history marked by extreme social and economic upheaval. By the 1870s and 1880s, the process of industrialization was well underway in most of France and, along with urbanization, this led both to new social strains and to an apparent rise in the severity of old ills: crime, madness, suicide, alcoholism, etc. All these problems were seen by contemporaries simultaneously as a pathological consequence of recent social developments and as a threat to the continuing existence of society itself. France's decisive defeat at the hands of the Prussians in 1870 sharply aggravated feelings of national insecurity and played a significant affective role in the endless late century debates over the decline in the quality and the quantity of the French population.

For all these reasons and more, Nye described the degeneration concept as a "metaphor of pathology," a sense of cultural crisis expressed in medical terms. I would like to suggest that the French model of the brain as a polarized organ spoke to at least some of the same psychological needs as the degeneration theory, though of course the cultural significance of the latter was to be far more profound.

How, precisely, did France's polarized brain serve the function I am claiming for it? Well, in the course of turning the left hemisphere into the educated, "human" side of the brain, Broca cast immediate suspicion on the right hemisphere, implying that it was permitted to remain in an uneducated, essentially animalistic state. From a variety of evidence, the idea began to take hold that the right hemisphere was responsible for various dark psycho-physiological processs out of human conscious control: sensibility, emotion (hysteria, passion, criminal impulsiveness), and trophic, instinctual life. These functions were seen as neatly complementing the supposed conscious, voluntary, intellectual activities of the left hemisphere.

With this development, the original rather static conception of hemisphere asymmetry as a measure of a person's relative rank on the phylogenetic ladder (Chapter Two) began to be rivaled by a more dynamic image of the human brain as an uneasy composite of (left-brain) human and (right-brain) beast. This image found expression in works claiming to have found that the right hemisphere was relatively more developed in women and so-called savages than in white men. It also found expression in the rise of a body of thought that argued for the right side of the brain as a natural breeding-ground for madness. The moral implicit in this last idea could hardly be clearer: the (left-brain) processes of civilization could be reversed; the (right-brain) beast out of which one had emerged still lurked within.

The fact that it was largely in France that a left/right polarized image of the brain developed and held sway may be explained to an important extent by the wider social and cultural developments mentioned earlier. However, it would be imprudent to suppose that this was the whole story. It is quite possible that we are also simply dealing here with differences in "national styles" of physiology and psychology (cf. Geison 1978, 331ff). In Germany, for example, everything suggests that interest in the comparative functions of the two brain-halves was relatively modest, despite the burgeoning of brain research in that country. This might be partly a consequence of the fact that German clinical neurology (aphasiology) did not really come into its own before the work of Carl Wernicke in 1874. Then, when it did, it tended to be dominated by the "geometric" orientation of the master, as briefly described in section 3-1. In contrast, French aphasiology remained loyal to Broca through the end of the century, never forgetting that, with his anatomical explanation for asymmetry, Broca had focused attention on hemisphere differences from the very beginning. The heightened nationalism and political animosity between the two countries, which followed in the wake of the 1870–1871 war, only would have tended to accentuate these differences in "national styles."

Nevertheless, even if the left/right model of the brain described in this chapter seems predominately to have been a French construction, the French were *not* unique in their general conception of the brain as an organ divided-against-itself. On the contrary, no less than three dual-opposition models of the human brain coexisted in European neurology at this time. In addition to the lateral model, one finds a hierarchical model, which divided the nervous system into higher (conscious, voluntary) and lower (unconscious, brutish) parts, and an older front-to-back model, which set up an opposition between the lofty intellectual workings of the frontal lobes and the crude, animalistic functions of the posterior lobes. This is quite a striking fact, and if the interpretation of the French work in this chapter is at all right, then one might begin to understand this wider phenomenon in neurology by looking at the way scientific and medical concepts can function in a society as metaphorical resources and by looking again at the specific social and cultural context that informed neurological research in different European countries at this time.

CHAPTER FOUR

The Post-Broca Case for "Duality of Mind": Basic Issues and Themes

Man is like those twin-births, that have two heads.
—W. Godwin, Caleb Williams, 1794

4-1. The Wider Context

MY ATTEMPT to make sense of the late nineteenth century effort to compare and characterize the two hemispheres of the brain has been complicated by the dramatic revival, in this same period, of interest in the older argument that the two hemispheres may function independently. The perceived structural and functional differences between the two sides of the brain (as outlined in Chapter Three), particularly the left-sided localization of language, were widely seized upon as evidence in favor of the duality of mind and brain. As Brown-Séquard put it in 1874, "the very fact that the loss of speech depends on a disease of the left side of the brain . . . is extremely important in showing that the two sides of the brain may act independently of each other" (Brown-Séquard 1874a, 5; cf. Descourtis 1882, 42).

At the same time, it is hard to say how far the discovery of asymmetry was directly responsible for the post-Broca revival of interest in "duality of mind." In studying the medical and scientific literature, one quickly realizes that ideas about duality and ideas about asymmetry tended to be grafted together in ways that, while sometimes mutually reinforcing, were just as likely to conceal contradictions. This suggests that one must look beyond Broca for an understanding of the late nineteenth century upsurge of interest in the explanatory potentials of the double brain.

At the end of Chapter Three, I proposed that the nineteenth century's left/right polarities of mind and brain in part were a reflection, within the scientific and medical discourse, of various social anxieties, loosely oriented around a nagging sense of the fragility of humanity's victory over its brutish past. The pleasing Judeo-Christian image of man as hardly lower than the angels had been badly shaken, but somehow not quite overturned, by the unwelcome revelation that he also stood uncomfortably close to the apes; and this seems to have encouraged considerable gloomy reflection upon the fundamental dividedness of human nature. Block indeed argued (1984, 474) that the fate of

Dr. Jekyll in Stevenson's classic tale of duality, *The Strange Case of Dr. Jekyll and Mr. Hyde*, represents the culmination of Stevenson's meditations upon the "somber ramifications of evolutionary psychology." As Jekyll declares in the story,

> It was on the moral side, and in my own person, that I learned to recognize the thorough and primitive duality of man. I saw that of the two natures that contended in the field of my consciousness, even if I could rightly be said to be either, it was only because I was radically both. (cited in Block 1984, 469)

Now, in searching for the sources of late nineteenth century fascination with brain-duality and duality of mind, I have been led back to last chapter's theme of human duplicity. It has become increasingly clear, however, that last chapter's emphasis on the specter of the man/beast raised by evolutionary theory and degeneration, while still very important, probably is not, by itself, sufficient to explain the pervasiveness of this theme. What other forces in the nineteenth century might have colluded to give birth to a widespread perception of man as a janus-faced soul, at war with himself?

In his influential *Science and the Modern World* (1926), A. N. Whitehead affirmed that, although every period of modern intellectual history has been marked by bitter feuds between opposing camps, the nineteenth century was distinctive for being an age in which each individual was "divided against himself." The deepest thinkers of this time, he felt, were "muddled thinkers"; their allegiance "claimed by incompatible doctrines," and their efforts at reconciliation tending to produce nothing but confusion (pp. 114–15). In the *The Divided Self* (1969), literary critic Masao Miyoshi concurs with Whitehead's historiography, arguing that men and women in the second half of the nineteenth century were faced with a bewildering spectrum of apparently incompatible choices quite specific to their own time: faith or agnosticism, traditional values or scientific progress, individualism or socialism, authority or freedom, common sense or poetic intuition. There is a value in possibilities, Miyoshi says, but all too often the late nineteenth century tended to see them in the form of rigid dualities: all or nothing, black or white (Miyoshi 1969, xv).

The nineteenth century's sense of moral conflict and dividedness, Miyoshi argues, was strikingly reflected in its imaginative literature. In his book, he goes on to develop this thesis through analysis of such literary conceits and themes as the Doppelgänger, the double, the polarization of evil and wholesomeness within the gothic villain/hero, the romantic "other self," the Jekyll-Hyde split personality, the identity crisis of the self-alienated hero, and the total "dissociation of sensibility" seen in many fin de siècle literary works. Although he focuses on the English scene, there is no doubt that his study could have been broadened significantly. The list of major nineteenth-century writers who, in one fashion or another, addressed themselves to the theme of self-dividedness and duality includes J. W. von Goethe, Jean Paul Richter (who invented the word *Doppelgänger*), Georg Büchner, E.T.A. Hoffmann, Guy de Maupassant, Franz Kafka, Fyodor Dostoevsky, James Hogg, Edgar Allan Poe, and Oscar Wilde (for more on the "double" in literature, see also Tyms 1949).

In *The Discovery of the Unconscious* (1970), Henri Ellenberger also pays considerable attention to the nineteenth century's fascination with the duplicity of human nature. Indeed, he goes so far as to assert flatly that "the entire nineteenth century was preoccupied with the problem of the coexistence of . . . two minds [within one person] and of their relationship to each other" (p. 145). However, he finds the chief source for this preoccupation neither in moral dilemmas nor in reflections on the implications of evolutionary theory but in a number of developments that overlapped popular and medico-scientific culture and that together conspired to reinforce the idea that the doctrine of unified consciousness was an exploded dogma, that the mind was far more extensive than the individual consciousness, that dwelling within each individual was a second unconscious "self," which "dreamed and prayed" (as von Hartmann put it in 1869) as the waking self labored to earn its "daily bread" (cf. Bastian 1869–70).

The advent of spiritualism in the 1850s had led the way, inspiring a vast spectrum of popular theories about the occult powers of the human mind. Lurking below waking consciousness, it was said, was "another mind" capable of tapping into various supernatural realms of reality. The rehabilitation of hypnosis, first by James Braid in Britain and later by Jean-Martin Charcot in France, further increased interest in the apparent duality of certain mental processes as it was shown how, for example, a hypnotized subject could be made to perform acts of which he had no memory upon awakening but would recall instantly upon being rehypnotized. Laymen and medical men alike also were fascinated to see the way in which a new, and generally more brilliant or interesting, personality could be made to manifest itself under trance, of which the primary "real" self would retain no memory. According to Ellenberger, both the rise of spiritualism and the new interest in hypnosis in turn must be understood as part of a larger neo-romantic reaction against positivism and scientific naturalism, which did not supersede the latter, but coexisted alongside it (1970, 278–79).

Michel Foucault has argued (1973, 116) that there is a link between an age's perception of human nature and its view of madness. If he is right, then the nineteenth century's belief that man was a chronically divided soul should have begun to change the way this era thought about mental disorders. Indeed, there is evidence that medical men of the time began to feel that even common forms of madness, which an earlier age might have seen quite differently, were marked by features pointing to a fundamental dividedness in the patient's intellectual and moral fiber. Andrew Wynter spoke in 1875 of certain patients who "whilst suffering from the choreic type of insanity alternately spit, bite, caress, kiss, vilify, and praise those near them, and . . . utter one moment sentiments that would do honour to the most orthodox divines, and immediately afterwards . . . use language only expected to proceed from the mouths of the most depraved of human beings" (Wynter 1875, 241). Chronic mental dividedness was also invoked to account for cases in which a patient suffered from "extravagantly insane delusions on some subject" while continuing to exhibit perfectly "sound reason and good judgment on all other subjects" (Maudsley 1889, 184) or cases in which the patient was actually aware that he was suffering from a delusion or a pathological impulse and struggled against his affliction (partial insanity, or *la folie lucide*).

Of course, the most spectacular manifestation of mental duality was the actual double, or divided, personality. Once regarded as a "very rare, if not legendary" happening, by 1880 or so, double personality had become one of the most widely discussed clinical disorders of the nineteenth century, particularly in France (Ellenberger 1970, 126; Binet 1889, 8–9). Two of the most famous cases, Mary Reynolds from the United States (Plumer 1860; Mitchell 1888) and Eugene Azam's Félida from France (Azam, 1887), were copied from book to book and circulated for decades.

The case of Mary Reynolds, widely known in France as *la dame de Macnish*, dated back to the early decades of the nineteenth century. The daughter of a Reverend William Reynolds, Mary's troubles began in 1811 when, at around the age of nineteen, she fell into a profound sleep from which she awoke like an infant, all memory and knowledge from her previous life having vanished. She quickly regained lost skills but remained unable to recall her past existence. Some weeks later, she reverted to her old state, and was then unable to remember anything of her recent transformation. Alternations between the two states continued for some fifteen to sixteen years, finally ceasing at the age of thirty-five, when she was left permanently in her second state. The differences between two personalities were quite remarkable. In her first, presumably normal state, Mary was taciturn, sober, a bit dull, with a tendency towards depression. In her second state, in contrast, she became gay, out-going, and even rather reckless. Her handwriting differed entirely from one state to the other.

The case of Félida was observed for many years by Bordeaux professor of surgery, Eugene Azam (1822–1891), who subsequently coined the term *dédoublement de la personnalité*. In her normal state, Félida had been a quiet, somewhat sullen girl, who at age thirteen began to suffer from a variety of hysterical ailments. These were soon supplemented by "crises," during which she would fall into a state of lethargy and awake a different person: vivacious, cheerful, and wholly free of illness. In her normal state, she was of only average intelligence, but Azam said that she became considerably more brilliant in her second state. When in the latter, she remembered all the events that had occurred in her previous secondary states and also had memory of her entire life. In contrast, Félida, in her first state, could remember nothing of events that occurred during her secondary phases. Both personalities regarded themselves as the "true" Félida and feared to fall back into the other state, which they considered pathological (Ellenberger 1970, 136).

Brown (1983) emphasized that neurologists in the late nineteenth century *had* to interest themselves in hypnosis, somnambulism, and dual personality cases, because the lay public was tending to favor a spiritualistic or supernatural interpretation of all these phenomena. As keepers of the new faith in science and the naturalistic worldview, neurologists saw that it was essential that such obscurantist ideas be promptly squelched. In the hard words of the American neurologist William Alexander Hammond,

> The real and fraudulent phenomena of what is called spiritualism are of such a character as to make a profound impression upon the credulous and the ignorant; and both these classes have accordingly been active in spreading the most exaggerated ideas relative to matters which are either absurdly false or not so very astonishing when viewed by the cold light of science. . . . Such persons . . . having very little knowledge of the forces inherent in their own bodies, have no difficulty in ascribing occurrences which do not accord with their experience to the agency of disembodied individuals whom they imagine to be circulating through the world. . . . Their minds are decidedly fetish-worshipping in character, and are scarcely, in

> this respect, of a more elevated type than that of the Congo negro who endows the rocks and trees with higher mental attributes than he claims for himself. (Hammond 1870, 234)

Thus, finally, we come to the view that at least some of the late nineteenth-century revival of interest in the double brain can be traced to this larger effort to fight the creeping tide of the occult. There was no denying that both the medium and the madman gave evidence of possessing "two minds"; it was also quite true, as the literary men never ceased to insist, that even the normal individual suffered from chronic internal conflicts and persistant dissociations of identity. Nevertheless, said the medical men, these phenomena all had their source in the lawful workings of the nervous system and were fully explicable in naturalistic terms. By grounding so-called double consciousness in the structural duality of a brain whose two hemispheres, as they saw them, were suffering from pathological independence of action, French medical men doubly undermined the phenomenological reality of the supernatural by at once *pathologizing* it and *materializing* it.

Although the double brain would prove to be an important explanatory, and ideological, resource for medical men, it naturally was not the only one. Frequent use was also made of the hierarchical model of the brain, briefly discussed at the end of the last chapter. Adherents of this model argued that cases of mental duality involved pathological "splitting" of the mind and brain along the horizontal axis, dividing higher (conscious and voluntary) centers from lower (unconscious and reflexive) ones. In the United States, for example, the alienist George Beard argued that "every human being lives two lives, the *voluntary*, in which he acts more or less under the control of the will, and the *involuntary*, in which he acts automatically, and over which the will has but limited power or none at all." This discovery, he was sure, would go far towards clearing up public confusion about "the great modern delusions of animal magnetism, spiritualism, and mind-reading" (cited in Brown 1983, 573–74).

Another American writer, Smith Baker, found the etiological origins of mental dividedness in individual heredity factors: "the suggested fact is, that certain personality characteristics of the two lines of ancestry do not always harmoniously blend in the personality of the individual under consideration." He spoke of the case of a fifty-year-old, well-bred, law-abiding man, who had been troubled all his life by a barely conscious impulse to pilfer and who now feared that, when senility overtook him, he would be unable to inhibit this impulse. "I find," said Baker, "that he was born of a parentage representing on the one hand, honor, probity, and trustworthiness; but on the other hand, just a generation back, a number of petty thieves, sexual perverts and dandified mediocres—giving a typical ground field for the Jekyll-Hyde personality" (Baker 1893, 667, 671).

4-2. *Anatomical and Physiological Considerations*

Before examining how the double brain was used as a resource in psychiatry, philosophy, and psychology, it would be helpful to review in a more systematic way how ideas about the double brain had changed and refined themselves, since the days of Wigan and the phrenologists (Chapter One); and what new evidence was being offered in support of belief in cerebral functional duality. Although the discovery of language's unilateral locus occupied an important place in the late nineteenth century's arsenal of evidence, it did not result systematically in so careful a readjustment of conceptual principles as one might have expected. And even as evidence, it did not occupy center stage so much as one might also have expected. In the late 1870s, a series of striking hysteria and hypnosis experiments would begin to excite far more interest in the duality of the human mind and brain than Broca's work had ever done. Because of its importance, this development is separately discussed in Chapter Six.

All the same, in very general terms, Broca's work undoubtedly was crucial to one important conceptual change in the idea of brain duality: the decline of the old views that asymmetry was inherently a sign of pathology and that simultaneous functioning of the hemispheres was absolutely necessary for all healthy mental functioning. In the wake of all the incredulity that had greeted the idea that only half the brain might serve speech, people now saw that Broca had simply called attention to one specialized dimension of what was really a quite basic principle of brain function. It has been said that people respond to every novel idea in science in three distinct stages: (1) "It's absurd!"; (2) "It's against the Bible!"; (3) "Oh, we've known *that* all along!" Suddenly, it was no more absurd that the two hemispheres should possess a certain amount of functional autonomy than it was that the two sides of the body should be able to move independently. Indeed, as English neurologist Samuel Wilks pointed out, the relative autonomy of the two lateral halves of the body was inexplicable unless one presupposed a degree of hemisphere functional independence. "The independence of the limbs on each side of the body seems to demand an independence of the ganglionic nerve centres which rule over them, and these, again, their own distinct hemispheres to govern them" (1872, 164–65).

In France, Jules Bernard Luys not only argued that the two brain-halves naturally possessed a degree of autonomy but believed that certain activities of civilized living artificially cultivated this capacity. The pianist, for example, learned to develop a distinct and separate melody within each of his two hemispheres simultaneously, while his two hands expressed them on the keyboard: "each lobe is capable . . . of acting separately and thus of generating a series of voluntary and conscious movements, inspired by a series of psychical operations, equally distinct on the right and left" (1879, 551). The pianist's alleged state of lateralized divided consciousness seems to have impressed Luys profoundly; he and his students harked back to this putative example of cerebral dualism for years (see, e.g., Luys 1889). Interestingly, the phenomenon of piano playing also had caught the attention of Carl Wernicke in Germany who, however, had taken quite a different view of its physiological significance:

> The novice piano player feels compelled to move the fingers of the left hand in synchrony with the right and dissociation between the hands is accomplished only through many years of practice. *Innervation of the action occurs only from the left hemisphere.* . . . In man, movements of the upper extremities may be practiced a thousand times, and yet innervation from *one hemisphere alone* is adequate to initiate movement of both extremities. (Wernicke 1874, 94–95, italics added)

Back in France, some neurologists began to emphasize that the relative functional independence of the two hemispheres was self-evident, not only in daily life but on simple anatomical grounds. As one of Luys' students, Gabriel Descourtis, stressed, hemisphere functional independence could be inferred, not only because the human brain was structurally dual but equally because recent careful investigation had shown ("chose bien remarquable") that the convolutions making up each of the secondary lobes (frontal, parietal, occipital, temporal) were never precisely identical on the left and right side. "Indeed," said Descourtis, "the dissimilarities [between] the two hemispheres constitute one of the best established anatomical points of our era" (Descourtis 1882, 6–7; cf. Luys 1879, 517). What had become of the equally "remarkable" anatomical symmetry between the hemispheres, so clearly seen by men like Bichat and Sir Charles Bell early in the century (Chapter One)? Again, one is struck by how theoretical expectations appear actually to change what people perceive.

The rise of the science of so-called cerebral thermometry gave a further important boost to the view that a certain degree of hemisphere functional autonomy was normal and desirable. Studies attempting to compare the relative temperature of different parts of the brain, including the two hemispheres, were carried out in France, Germany, Italy, and the United States beginning in the early 1870s. All of them took their point of origin from the supposition that the brain, somewhat like a muscle in action, manifested its dynamic power by a local increase of heat, which could be detected by laboratory instruments (Luys 1881a, 76). The subsequent discovery that brain temperature varied between the two hemispheres and their lobes seemed an important new confirmation of the relative functional independence of those areas of the brain. The findings also confirmed what everyone already "knew" about the differential intellectual capacities of the different parts of the brain: the left hemisphere tended to be hotter than the right; and the frontal lobes tended to be hotter than the parietal lobes; which were, in turn, somewhat hotter than the occipital lobes.

Although Broca had been satisfied simply to strap thermometers onto the scalps of his subjects, others strove to eliminate any possible contamination of the data caused by the presence of scalp vessels and their temperatures. An American, Josiah Stickney Lombard, former assistant professor of physiology at Harvard, developed a device he called a thermoelectric apparatus, which exerted pressure on the part of the scalp where measurement was to be made, leaving it pale and bloodless so that a true reading could be made of the brain, rather than the blood, temperature. Lombard hoped that by examining "the relative temperatures of different portions of the surface of the head during increased mental activity . . . information could be obtained as to the comparative importance of the parts played by different portions of the brain in the evolution of thought and the different emotions." He was unable, however, to obtain any consistent results with his six human subjects, in spite of extensive trials (Lombard 1878). Meanwhile, German neurophysiologist Moritz Schiff had already gone one step beyond Lombard, actually embedding thermometers in symmetrical points of (animal) brains to ensure that he was recording truly "evoked" (his term) temperature changes, free from vascular influences. It was through Schiff's pupil, Angelo Mosso, that Hans Berger was inspired to replace thermometers with electrodes in 1924, giving rise to the concept of the EEG (Schiller 1979, 218–19).

By the end of the century, the relative functional independence of the two hemispheres had become so widely accepted that one neurologist, Pierret, not only saw no particular need to defend the thesis, but went on to suggest that the nervous system's contralateral organization, long puzzling to physiologists, had evolved as a compensatory strategy against lateral autonomy.

> If things remained in a state of noncontralaterality, nervous incitations, proceeding from one hemisphere, would not be able to pass from one side of the spinal cord to the other. . . . Passage [locomotion?] would be more complicated and absolute synchrony of bilateral movements virtually impossible. Each half of the cord would have too great an individuality.
>
> Consequently, there arose, for superior vertebrates, a new characteristic of improvement, that is the contralaterality of the long commissures. (Pierret 1895, 270)

Belief in the *relative* functional independence of the two hemispheres in health, then, represented one important conceptual development in the idea of brain duality since the days of Wigan. I must at this point call attention to a second, and far more controversial, new twist in the old story: the development of the idea that the functional duality of the brain posed a serious challenge to belief in cerebral localization.

The man who became most closely associated with this argument was Edouard Brown-Séquard. In Chapter Three, attention was paid to Brown-Séquard's views on the respective functions of the right and left sides of the brain. It must now be pointed out—and stressed—that he did not accept the idea that these functional differences were a consequence of innate structural differences between the hemispheres resulting from differential growth, blood-supply differences or whatever. Nor did he accept, as Broca had argued, that functional asymmetry, was an inevitable and laudable reflection of the effects of education and civilized life upon the individual. For Brown-Séquard, functional asymmetry was caused by a *failing* in the educational system; a result of "the fact that, through the fault of our fathers and mothers, . . . [w]e make use of only one-half of our body for certain acts, and one-half of our brain for certain other acts—we find that it is owing to that defect in our education that one-half of our brain is developed for certain things, while the other half of the brain is developed for other things" (Brown-Séquard 1874a, 10). "If children were . . . trained [to develop both brain halves], we would have a sturdier and healthier race, both mentally and physically" (1874b, 333; cf. the arguments of the ambidextral culture movement, discussed in section 4-5).

But Brown-Séquard went still further. Not only was each hemisphere a potentially complete brain for the activities of intellectual life, each was also fully capable of carrying out *all* sensory-motor activities, that is to say, sensory-motor activities for both sides of the body. He flatly denied, then, what all the recent animal experimentation by men like David Ferrier and all the older data from human pathology seemed to show: that each half of the cortex essentially possessed sensory-motor centers only for the opposite half of the body, with some limited power over the ipsilateral half. And the chief reason Brown-Séquard denied this virtually unquestioned tenet of nineteenth century neurology is that he believed that in fact there were no sensory-motor "centers" in the brain at all; the whole localization enterprise was misguided. In an 1876 lecture in Boston entitled "On Localization of Functions in the Brain," Brown-Séquard declared,

> The character of symptoms in brain diseases is not in the least dependent on the seat of the lesion, so that a lesion of the same point may produce a great variety of symptoms, while on the other hand, the same symptoms may be due to the most various causes, various not only as regards the kind, but also the seat of the organic alteration. (cited in Olmsted 1946, 152)

The tendency for unilateral motor and sensory disorders to result from damage to the contralateral side of the brain was simply due, in Brown-Séquard's opinion, to "an inhibitory influence, exerting itself on other portions of the nervous centers (encephalon or spinal cord)" (Brown-Séquard 1887). In other words, activity in, or damage to, almost any part of the central nervous system could set off, or restrain, action in some other part. No region of the brain in particular, therefore, could be said to have a definite and invariable function, since every region was functionally dependent upon the action of various other, often distant regions (Olmsted 1946, 118).

In an 1890 paper published in the popular American journal, *The Forum*, Brown-Séquard summed up the evidence in favor of his belief in sensory-motor cerebral dualism. To begin, it was well known that minor damage to half the brain sometimes resulted in extensive or even total loss of function while, contrariwise, severe damage to the brain not rarely caused little or no disability. "I need not say that such facts are quite decisive against the universally admitted view that each half of the encephalon . . . serves only for the half of the body on the opposite side" (1890, 627–28). Although unilateral paralysis and anaesthesia often did occur from surface damage to the contralateral side of the brain, the neurological community had completely misinterpreted the significance of this fact. This was demonstrated by certain experiments Brown-Séquard had performed on frogs.

> It is known that when both cerebral lobes are taken away from these animals, no paralysis appears. I have found, however, that when one of those lobes is cut transversely, hemiplegia is caused immediately on the opposite side, just as would be the case in man from analogous lesions. If, then, I divide at the same level and transversely also, the other cerebral lobe, the paralysis produced by the first operation disappears, and the frog seems to move all its limbs just as well as if nothing had been done to the brain. (1890, 631–32)

What had happened, Brown-Séquard felt, was that the first lesion had brought about a dynamic imbalance between the two brain-halves, causing paralysis. By inflicting similar damage on the other lobe, equilibrium had been restored. Similar principles, Brown-Séquard was sure, were at work in cases of human hemiplegia.

For Brown-Séquard, however, the most powerful evidence for his view that one hemisphere is sufficient "to originate all the voluntary movements of the two sides of the body" came from the clinic not from animal experiments. Over the years, he had continually been calling attention to certain clinical cases involving "considerable alteration, if not destruction of a whole cerebral hemisphere, without paralysis, or with only a slight paresis" (1890, 638). Clearly, this would only be possible if one hemisphere could serve both sides of the body.

Although not mentioned by Brown-Séquard in his 1890 paper, the German neurologist Friedrich Goltz, another critic of cerebral localization, had recently performed a "unique amputation" that offered apparent experimental confirmation of this last argument. In his 1888 paper, "Ueber die Verrichtungen des Grosshirns," Goltz declared,

> I will begin by relating an experiment which I hope will be acclaimed by all true friends of science. I succeeded in observing for fifteen months an animal in which I had taken away the whole left hemisphere. . . . [A] dog without a left hemisphere can still move voluntarily all parts of his body, and . . . from all parts of his body, actions can be induced which can only be the consequence of conscious sensation. This is incompatible with that construction of centers which assumes that each side of the brain can serve only those conscious movements and sensations that concern the opposite half of the body. (Goltz 1888, 119, 130)

The idea that both brain-halves might be capable of bilateral sensory-motor functioning was received with incredulity and general hostility, even by writers convinced of the functional duality of the brain for *mental* processes. In his own (1889) article on brain duality, Henry Maudsley opined in his usual blunt way that "one would receive this opinion of so practised a vivisectionist [as Brown-Séquard] with more assurance had any adequate result ever come from his multitude of cruel experiments" (p. 164n). David Ferrier, although he believed that the brain "as an organ of ideation" was dual, had no doubts of its single nature so far as sensory-motor functioning was concerned, and attacked Brown-Séquard for his "peculiar views that symptoms of cerebral disease [such as hemiplegia] are due to some dynamic influence exercised by the lesion on parts situated at a distance (and always apparently out of reach) which are credited with the functions lost or otherwise disturbed" (cited in Olmsted 1946, 146–47). At the time of this remark, Ferrier had already scored a rather decisive victory over Goltz's antilocalization views. During the London International Medical Congress of 1881, Goltz had managed to impress his colleagues with his decorticate dogs ("Hunde ohne Grosshirn"), who retained reasonably good sensory-motor coordination. Ferrier's masterly counter-demonstration with monkeys, however, brought the verdict more or less unanimously down on his side (Haymaker 1970; for Goltz's own assessment of the meaning of his experimental work with dogs, see Goltz 1892).

At the time, Ferrier had been forced to admit that he did not dispute Goltz's "facts," but only rejected 'his conclusions." It is worth realizing that after his 1881 triumph, the facts remained, undigested anomalies. Only with the rise of a generation of neurologists increasingly disillusioned with the rigid localizationist approach of their forebears (see Chapter Nine), would such facts begin to be examined anew. This development, in turn, would result in a reassessment and partial rehabilitation of such early critics of localization as Goltz and Brown-Séquard (see, e.g., Walther Riese's assessment of these men in his 1959 *A History of Neurology*).

4-3. Psychiatric Applications

The increasing acceptance of the idea that the brain's two hemispheres possessed in health a certain amount of functional autonomy gave a slightly different cast to late nineteenth-century attempts to understand various forms of mental pathology in terms of hemisphere functional independence. The view, long championed by Claude Bernard, that there was a continuity between normal and pathological physiological functioning served as the new point of reference. Thus, in his 1879 paper on the double brain, Jules Bernard Luys argued that in cases of mental pathology man's natural aptitude for hemispheric autonomy was grossly exaggerated. Anatomical evidence for this was found in the fact that weight asymmetries between the two sides of the brain were much greater in the insane than in the sane (25–35 grams instead of the normal 5–6). Not only that, but the normal left-sided predominance found in the sane was frequently reversed in favor of the right side, further disrupting normal dynamic relations. As mentioned in Chapter Three, Luys believed that this reversed asymmetry pointed to a particular participation of the right hemisphere in madness. It *also* meant, in his view, that many madmen might retain a certain lucidity on the nondiseased, presumably left, side of their brains. If morbid hypertrophy of one hemisphere could coexist with the integrity of the other, then one had a possible explanation for cases of lucid insanity, struggles against obsessions, contradictory impulses, even cases of full-blown double personality, like Azam's Félida:[1] all those "manifestations various in appearance, yet always identical at bottom, that express, as it were, the internal conflict of the human being divided into two independent and insubordinate sub-individualities" (Luys 1879, 518; see also Urquhart 1879–80b).

In a similar way, Gabriel Descourtis argued in his 1882 medical dissertation that, although the brain was an amalgamation of structurally and functionally diverse parts, unity normally managed to

[1] At least one attempt also was made to account explicitly for the "other" great nineteenth century case of double personality, that of Mary Reynolds, in terms of the double brain. In 1888, the neurologist Silas Weir Mitchell, famous for his diet and rest approach to hysteria, presented a paper on the Reynolds case before the College of Physicians of Philadelphia. During the discussion period following, Henry Hartshorne said, "It may be not entirely without interest to speak in this connection of the possible separable junctions of the two hemispheres of the brain. Brown-Séquard in a paper published some years ago, maintained that each hemisphere is capable of performing all the functions of the brain. If, then, we could suppose that by some interruption of the function of the corpus callosum, or in some other manner, there should be a totally separate action of one hemisphere from that of the other, I think there would be some suggestion of an explanation."

[Hartshorne's distinguished colleague, Charles K. Mill, disagreed,] "Instead of the idea of the two hemispheres of the brain acting separately being of service in the explanation of these cases, its acceptation might, in some respects, interfere with a satisfactory explanation. With our present physiological knowledge the different levels of the general cerebro-spinal axis would better explain this condition" (Mitchell 1888, 386–87; see also Carlson 1984, to which I am indebted for this reference).

emerge from multiplicity. Under certain conditions, however, there could be a pathological amplification of the capacity for independence that interfered with normal cooperative efforts, leading to a "rupture of the psychic unity" (1882, 62).

The most common sort of psychological "rupture," Descourtis believed, was the simple division into "two minds" caused by a breach of communication between the two hemispheres. He cited a case, originally recorded by Dr. Jaffé back in the 1850s, which he felt clearly illustrated this condition:

> D . . . always speaks using the pronoun *we*; *"we will go," "we have walked a great deal."* He says that he speaks like this *"because there is someone with him"*; at meals, he says: *"I've had enough but the other has not."* He begins to run; he is asked why, and he replies that he would prefer to stay [where he is], but it is the *"other"* who forces him on, even though he holds him back by his coat. One day, he hurls himself upon a child in order to strangle it, saying that it is not he but the *"other."* Finally he tries to commit suicide, *"to kill the other,"* whom he believes to be hidden in the left part of his body; he also calls him *"left* D . . ." and names himself *"right* D . . . ," bad D . . . and good D . . . , etc. The patient falls little by little into dementia.—Autopsy reveals a considerable difference between the two halves of the brain; . . . It is evident that the unilateral seat of these lesions was the cause, if not unique, at least essential of the double personality delirium; the individual was different on each side, he felt himself double. (cited in Descourtis 1882, 38; also recounted in Luys 1888, 524; Lyon 1895, 108; Ribot 1891, 127–28)

Other historians have argued that the positivistic orientation of nineteenth-century alienists meant that the symptoms of mental pathology were regarded simply as signs that something was amiss in the brain; the actual content of the madman's delusions was deemed irrelevant to the diagnostic enterprise. "As 'subjective' data . . . [such delusions] could form no part of the 'objective' sciences of physiology and medicine" (Clark 1981, 284–85; cf. Foucault 1973, 198). The late nineteenth-century literature implicating the double brain in madness suggests that this interpretation may be rather oversimplified. Certainly, the new breed of "scientific" alienists *claimed* not to be interested in the madman's psychological reality, but in actual fact, they could not easily avoid taking some notice of what their patients had to say and these patient reports, these "subjective" data, could and did influence the allegedly objective physiological theories formulated to account for each case.

The interpretation of the case of D is a case in point, and it is quite typical of a certain form of reasoning that runs through much of the nineteenth century duality of mind literature. A patient's (subjective) sense that he was "two people" was seized upon by the doctor as a key piece of evidence in support of the physiological (objective) hypothesis that his two hemispheres had fallen into a state of pathological independence or disequilibrium. The double-brain hypothesis was then used by the doctor to explain away the patient's subjective state and provide an allegedly deeper-level, objective account of the patient's condition. It is a fascinating sort of mirror-logic.

One sometimes, indeed, has the impression that certain doctors and patients were unwittingly caught up in a sort of *folie à deux*; the former shaping their theories in accordance with the latter's testimony; the latter's testimony being shaped by the former's theories. This point is nicely illustrated in one of the more popular late nineteenth century examples of a supposed case of pathological independent hemisphere action: the case of "Monsieur Gabbage," apparently first recounted by the French alienist and professor Benjamin Ball, in the opening lecture of a course on mental disorders that he offered in 1884. In this case, the patient *himself* urged the double-brain thesis upon his doctor and his assessment of the nature of his own delusion was considered compelling evidence in favor of this particular physiological thesis.

Ball's patient was a young man who had been overcome by sun and heat during a trip to South America and had lain unconscious for a month. Soon after recovering his senses, he heard a voice inquiring after his health, but was unable to see anyone else in the room. In response to his questions, the voice identified itself as "M. Gabbage," and shortly after that, the young man was able to see him, but only from the bust upwards. Gabbage turned out to be a dark, strongly built individual with a beard and heavy brows, who always dressed in a hunting outfit. He fell into the habit of tormenting the young man with incessant questions and later forced him to commit senseless and violent acts. Ball explained how

> One day, when conversing with him on the subject of his impulsions, he said to me: "You are not up to date in science; you seem to be unaware that one often has two brains in one's head. This is precisely the case with me. Gabbage has the left brain and I possess the right brain. Unfortunately, it's always the left side that gets the better of me, and this is why I cannot resist the advice of this man who

> appears to be an evil spirit or at least a malevolent fellow." . . . Here we have then [continued Ball] a brain whose functions seem to be quite clearly doubled, and one can readily believe, following the theory of the patient himself, that one of the hemispheres is fully delirious, while the other regards it with compassion. (Ball 1884, 37)[2]

Naturally, not every alienist and neurologist was willing to go as far in his claims as Luys, Descourtis, Ball, and their followers. A good number of the nineteenth-century medical men interested in the explanatory potentials of the double brain made considerably more circumscribed arguments. In 1868, for example, a German physician, Dr. Jensen, ventured to suggest before his psychiatric colleagues in Dresden that the phenomenon which would today be called *déjà vu* or *paramnesia*, the sense of having lived through an experience before, might be due to a disruption of the normal cooperative actions between the two hemispheres in perception. Such an argument was not wholly original; it appeared, for example, in a different form in Wigan's 1844 *Duality of Mind*. It was Jensen's paper, however, that seems to have given the idea its first significant impetus, at least on the continent.

> just as two images take shape in the eyes, so do two perceptions in the hemispheres [independently] take shape, which however under normal circumstances mostly overlap, are almost wholly congruent and so are perceived as just single. What, though, when circumstances are not normal? when the two halves [of the brain] do not function in a congruent manner but, like a squinting eye, there occurs instead an incongruence between the functioning of the brain's two hemispheres? In the case of eyes, there is a doubling of images—in the case of the brain there would correspondingly be a doubling of perception. . . . We project double images outwards beside each

[2] It is not clear how unusual Ball's patient was in diagnosing his own case in terms of the independent functioning of his two hemispheres. Certainly, there is no reason to see anything extraordinary in this man's spontaneously volunteering such an interpretation of his condition. After all, as William Ireland remarked in 1886, "The insane are quick to catch at new scientific notions to explain their delusions. Complaints of being electrified and being magnetised against their will have long been common; and since the invention of the telephone, they have said that there are telephones in their rooms, or that people use this instrument to torment them. In a similar fashion, the medical superintendents of asylums will hear many whimsical applications of the conception of the dual functions of the brain should it become popularised" (p. 345).

Obviously, as Ball's response to his patient shows, not *all* "medical superintendents" necessarily would consider it "whimsical" that patients should attempt to explain their own deluded condition in terms of the dual functioning of the brain.

> other in space. Would it not be possible that we project double perceptions outwards beside each other in time? (Jensen 1868, 50–51)

An asylum doctor from Colditz, Max Huppert, sought to extend Jensen's hypothesis by proposing (1869; 1872) that a similar sort of cerebral disruption might be responsible for a variety of cases he had observed involving patients who, in one way or another, believed that someone else in their head was forever anticipating their private thoughts and actions ("Doppeldenken, Doppelvorstellungen"), or who were tortured by voices that echoed back at them everything they said ("Nachreden"). Huppert was particularly interested in the fact that all of his patients suffered very little or not at all from their disorder when asleep and dreaming. To explain why this should be so, he appealed to the authority of Schroeder van der Kolk, who had felt that generally only *one* of the brain's two hemispheres (the one lying closer to the pillow) was active during sleep (Huppert 1869, 544; 1872, 103–104; see also Chapter Three, note 5).

Other alienists began to argue that some theory of hemisphere functional duality was necessary to account for the phenomenon of unilateral hallucinations (but see the rival theory of Régis 1881). In Scotland, Alexander Robertson had been an early and influential advocate of this view, reporting on a number of cases in which hallucinated voices had been heard in only one ear (Robertson 1875). In France, Valentin Magnan came to a similar opinion after studying several cases in which two distinct auditory hallucinations were referred to the left and right ear, respectively (Magnan 1883). Likewise, America's leading neurologist, William Alexander Hammond, declared in 1885 that "unilateral false perceptions . . . afford strong evidence of the correctness of the theory originally proposed by Wigan, but since his time adopted by many others, that there are in fact two brains, [with] the two hemispheres . . . capable of isolated and even different action" (Hammond 1885). In Britain, medical superintendent William Ireland, one of that country's most prominent popularizers of the double brain theory, offered the following illustration of such a "unilateral false perception":

> I knew of a patient, a single woman aged thirty-six, who imagined that a large green fish followed her about and bit her between the shoulders. It used to look over her right shoulder at her with its large round eyes. When asked if she heard voices, she said, "It seems to come over my shoulder (right). . . . It does not say anything, but I don't like the look of it." On glancing over her left shoulder she

could not see the green fish. (Ireland 1891, 1168; for Ireland's other publications on the double brain, see Ireland 1881, 1886, 1892)

4-4. Philosophical Implications

The years immediately before Broca saw one of the first substantive meditations on the philosophical implications of brain-duality, offered up in this case from the pen of a German writer. It needs to be understood, though, that Gustav Fechner (1801–1887) was working within a distinctly different metaphysical context than any of the other individuals examined so far. A panpsychist strongly influenced by the Naturphilosophie of his country, Fechner was interested in the possibility of refuting materialistic metaphysics using the methodology and data of modern science. In his 1860 *Elemente der Psychophysik*, he set out to prove that there existed a lawfully determined relationship between the physical and the mental, the noumenon and the phenomenon, which could be expressed in precise mathematical terms. If such a functional relation could be empirically substantiated, he believed, then the conclusion would follow that mind and matter were simply alternative ways of perceiving the same reality.

Fechner's specific reflections on the implications of brain duality emerged out of his consideration of certain experiments that the Swiss naturalist Charles Bonnet long ago had carried out on segmented worms. The animals had been cut into two or more pieces along the line of segmentation, and it was found that each piece continued to maintain an independent existence, with the full range of behavioral responses appropriate to its species. According to Fechner, these studies did *not* mean, as the materialists would be inclined to argue, that Bonnet had "split" a single mind into two halves. Rather, he had created two physical conditions conducive to the rise of two independent minds. The result was a doubling of observable psychic activity (activity above the mind-body "threshold"); an "awakening of a new mind [Erwachen einer neuen Seele]" within each of the worm's segmented halves.

A human being, of course, could not survive a complete splitting of his nervous system, since the principle of "solidarity" between his organic units was much stronger than in a lowly animal like the worm. Nevertheless, in theory, the same psycho-physical laws that were true for Bonnet's worms were applicable to the mind-body relationship in man:

> Were it possible for the two halves of a man divided lengthwise to carry on living at all, that is, for the psycho-physical activity in both

> halves to remain above the threshold, then we would undoubtedly see something equivalent to the doubling of a man. . . . Undoubtedly, the two halves would begin with the same states of mind, the same dispositions, knowledge, memories, the same consciousness in general; by degrees, however, as they passed through different circumstances, they would develop differently. (Fechner 1860, 536–37)

One of the most important pieces of evidence in favor of this argument, Fechner felt, came from clinical reports of individuals who had lost an entire cerebral hemisphere with little or no loss of mental capacity. In such cases, the remaining healthy hemisphere had clearly provided the conditions for the "awakening of a new mind" in a manner essentially identical to that of a segmented unit of a worm. "Whether equivalent segments of the psycho-physical system lie behind or beside each other is only an insignificant distinction, and whether the separated piece is destroyed or not is likewise unimportant, so far as the surviving one is concerned" (p. 536). It could now be understood, Fechner said, why many insane people regained their wits just before death: the diseased half of their brain responsible for the madness had finally slipped below the psycho-physical "threshold," setting free the healthy half to give birth to a new mind.

Admittedly, because of the high level of "solidarity" between the two halves of the human brain, a single hemisphere supporting a mind on its own was less effective and tired more readily than two hemispheres working in tandem. Fechner used a "horse and wagon" analogy to bring this point home:

> The two cerebral hemispheres can actually be compared to two horses who are attached to one and the same wagon. . . . During the time both are on the wagon, they are not just there to stand in for each other, but also to support each other along the way; for if one of them is released, then the wagon goes more sluggishly or, if it still goes at about the same speed because the one horse pulls more strongly, it cannot carry on so long. (Fechner 1860, 536)

Although the lines of influence cannot be traced with any certainty, it is not unlikely that Fechner's 1860 discussion of the double brain inspired his countryman Eduard von Hartmann (1842–1906) to address himself to the same topic several years later. Like Fechner, Hartmann was an idealist who rejected the mechanistic assumptions of much of the science of his time. Nevertheless, again like Fechner, he professed to believe that any viable alternative philosophy must derive its conclusions from data provided by the natural and historical sciences, according to the scientific method. In his immensely successful 1869 *Die Philosophie des Unbewussten*, he set out to explain the dialectical interaction of will and ideas through which self-consciousness was able to come into being out of unconscious potencies. In its metaphysical dimension, von Hartmann's unconscious need not detain us; physiologically considered, we can simply note that he believed it served as the ground for the total consciousness of an organism. That is to say, he held that conscious mind was an emergent product of a myriad of unconscious processes, which in turn were dependent upon cerebral activity: "*There is no conscious mental activity outside or behind the cerebral function*" (von Hartmann 1869, 73). This given, one could further lay down "as a principle" that "*separate material parts give separate consciousness*" (p. 117). It followed that the human brain, which consisted of two fully distinct halves, necessarily produced two separate streams of conscious mental activity. The apparent unity of our identity was simply due to the fact that the commissures connecting the two hemispheres mingled the two streams of consciousness into one. As von Hartmann explained it,

> If one imagined the union of the *brains* of two men possible by a bridge as capable of conduction as is that between the two hemispheres of the same brain, a mutual and indivisible consciousness, including the thoughts of both brains, would immediately embrace the hitherto separate consciousnesses of both persons; each would no longer be able to distinguish his own thoughts from those of the other; *i.e.* they would no longer know themselves as two Egos, but only as one Ego, as my two cerebral hemispheres also only know themselves as one Ego." (p. 118)

The reflective strain of thought in the double-brain literature introduced in the 1860s by Fechner and von Hartmann persisted into the later decades of the century, although it often tended to be intermingled with more pragmatic medical concerns. For example, Henry Maudsley (1889), in a paper on the double-brain published in the British quarterly review of philosophy and psychology, *Mind*, asked the question, How do the two hemispheres, each possessing, as he supposed, a certain amount of functional autonomy, "act towards one another in thinking"? Concealed within this laconic sentence lies an emotionally charged philosophic dilemma that had been eloquently expressed several years earlier by William Ireland:

> How can one side of the brain compare its own impressions with those of the other side? How can it know anything save its own

> impressions? How can unity come out of a double set of impressions? Is there not something behind which reads off both sets of impressions and fuses them into one? Is there an organ of the Me—an Ichheits-organ? or have we reached the mind itself? (Ireland 1886, 340)

The first point to notice, Maudsley proposed, was that a person performing two separate acts employing his two hands, and thus his two hemispheres, was unable to concentrate on both simultaneously. He was forced, rather to "pass in thought from one to the other, a rapid alternation of consciousness taking place." It was logical that this should be so, since a "simultaneous consciousness in such a case would necessarily be a distracted or dual consciousness." What was happening, then, in the brain during this experience of alternating consciousness? Maudsley offered three possibilities: (1) consciousness was jumping from one hemisphere to the other; (2) both hemispheres, acting together "by a sort of immediate sympathy," became conscious at the same instant of the activity going on in one or the other side of the brain; or (3) each hemisphere was uniquely conscious at all times of the activity going on within it. The last of these options seemed logically impossible; the second unparsimonious and, hence, implausible; the first, "an alternating action of the hemispheres corresponding to the alternating consciousness," struck Maudsley as the most likely of the three.

Now, as one of the more hard-headed advocates of the new naturalist philosophy (see Chapter One), Maudsley was scornful of the notion that consciousness played an effective role in the workings of the brain. The fact, therefore, that he thought concentration alternated between the two hemispheres does not mean that he believed it somehow to be directing their respective activities. Given this, how did Maudsley explain that, when the two hemispheres were engaged in simultaneous but different actions, they were able to cooperate with each other? How came the left hand to know what the right was doing? Or, as Maudsley put it, "from what higher source do the hemispheres of the brain obtain their governing principle of unity?" His answer was that unity of purpose "does not come from above but from below." The two hemispheres were able to cooperate because both had become infected by the same *desire* or sense of purpose welling up from the lower-order affective organs, the organs of organic life or "feeling" that, unlike those of intellectual life, were unified. Once thus emotionally united, the two hemispheres then had slowly and painfully to *learn* how to work together in order to accomplish "the [common] end or aim in view." In this respect, they could be compared to "two lithe and supple acrobats who can writhe their bodies in a conjunction and succession of the most rapid, nice and complicated movements . . . which they have thoroughly learned to do by practice but cannot do until they have learned them by much travail and pains." In sum,

> It appears, then, that the unity of the intellectual life which, so far as the division of the cerebral hemispheres is concerned, might apparently be almost dual, is based upon the unity of feeling, and this again upon the unity of the organic life. For although there is symmetry of the organs of animal life on the two sides of the body, there is not a like symmetry of those of organic life throughout the body. . . . It is from the life of the *whole* body that the constituents of the mental life are derived, and inasmuch as the cerebral hemispheres are organs ministering to this life, they must necessarily have its fundamental unity. (pp. 176–78).

If "the organic life supplies a basal unity to the action of the cerebral hemispheres," then "failure of the organic driving force" could lead to "incomplete union or actual disunion" between the two sides of the brain. One would then be witness to "a self divided against itself . . . from the weak and almost independent action of the disunited halves." Disunion between the hemispheres was also likely to arise, Maudsley felt, when only *one* of the two half-brains fell prey to disease, leading, again, to a conflict between a true self and a false one, "a distracted or double self." Many different types of mental pathology, Maudsley believed, would "repay examination in the light of the theory of a discordant action of the hemispheres" (p. 185).

The adequacy of Maudsley's "organic" solution to the problem of what united the two halves of intellectual life was challenged by J. M. Baldwin in the next issue of *Mind* (Baldwin 1889). The unity of feeling, which Maudsley had attributed to the unity of its generative organs, "could not be apprehended as such," Baldwin pointed out sharply, "in the absence of a circumscribed consciousness which, through its own unity, takes it to be what is." The unity of consciousness, in other words, far from arising out of the unity of feeling life, was actually prior to and responsible for the perceived underlying unity of one's states of feeling. Given this, perhaps it was necessary, Baldwin suggested, to admit that, in addition to feeling and intellect, there existed in each individual "a unitary subjectivity additional to the unity of the sensory content" and possibly inexplicable in terms of physical organization alone. In attempting to reconcile the duality of the brain with the unity of the known, it seems Maudsley had failed to account for the knower.

Baldwin's metaphysical alternative, however, might have been a little premature. Was the duality of consciousness into knower and known, subject and object, really irreconcilable with the duality of the brain? One could adopt a radically contrary position and argue, as the now forgotten British philosopher Robert Verity did in 1870, that dualism actually found its *explanation* in the duality of the brain. Verity argued his case on somewhat precarious analogical grounds, emphasizing the formal correspondence between subject and object and the "anatomical distribution of the brain" into two hemispheres, separated from each other but communicating "through central organs of unity" (1870, 44). He proposed that the self (the subject) had its anatomical focus in one hemisphere (on p. 46, Verity identified the "subjective" hemisphere as the right one) while the object of every individual's thoughts had its focus in the other, presumably left hemisphere (cf. Watson 1836). Once it was realized, Verity affirmed, that the age-old distinction between subjective and objective reality was simply a function of the duality of the brain, the way was opened for nothing less than a "new theory of Causation" sufficient to "solve the scepticisms of Locke, Berkeley, Hume, and Kant, and to reconcile and identify Idealism and Realism in one common synthesis" (Verity 1870, 31–33). The nineteenth century has never been notable for its humility.

From the point of view of this study, the chief interest of Verity's thesis lies in the way it brings into focus the sort of logic to which nineteenth-century materialists were forced to resort when confronted with the problem of self-consciousness. If all thought processes were ultimately reducible to brain-processes, then the "self" that came to know those processes must also be traceable back to a cerebral agency. It might seem, then, that a *second* brain was needed to "reflect" the processes going on in the first. This was Verity's conclusion. He failed to ask, though, how the second brain, housing the "self," came to know of its own existence. Was a *third* brain required to "reflect" the processes going on in the second? As Riese and Hoff point out, once one began this sort of reasoning, there was no logical end-point, "no end of brains and matter" (Riese & Hoff 1950, 62). Hughlings Jackson, writing in the years just after Verity, was probably the only double-brain theorist fully to recognize the problems inherent in the latter's approach. His own ingenious, if somewhat contrived, alternative vision of how the subjective and objective halves of consciousness interact across the hemispheres is discussed in Chapter Seven.

We see also that Verity was trying to *reconcile* brain duality with a traditional reified conception of the self. This, though, was not the only possible recourse. Early in the twentieth century, one writer declared that the new insights into the functional duality of the brain required a dramatically new perspective on the nature of personal identity. In a 1908 article, "L'Homme, est-il simple, double, ou multiple?" Camille Sabatier asserted that "Man is formed of two cobeings subjected to a coupled existence" (p. 358), and his 1918 book, *Le Duplicisme Humain* was dedicated to working out the implications of this discovery.

Sabatier focused on "the undeniable sentiment of the Unity of the Self" and heaped scorn on Wigan (whom he called Wagan) and others whose theories of mental duality failed to take into account the coordinating role of this self (1918, 178–79). Sabatier admitted that, "made of two cobeings, both *apersonal*, the *personal* Self naturally has the same faculties as each of its components." Nevertheless, the whole was greater than the sum of its parts, finding an overarching unity *not* in the static dual structure from which it sprang but via an endless dynamic process of coordination: "In the end, it is precisely from this definitive coordination, from this *conjugation* of the two coentities [co-êtrales] whereby a lasting Memory is created, that the sense of the Personality arises and becomes established; while oblivious of themselves, the cobeings underneath the Self continue to carry out the penumbra life of the *Subsconscious*" (1918, 89–90).

In the end, though, Sabatier rather compromised his novel emphasis on process rather than structure, unable to resist trying to root the "Moi personnel" in one more or less circumscribed part of the brain: the corpus callosum. The metaphysical craving that had set the old Cartesian physiologists off on their search for the seat of the soul seems to have lost less of its force than most writers of the time probably would have liked to admit. Indeed, Sabatier's defense of his decision to designate the corpus callosum the chief site of the self's activities would not have been wholly out of place in one of Descartes' works:

> it is difficult for us to avoid seeing in the Corpus Callosum the seat and the organ of the functions of Coordination . . . which amounts to saying that the Corpus Callosum appears to us like the seat, like the principal organ of the Self. And in support of this judgment there also tells a circumstance of its anatomical constitution: viz. that it belongs neither to the right cobeing nor the left cobeing; but . . . lies equally on one and the other upper rim of the two cerebral hemispheres. (1918, 117)

4-5. Popularization and Further Extensions of the Hypothesis

Brown-Séquard's 1874 argument that parents should train their children to develop *both* sides of their brain (section 4-2), anticipated what was to become a relatively vocal point of view at the turn of the century. In America, James Liberty Tadd, director of the Philadelphia Public School of Industrial Art, promoted ambidextral training as part of his effort to introduce comprehensive "real manual training" into the public school system. In his impressive and lavishly illustrated manual (the cover of the first edition, shown in Figure 3, shows a young girl demonstrating "ambidextrous coordination in four directions"), he declared,

> Biology teaches us that the more the senses are coordinated to work in harmony in the individual, the better. If I work with the right hand, I use the left side of the brain; if I employ the left hand, I use the right side of the brain. . . . I am firmly convinced that the better and firmer the union of each hand with its proper hemisphere of the brain, and the more facility we have of working each together, and also independently, the better the brain and mind and the better the thought, the reason and the imagination will be. (Tadd 1899, 48)

By 1900, according to British osteologist E. Noble Smith, some 2,000 Philadelphian schoolchildren were regularly undergoing training and were said to have become "relatively sharper and more intelligent than others" (Smith 1900, 580).

Although Smith had urged in his 1900 article that Britain follow America's example and adopt ambidextral education methods, it seems that the 1903 founding the the British Ambidextral Culture Society by Belfast grammar school principal John Jackson owed little, if anything, to Tadd's methods (see Harris 1980 for a discussion of its origins). John Jackson (not to be confused with the neurologist John Hughlings Jackson) had begun his career as an educational reformer with a vigorous campaign in the 1890s for the introduction of a "vertical" style of writing in the schools. His pamphlet, *Upright Versus Sloping Writing* (Jackson 1895), attacked the sloping style of penmanship for "a tithe of evils": twisted spines, cramped hands, compressed chests, unequal strain on the eyes. In his later call for teaching ambidextral writing in the schools, he always had in mind *vertical* ambidextral writing.

Denying any insurmountable organicist basis for handedness and speech asymmetry and pointing to Wigan and Holland's old claims

Figure 3. James Liberty Tadd's program for two-brained education: a student illustrating "ambidextrous coordination in four directions." (Source: Tadd 1899, 103. Reprinted with the permission of the Bodleian Library, Oxford University.)

that each hemisphere is a distinct and perfect organ of thought (see, e.g., Jackson 1903–04), the ambidextral culturists denounced the folly that wasted half the educational potential of future generations. As one member of the society, W. A. Hollis, put it, "The time has now arrived when our posterity must utilize to the utmost every cubic line of brain substance; and this can only be done by a system of education which will enforce an equal prominence to both sides of the brain in all intellectual operations" (cited in Jackson 1910, 20). By the systematic teaching of ambidexterity, Jackson and his followers promised,

Britain could look forward to a brave new world of two-handed, two-brained citizens, with untold benefits for health, intelligence, handicrafts, sport, schoolwork, industry, and the military.

> each hand shall be absolutely independent of the other in the production of ANY KIND OF WORK whatever; . . . if required, one hand shall be writing an original letter, and the other shall be playing the piano . . . with no diminution in the power of concentration. (Jackson 1905, 225)

The ambidexterity movement was bitterly attacked in the United States by the Boston physician George Gould and in England by Sir James Crichton-Browne, whom Jackson dubbed "the English Goliath of lopsidedness." In a 1907 lecture, "Dexterity and the Bend Sinister," delivered before the Royal Institution of Great Britain and later at Cambridge University, Crichton-Browne reaffirmed the old evolutionary arguments of the Broca school that "it is by the superior skill of his right hand that man has gotten himself the victory," and "to try to undo his dextral pre-eminence is simply to fly in the face of evolution." The ambidextral movement was "foolish, visionary, and pernicious," and included among its advocates those who "are addicted to vegetarianism, hatlessness, or anti-vaccination and other forms of aberrant belief" (cited in Jackson 1909, 4, 8). In Boston, George Gould took a similar tack. "No attempt [to eliminate handedness] can wholly succeed; none should, and the partial successes produce cripples and awkwards, if not disease and tragedy. The most foolish, impertinent, ignorant, expensive, resultless and maiming fad is that of the ambidexterity mongers" (Gould 1907, 598; cf. Gould 1908).

The shrillness of this attack by official science is quite striking, and says something about the extent to which the post-Broca interpretation of asymmetry had gotten under neurology's skin. When looked at dispassionately, Jackson's arguments for ambidextral cultivation do not seem entirely absurd. That there existed in man a natural tendency towards "unidexterity" (handedness) was not denied, and the Ambidextral Culture Society had no intention of trying to undo "man's dextral pre-eminence." Ambidextral culture was perfectly compatible, Jackson claimed, with "the inherent and universal tendency to give a preferential use to one hand," one simply developed the other hand "to the nearest attainable equality" with its fellow.

> who is Sir James Crichton-Browne that he shall fix the maximum limit to which [the left hand's] . . . dexterity shall go? Has evolution given him a power of attorney and instructed him to say to this

> specific limb, "Thus far shalt thou go and no farther" in the perfecting of thy manual and digital capacities? Is he under the impression that Dame Nature has invested him with plenary powers to dissolve the Ambidextral Culture Society . . . ? (Jackson 1909, 12)

In the end, rightly or wrongly, official science prevailed, and ambidexterity advocates were forced to retreat to the margins of scientific discourse where they lingered on in various more or less sensationalist and cranky guises. One finds, for example, in the 1920s, a certain writer in the United States claiming to have brought his bimanual powers

> to such a pitch as to be able to exist on only three hours sleep out of the twenty-four. Thus from the hour of awakening until 4:30 PM he would spend his time in the ordinary right-handed activities of writing, reading, etc. From 4:30 PM until 5 PM he would perform 'left-handed' exercises, at the conclusion of which he would be ready for his left-handed existence. Until 3 o'clock in the early morning he would spend his time writing mirror fashion with the left hand or reading with the aid of a mirror and a table lamp. . . . By thus employing alternately as it were the left and right hemispheres, the writer claims to have twice the capacity for work of a completely right-handed individual. (Critchley 1928, 25–26)

As the case for duality of mind became increasingly popularized, the double brain's versatility as an explanatory resource seem to have grown correspondingly. Brain duality began to be invoked to account for a whole grab-bag of everyday psychological curiosities: reverie, castle building, counting steps absentmindedly, chess playing, struggles with temptation, the half identification with their roles by actors (Hall & Hartwell 1884). Wigan and Jensen's old theory explaining paramnesia in terms of discordant perceptions between the two hemispheres continued to find favor until the end of the century, especially in England; Maudsley was among its more influential advocates (Maudsley 1889, 187). It also was felt widely that some theory of hemisphere independent action was required to account for certain features of dreaming. One author, for example, pointed to cases where the dreamer creates a drama in which his whole conscious being is made to play a part "and yet who has to follow the developments of the story in . . . striving and baffled ignorance," even though these developments are all his own invention (Greenwood 1892, 177). Another drew attention to those situations in which one dreams and yet knows that one dreams, "as though one half of the mind watched the other half from a more or less external and separate standing ground" ("Dreams" 1892, 190). Still other authors were more impressed with cases in which a dreamer carries on a long conversation with another person and two distinct trains of thought are evolved (Bastian 1880, 492) or in which "the dreamer makes his own points and solves his own riddle" (Maudsley 1895, 220).

In America, R. C. Word, managing editor of *The Southern Medical Record* and professor of physiology at Southern Medical College in Georgia, was inspired by Brown-Séquard's (1874a) American lecture on brain duality. Word proposed that the duality of the brain might be able to account for "certain mysterious phenomena . . . which hitherto have baffled every effort at explanation": mind-reading and spiritualistic slate-writing. He felt there was good reason to suppose that "under certain peculiar circumstances one side of the brain may converse with the other side" (Word 1888, 85).

For a long time, Word observed, philosophers had divided human mental life into two distinct sides (conscious/unconscious, voluntary/involuntary, etc.) but "no one, so far as I am aware, has thought of their possible dependence upon the two separate and independent brains—the one ordinarily passive, the other active; the one ordinarily automatic, the other active or conscious." Mind-readers, mediumistic slate-writers, and people who were easily mesmerized all had a gift for focusing mental energy within the passive side of their brains. Drawing upon the language of that peculiarly American interpretation of the idea of animal magnetism, so-called electro-biology, Word declared that one might consider all such people constitutionally "electro-negative." "In a manner akin to electric induction," all "electro-negative" individuals were extremely susceptible to influences and mental impressions emanating from an "electro-positive" outside party; i.e., someone whose mental energies were focused in the *active* side of his brain. This explained, according to Word, the mesmeric *rapport* and the phenomenon of mind-reading. The case of spirit slate-writing was slightly more complicated.[3] Word argued that

> it is possible, in certain exceptional instances, for one side of the brain to be electro-positive and the other side electro-negative in

[3] By *slate-writing*, Word was unclear as to whether he was referring to mediumistic automatic writing or to so-called direct writing, in which letters were supposed mysteriously to appear on a designated surface without any physical assistance from the medium. I am indebted to Carlos Alvarado's paper (1986) for drawing my attention to this ambiguity and making the distinction between the two types of writing clear to me.

> the same individual. Under these circumstances the link or connection which ordinarily ties the two brains together is in some mysterious way severed or, for the time being deprived of its co-ordinating influence. . . . In this condition the electro-positive side may ask questions which may be automatically answered by the electro-negative side. Herein we find an explanation of what is called slate-writing, practiced by what are called slate-writing mediums or spirit-writers. (Word 1888, 87)

I have argued that the hypothesis of dual brain functioning was part of medicine's materialistic reply to the various supernaturally inclined interpretations of the abnormal that saw such wide currency in the last decades of the nineteenth century. At this point, I must qualify myself somewhat. In Word's paper, spirit-writing was not "explained away" by reference to the double brain; on the contrary, the double brain was made an accomplice to the workings of the supernatural. Nor can Word's approach be dismissed as an isolated exception; as will be seen, the double brain was appropriated into the controversial borderland fields of the paranormal or occult on a number of occasions during the last decades of the nineteenth century. How can we explain this? Perhaps the aura of the extraordinary surrounding the idea of two thinking brains inside one head was in the end almost as attractive to frank devotees of the marvelous, as it seems to have been to those medical men, devoted in their own way to the elucidation of the marvelous, but bound by philosophical and professional obligation to make it conform to the "facts-and-dust" laws of nineteenth-century neurological reality.

CHAPTER FIVE

Left-Brain versus Right-Brain Selves and the Problem of the Corpus Callosum

5-1. Two Brains/Two Opposing Personalities?

AT THE BEGINNING of Chapter Four, I mentioned that Broca's asymmetrical localization of language was seen as important new evidence for the theory of hemisphere independent action. At the same time, Broca's work should have suggested that an important revision of Wigan's old theory was in order. So long as the "laws of symmetry" reigned undisturbed in physiology, physicians attempting to account for conflict or doubling in the human personality were free to delegate a "self" indifferently to one or the other hemisphere. After 1865, however, to claim that independent action of a man's two hemispheres could turn him into some sort of Dr. Jekyll and Mr. Hyde, one would have to argue further that Jekyll would tend to focus his personality in the civilized, rational left hemisphere, while Hyde would give vent to his criminal instincts from somewhere in the recesses of the uneducated, evolutionarily backward right hemisphere.

Luys came close to making an argument of this sort when he proposed, in 1879, that asymmetry in the "wrong" direction (i.e., in favor of the right hemisphere) tended to upset the normal dynamic balance between the two sides of the brain and led to pathological independent action. He failed, however, clearly to state his position on the relationship in psychopathology between hemisphere independent action and hemisphere functional differences. This is a surprisingly common failing in the late nineteenth-century duality of mind literature. Although authors made use of the post-Broca data on asymmetry to buttress their case for hemisphere functional independence, the moment the case was considered proved, they tended to launch into arguments about "two selves" in two brains without, for example, seeming at all concerned that both their selves could almost always talk.

How could they have failed to address themselves to such an obvious objection? Perhaps, the reason harks back to Chapter Four's argument that neurophysiological explanations for abnormal phenomena, like double personality, represented part of medical science's ideological arsenal against the varieties of superstitious flimflammery that continued to delude the public mind. If true, this suggests that, in the final analysis, some of the double brain theorists might have been disinclined to equivocate over the nicer details of brain physiology and anatomy should such details threaten a larger argument they had an investment in defending.

It would be a mistake, though, to think that this is the whole story. There seems every reason to think that, in addition to its implicit propaganda function, the double brain also served a significant cognitive function for a profession doing its best to come to terms with some of the wilder anomalies of human experience in a hard-nosed, scientific fashion. It seems likely that the notion of a link between split or duplex consciousness and the double brain was partly so attractive because it permitted medical men to address themselves with good conscience to a provocative *psychological* theory, that man really has "two minds," while simultaneously seeming to eschew psychological categories of explanation in accordance with the somatic avowals of their profession. The double brain, in short, allowed these French alienists to act as "crypto-psychologists," harnessing the language of the brain to lively discussions about the subjective world of the divided person. So seductive was the metaphor, so neat the isomorphic mapping of selves onto brain-halves that absence of, or even blatant contradictions in, evidence on the physiological front seem largely to have failed to deter.

At the same time, there *were* a few exceptions to the general trend, a few discussions of divided consciousness or double personality that were backed by a certain amount of physiological evidence seemingly consistent with the nineteenth-century post-Broca data on hemisphere differences. In this chapter, which continues along the theme of Chapter Four, some attention will be given to salient examples of this countercurrent in the double brain literature. I then take a broad look at contemporary objections to the post-Broca case for a link between duplex personality and the double brain, with particular attention to the problem of the corpus callosum.

5-2. Frederic Myers on Brain Duality and the "Subliminal Self"

Frederic Myers was one of the founders, in 1882, of the British Society for Psychical Research. This society was devoted to the classification and study of the so-called unclassified residium of alleged exceptional psychical events: telepathy, clairvoyance, altered states of consciousness, hauntings and apparitions, apparent communications from the "spirit world," etc. Myers, a classicist and minor poet, had been led to take a leading role in the society's activities because, to his mind, the apparent inexplicability of psychical phenomena in terms of known laws of matter and motion challenged the stern materialism of Victorian science, especially as presented by the aggressive young field of psycho-physiology. More important, in Myers' view, these phenomena opened up new grounds for believing that the human personality might survive physical death. If the immortality of the human soul could be scientifically confirmed, then one would be in a position to give a resounding affirmative reply to what Myers considered the "central question" of life: *Is the universe friendly* (James 1973, 211)?

In his effort to establish human immortality as a "structural fact of the Universe," Myers not only did tireless research into psychical phenomena but also began, over a period of about fifteen years, to develop a theory of the human subconscious mind or, as he would ultimately call it, the subliminal self. This theory was intended to offer a coherent explanation for all manner of paranormal psychological events and simultaneously to set up a framework in which the survival of the human personality after death was made to seem quite plausible. As Myers saw it, normal, everyday consciousness represented only a tiny, focused portion of a total self whose consciousness extended far beyond the physical or terrene world. Various extraordinary events of psychological life, then, were to be understood as "subliminal uprushes" from this enveloping mother-consciousness, which, emerging as they did from a nonphysical realm into the "supraliminal" or terrestrial consciousness of mundane (physical) life, naturally eluded all explanation in terms of physical laws (see Myers 1915).

Given the limitations of existence on the physical plane, the subliminal self was capable of communicating with normal consciousness only indirectly, and imperfectly, manifesting its presence through "*makeshift* usage" of the human brain (Myers 1898–99, 386). In 1885, Myers published what William James called a "highly important" article (James 1890, I:400) in which he argued that this secondary self relied primarily upon the right hemisphere for its activities. This was because it could "appropriate the energies" of that side of the brain more readily than those of the left hemisphere, "which is more immediately at the disposal of the waking mind" (Myers 1885, 43).

Discussing the "automatic writing" produced by subjects using a planchette, the forerunner of today's ouija board, Myers quoted from a communication by the Reverend P. H. Newnham, who had "very urgently" called attention to the "low *moral* character" of the personality that appeared in the course of the planchette writing. Stressing that this personality was "distinct from the *conscious* intelligence and character of either of the two parties engaged in the experiments" (pp. 21–22), Newnham had gone on to ask,

> 1. Is this "third intelligence" analogous to the "dual state," the existence of which, in a few extreme and most interesting cases, is now well established? Is there a latent potentiality of a "dual state" existing in *every* brain . . . ?
>
> 2. Is it possible that this "dual state" arises from the fact that we habitually use only one of the cerebral hemispheres for the transaction of our ordinary brain-work; leaving the other, so to speak, untrained and undisciplined? and so, if the untrained side of the brain be suddenly stimulated to action, its behaviour is apt to resemble that of a child, whose education has not been properly attended to [and for whom] . . . morality is simply a matter of convenience. (Myers 1885, 22)

Was there any physiological evidence in favor of Newnham's proposal? Myers believed that there was. He began by acknowledging his indebtedness to papers by Pitres and Claude Bernard on the neurophysiology of agraphia and aphasia, to the "much pertinent matter" in Bérillon's 1884 thesis on the double brain (see Chapter Six), and especially to the work of Hughlings Jackson on affections of speech and the "evolution and dissolution" of the nervous system (for Jackson, see Chapter Seven). Myers then pointed out that the writing produced automatically by subjects using a planchette was filled with errors remarkably similar to those found in the work of aphasic patients, presumed by most pathologists to be relying upon their "partially untrained" right hemisphere. In both cases, words came out transposed, backwards, in mirror-image, or otherwise distorted. The fact, as the embarrassed spiritualists were forced to admit, that "planchette . . . is sadly given to swear" was seen by Myers as additional evidence for particular right-lobe involvement in planchette writing. He turned here to Hughlings Jackson's demonstration that otherwise speechless patients often retained an ability to curse as a means of "automatic" emotional release, an action Jackson believed to be mediated by the right hemisphere.

In many ways, Myers followed the strict post-Broca line in regard to the right hemisphere, shaking his head gloomily over the "inferior evolution" of that side of the brain, with its "traces of that savage ancestry which forms the sombre background of the refinements and felicities of civilized man" ("Report on the general meeting" 1886, 228). At the same time, it is important to realize that, for Myers, the right brain's coarse "supraliminal" or terrestrial personality in no way reflected unfavorably upon the excellent character of the subliminal self, forced to *utilize* that side of the brain but not physically dependent upon it. This is why he *also* could suggest that the right brain-half might play a special role in certain forms of genius (Myers 1892) and might even serve as a sort of neurophysiological pipeline to the spirit world (Myers 1898–99). His view of the right hemisphere as a kind of meeting-ground for the living and the dead accounts for his "curious suggestion," in the puzzled words of one historian, that "the errors and confusions which deceased persons fall into when endeavouring to control a medium's brain may sometimes resemble the aphasias and agraphias which may be produced by brain damage" (Gauld 1968, 292n). As Myers explained it,

> Now, I think it possible that our left hemispheres, having been more constantly used than our right hemispheres, may be more crowded and blocked (so to say) with our already fixed ideas. An external intelligence, wishing to use my brain, might find it convenient to leave alone those more educated but also more preoccupied tracts, and to use the less elaborated, but less engrossed, mechanism of my right hemisphere. (Myers 1898–99, 386)

In the United States, William James was sufficiently impressed by Myers' automatic writing theory to suggest in his *Principles of Psychology* that, in at least some cases of personality dissociation, "the systems thrown out of gear with each other are contained one in the right and the other in the left hemisphere" (James 1890, I:399–400). In floating this idea, James referred not only to Myers' work, but to Maudsley's "instructive" 1889 essay in *Mind*, and Luys' 1888 article in *l'Encéphale*. He had earlier (p. 390) called attention to Wigan's book but does not seem to have been terribly impressed by it.[1]

In contrast to James, Pierre Janet, in his 1889 dissertation, *L'Automatisme Psychologique*, was highly critical of Myers' attempt to implicate the double brain in mediumistic phenomena. To begin, there was no good reason to suppose that, just because spirit writing tended

[1] Although it is not cited in *Principles*, Eugene Taylor has recently shown that James was also familiar with Gabriel Descourtis' 1882 dissertation, *Du fractionnement des opérations cérébrales et en particulier de leur dédoublement dans les psychopathies*. One of Luys' students, Descourtis became one of the better-known and more active advocates of the double brain hypothesis. Taylor reports that James checked out and marked up a Harvard College Library edition of Descourtis' dissertation in preparation for his own 1896 Lowell Lectures on "Exceptional Mental States." James also personally owned a special edition of this work (Taylor 1984, 201, 146; see as well the review of Descourtis' thesis in *L'Encéphale* praising it as an exceptional contribution to a difficult subject that few young doctors would have had the courage to tackle [Duval 1882]).

to be garbled and awkward, it must be a product of the right side of the brain: "The writing is more unskilled because it occurs in novel conditions, without the subject seeing the paper, without his using visual images, etc." Similarly, Janet was not convinced that the immoral and obscene language frequently encountered in automatic writing pointed to any necessary connection with the right side of the brain. "What, oaths, obscenities and insults can only come from the right hemisphere?" Such reasoning, he felt, recalled Gaëtan Delaunay's wild style of speculation, in which dreams produced by the right hemisphere were said always to be absurd and those produced by the left always rational (see Chapter Three). The claim that automatic writing often came out backward or in mirror-image was more interesting, since such distortions were not rare in the writing of left-handed children and aphasic patients. Nevertheless, Janet asserted that he himself had never observed the phenomenon in samples of mediumistic writing and thought it too uncommon to serve as the basis for a general theory. Finally, he focused on the fact that Myers' argument for a link between the right hemisphere and mediumistic automatic writing hinged on the supposition that recovery of speech in cases of left hemisphere damage always depended on action of the right hemisphere. Was this absolutely certain?

> M. Charcot himself, through his theory of the different sensorial types of language, has indicated to us another possible hypothesis. The patient can re-establish his language [capacity] by developing another faculty of representation, the faculty of auditory representation for example, to substitute for the obliteration of visual images; and one would witness then a new education of language or of writing capable of showing all the phases that M. Myers has indicated, without the right brain having to intervene more particularly than usual.
>
> It is a difference of this type, psychological rather than anatomical, which seems to exist between the diverse simultaneous languages of the medium. (Janet 1889, 417–18)

Myers' reply to Janet, published soon afterwards in the proceedings of the Society for Psychical Research, was unrepentent. The British researcher urged his French colleague to keep an open mind on the automatic writing question and not to "lose the true independence of each experiment by falling prematurely under the *power of suggestion* of any one theory" (Myers 1889–90). Myers also remarked pointedly that, so far as he could see, much of Janet's *own* work actually seemed to offer significant support to the various theories in the literature linking brain duality to double personality and alterations in consciousness. "I can scarcely understand why M. Janet disapproves of this view" (p. 189). There was, for example, Janet's study of a certain somnambulistic patient named Rose. As Myers noted,

> [Janet] found that,—whereas in ordinary life and in all previous somnambulisms she was wholly anaesthetic,—yet both in this new somnambulism and in those blank periods of life she was only hemianaesthetic,—having recovered tactile and muscular sensibility on the right side. . . . [These findings] seem to me to add confirmation to my own view . . . that alterations in the predominance of one or the other cerebral hemisphere have something to do with these changes of personality, of which automatic writing is now recognised as one of the most instructive manifestations. (Myers 1889–90, 188–89)

5-3. The Strange Case of Louis Vivé

Not long after the publication of Frederic Myers' article on automatic writing, he and his brother, Arthur Myers, became aware of a bizarre and complicated case of multiple personality being studied in France by two physicians, Henri Bourru and Ferdinand Burot at the naval hospital of Rochefort. To the two British investigators, this case offered further evidence in favor of the thesis that at least some forms of divided consciousness were caused by a functional dissociation between the two hemispheres. Louis Vivé (Louis V.) was a young male hysteric with a turbulent childhood history. Left to fend for himself on the streets by a degenerate mother and forced into a life of petty crime to support himself, he was ultimately sent to a reformatory. There his nervous system was first thrown out of balance at the age of fourteen when, working in a vineyard, he narrowly escaped being bitten by a viper. His terror from this episode was said to be such that he began to suffer from hystero-epileptic attacks, each of which left him paralyzed and anaesthetic in different parts of his body, with corresponding alterations of memory and personality. Over the years, he had passed through a series of hospitals, finally coming under the care of MM. Bourru and Burot at Rochefort. On admittance, the two doctors noted that their patient was hemiplegic and anaesthetic on the right side of his body, suggesting a "functional" disorder of the left hemisphere. He seemed to have considerable difficulty speaking distinctly, but this did not prevent him from haranguing the hospital staff "with monkey-like impudence," preaching "radicalism in politics and atheism in religion," trying to "fondle" people who pleased him, and generally behaving in a "violent, greedy, and quarrelsome" manner (Myers 1886, 650; also A. T. Myers 1885–86; Bourru & Burot 1886–87).

It was then found, using magnets and metals, that the patient's hemiplegia and anaesthesia could be "transferred" from the right side of his body to the left, a procedure that will be described in detail in the next chapter. A remarkable transformation in personality occurred. "The restless insolence, the savage impulsiveness have wholly disappeared. The patient is now gentle, respectful, and modest. He can speak clearly now, but he only speaks when he is spoken to. If he is asked his views on religion and politics, he prefers to leave such matters to wiser heads than his own" (Myers 1886, 650). Other magnetic and metallic manipulations (on the neck, the top of the head, the right thigh, etc.) resulted in more subtle variations of character, but the case was made that these were simply intermediate forms of the basic personality dichotomy observed in conjunction with the left and right hemiplegia, respectively (see the Table 3 clarifying the phenomena associated with six of Louis V.'s states; eventually eight states would be discovered).

Given this, it was not unreasonable to suppose that Louis V.'s personality changes were caused by some sort of dual, alternating action of the brain's two hemispheres. In the full-length book they eventually wrote about their patient, *Variations de la Personnalité*, Bourru and Burot declared,

> If one reduces this case to the patient's two states of principal consciousness, one can assuredly see in it a most striking example of the dualism of the nervous centers. . . . It seems that Louis V. . . . directed by the right hemisphere of his brain, is a different individual from the Louis V. . . . who corresponds to the left hemisphere. The right-sided paralysis [inhibiting the left hemisphere] only lets the violent and brutal aspects of his character appear; the left-sided paralysis [inhibiting his right hemisphere] transforms him into a quiet and well-bred lad. (Bourru & Burot 1888, 127)

The assumption was that Louis' childhood traumas had knocked his brain out of kilter, plunging the two brain-halves into a state of disequilibrium and pathological independence. This view was strongly reinforced by certain experiments showing that more or less equal functioning could be temporarily restored to both sides of Louis' brain by applying a magnet to his head or putting him in an electric bath. When this was done, the patient would abruptly adopt all the mannerisms of a young child, set back in memory, character, and ability

Table 3. Louis V's six states at Rochefort, 1885

	1	2	3	4	5	6
Paralysis	Right hemiplegia.	Left hemiplegia, affecting face.	Left hemiplegia, not affecting face.	Paraplegia.	Paresis of left leg.	No paralysis.
Anaesthesia	Right side.	Left side.	Left side.	Of lower half.	Of left leg.	Hyperaesthesia of left leg.
Character	Violent.	Quiet.	Quiet.	Shy, speech childish; tailor.	Obedient; boyish.	Respectable.
Education	Fair.	Good.	Good.	Bad.	Good.	Moderate.
Esthésiogènes	—	Steel on right arm.	Magnet, &c., on right arm.	Magnet on back of neck.	Magnet on top of head.	Soft iron on right thigh.
Dynamometer	Rt. = 0 Lt. = 80lbs.	Rt. = 80lbs. Lt. = 0	Rt. = 80lbs. Lt. = 0	Rt. = 45lbs. Lt. = 44lbs.	Rt. = 40lbs. Lt. = 44lbs.	Rt. = 66lbs. Lt. = 70lbs.

SOURCE: A. Myers, 1885–86.

to the days just before the viper incident and the onset of the hysteria; i.e., just before, as Myers put it, "his *Wesen* was . . . cloven in twain" (1886, 652).

In his time, Louis V. was quite a cause célébre, and was discussed by Ribot, Camuset, Richer, Legrand du Saulle, Jules Voisin, William James, and doubtless others as well. His extreme sensitivity, even "at a distance," to metals, magnets, and as was soon discovered, medicines and toxic substances in sealed tubes brought him still greater fame, as he became a prime subject in a series of works on the borders of mainstream neurology, dealing with the alleged capacity of the human nervous "fluid" to interact with cosmic forces and radiations (Bourru & Burot 1887; Berjon 1886; Alliot 1886). Given this notoriety, it is a bit surprising that his case is so little known today; if nothing else, it merits recognition as probably the most extensively studied example of male hysteria in the nineteenth century. Most of the later nineteenth-century commentators seem to have accepted, or at least did not explicitly challenge, the "double brain" interpretation of Louis V.'s personality changes championed by Frederic Myers and, with slightly more equivocations, the patient's own doctors. One prominent exception, again, was Pierre Janet, whose 1899 study comparing hysterical symptoms associated with left and right hemiplegia (see Chapter Three) had given no credence to the idea that intelligence was particularly associated with left hemisphere functioning. This, Janet felt, argued decidedly against Myers' claim that a coarse, uneducated personality will inevitably be associated with right-brain action and a more refined personality with left-brain action (Janet 1899, 855).

5-4. Lombroso on Mediumistic Consciousness and the Right Hemisphere

Like Myers, the Italian criminologist and psychiatrist Cesare Lombroso was a whole-hearted convert to the post-Broca view of the right hemisphere as evolutionarily retarded, neurotic, and generally pernicious (cf. the discussion of Lombroso's work on "born criminals" and left-handedness in Chapter Two, section 2-6). Yet he also had no difficulty agreeing with Myers that the right side of the brain might play a special role in the production of certain types of paranormal phenomena.[2] His position on this issue is in many ways characteristic of nineteenth-century medical assumptions of an essential or under-

[2] I am indebted to Carlos Alvarado's 1986 paper "Early speculations about psychic phenomena and the brain's hemispheres' for bringing this aspect of Lombroso's thinking to my attention and guiding me to the relevant primary sources.

lying symmetry between the pathological and the supernormal. There was a fundamental link, Lombroso believed, between the morbid antics of the hysteric and the inspired abilities of the medium, the genuinely supernormal nature of which he had no aim of discrediting. He compared this putative relationship to the intimate affinity that, in his view, existed between the "creative frenzy" of genius and the "psycho-epileptic paroxysm" of the madman (Lombroso 1909, 119).

The "abnormal state" of the medium was at least partly proved, according to Lombroso, by the fact that many of the "spiritistic phenomena" she produced (apparitions, rappings, etc.) seemed to take their point of origin from the left side of her body (cf. the discussion of hysteria, Chapter Six, section 6-1). In a reference to this alleged finding, Lombroso went so far as to speak metaphorically of "the usual spiritistic left-handedness" (1909, 117). He also called attention to the supposedly high incidence of actual left-handedness among mediums, and their tendency in the trance state, as of course discussed by Myers, to write automatically in mirror-image or reversed. All this seemed "to indicate the increased participation of the right lobe of the brain in mediumistic states, as occurs with hypnotised persons, and would explain the concomitant unconsciousness" (Lombroso 1908, 378).

Some of the most novel evidence in support of this hypothesis had come, Lombroso believed, from the recent investigations by the Italian psychiatrist Enrico Morselli of the medium Eusapia Palladino, whose talents Lombroso himself had studied and found quite impressive. Born into charity in Murge, Italy, in 1854, Palladino had been subject since childhood to various mediumistic or hallucinatory manifestations: hearing mysterious raps coming from pieces of furniture on which she was leaning, having her bedclothes stripped from her at night by an invisible hand, seeing ghosts or apparitions (Lombroso 1909, 39–40). It was not long before her way with the spirits was discovered, and she settled down to a life as a professional medium. From a clinical point of view, it was clear to Lombroso that this woman possessed a number of traits characteristic of the hysterical or epileptic character: these included a pronounced asymmetry of the cranium in favor of the right side (suggestive of greater right-brain development), asymmetry in arterial pressure in favor of the left side of the body, marked tactual left-handedness coexisting with a general weakness or paresis in the limbs on the left side (pp. 105–107).

In discussing Palladino, Lombroso called attention to Morselli's habit of using a dynamometer to measure the medium's hand grip strength, both before and after a séance, on the working assumption that any changes might in some sense reflect changes in her supply of "psychic force." Morselli had been particularly struck by differences in energy loss between the medium's right and left hands, the left hand always suffering markedly more after a séance than the right. In one particularly notable session, held in 1901, Palladino had actually *gained* strength in her right hand, relative to the reading taken before the session, while losing it in her left (usually dominant) hand. At the same time, Morselli (usually right-handed) had become temporarily left-handed, losing significant strength in his right hand and gaining some in his left, relative to the preséance readings (Alvarado 1986, 20–21). It was quite common, Lombroso believed, for the abnormalities of the medium to be transferred to one or more of the sitters during a séance. Morselli's transitory left-handedness during that 1901 session nicely confirmed, in his view, the provocative hypothesis put forward by E. Audenino: that normally right-handed individuals may turn briefly left-handed during abnormal or reduced states of consciousness, such as trance, somnambulism, hysterical double consciousness, intoxication, etc. (Lombroso 1908, 378; see Audenino 1908, 301).

Certainly, Lombroso admitted that the hypothesis linking mediumistic consciousness to right hemisphere action was not without its limitations. For example, though useful for explaining automatic writing and other mediumistic automatisms, "it yet does not serve in the case of those [mediums] who write at one and the same time two communications, or even three. My readers will recall, as making against this hypothesis, the simultaneity and contemporaneity of certain phenomena in mediumistic seances." And Lombroso singled out for attention the medium Mansfield, who had the impressive capacity of writing "simultaneously with both hands in two different languages, while speaking other subjects with persons present" (Lombroso 1909, 158, 160). Clearly, there were still more things on heaven and earth than the two lobes of the brain alone could encompass.

5-5. Lewis Bruce and the "Welsh" Case

Toward the end of the century, the Scottish physician Lewis Bruce would revive lagging interest in the argument for hemisphere independent action with his report of a patient suffering from a variety of double personality. The patient alternated between periods in which he spoke English and was right-handed and periods in which he spoke a rather incoherent form of Welsh and only used his left hand.

> When in the English stage, right-handed and presumably using the left cerebral hemisphere, he is the subject of chronic mania. He speaks in English, but understands and will converse in Welsh. He is restless, destructive, thievish, and is constantly playing practical jokes on his fellow patients. He exhibits a fair amount of intelligence. He writes, draws pictures of ships, and relates incidents in his past life. . . . His memory is, however, a blank to anything that occurred during the Welsh stage. . . .
>
> When in the Welsh stage, left-handed and presumably using the right cerebral hemisphere, he is the subject of dementia.
>
> His speech is almost unintelligible; I have got Welsh patients to act as interpreters; they tell me that they cannot understand much of what he says, but what they do understand is spoken in the Welsh language. . . . He has no idea of English, does not understand the simplest remarks, but stares vacantly at the questioner. . . .
>
> He sits doubled up in a chair for hours . . . is now shy and suspicious, does not recognize the doctors or attendants, and appears to be constantly on the lookout for unseen danger. (Bruce 1895, 60–62)

Bruce placed a good deal of emphasis on the shifts in motor capability associated with his patient's transition from one personality to the other and, in a later publication, harshly criticized the "many cases of spurious duality" published in the medical journals "as cases of dual brain action" without an "iota of proof" (Bruce 1897). An anonymous reviewer of the Bruce case agreed that, "on reflection, the main symptom of uncontrovertible value in this case seems to be the use of the hands" ("Notes of a Case of Dual Brain Action" 1896, 217). Another commentator noted that the patient's speech was not altogether lost during the periods when he was presumed to be dominated by his right hemisphere, but was limited to Welsh, his native tongue. "This, as far as it goes, would indicate that either the left brain was not altogether inactive, or that there were bilateral speech centers" (Bannister, cited in Kiernan 1896, 32). Certainly, as we have seen (Chapter Two), Broca had supposed from the very beginning that the right hemisphere possessed a certain amount of latent language capacity, which could express itself under certain, usually pathological, conditions.

Even though, in his case of dual brain action, both brain-halves seemed able to "talk" to a limited extent, Bruce had been greatly impressed by the pronounced difference in "mental power" between them, the right standing "at a much lower level than the left." Was this due, he asked (p. 64), "to the unequal ravages of disease or to the unequal development of education?" In other words, had madness *caused* the right-brain's dementia, animal wariness, and incapacity to respond in a human manner to other people; or had it simply *set free* that hemisphere's innate brutishness, normally damped down by the controlling power of the educated left hemisphere? We know what Broca is likely to have said, and Bruce makes it clear that he was inclined to favor the same alternative.

Two years later, Bruce extended the argument for dual brain action to certain phenomena associated with epilepsy. It seemed to him that "the peculiar mental state known as epileptic automatism might be due in some instances to the paralysing effect of the seizure in one hemisphere [the left], while the other hemisphere was capable of carrying on a separate mental existence" (Bruce 1897, 115–16). He described two patients who, following right-sided epileptic attacks, gave evidence of heightened right hemisphere activity (unnatural left-handedness or fugitive right-sided paralysis) and simultaneously underwent significant personality changes. The expression of the first "was completely altered, she recognised neither the nurse nor myself, and she jabbered in an angry way like a monkey." In short, "her mental state, never very high, was then practically that of an uneducated savage." The other patient became extremely confused and disoriented: "He does not recognise the attendants who are usually about him and when spoken to he makes peculiar noises but cannot speak. He is extremely stubborn, resists if held, and all his actions show that although conscious he is not cognisant of his surroundings" (pp. 116–18).

5-6. Bleuler on "Unilateral Delirium"

Not long after Bruce published his views on the double brain, the Swiss psychiatrist Eugen Bleuler, who would soon become internationally renowned for his dynamically oriented studies on schizophrenia, also looked to the functional duality of the brain to account for a remarkable case of unilateral delirium he had had the opportunity to observe (Bleuler 1902–03; for an English-language summary, see also Meyer 1904b). The patient, suffering from general paralysis (insanity resulting from syphilis) had lain in his bed at Bleuler's psychiatric hospital in Zurich, perfectly quiet and relaxed on the left side of his body but in a state of violent delirium on the right side. The right hand grabbed at ropes, chopped things with an axe, sowed seeds, and slung away unwanted invisible objects with great vigor. Sometimes it would seize hold of the blankets or the pillow and try to yank them away, and once it upset the patient's dinner. When this happened, the left hand readjusted the bedclothes, wiped the patient's mouth, and gave every appearance of remaining in contact with reality (Bleuler 1902–03, 361). The consciousness associated with the delirious right-handed activities of the patient had full command of language; the rational consciousness corresponding to the patient's left-handed activities seemed occasionally to be able to speak but was considerably more limited in this respect (p. 366). What was one to make of this? Bleuler did not hesitate to volunteer the following interpretations:

> We know that one brain-half is sufficient to represent a psychical personality. [Bleuler here offers examples of individuals who had lost half a brain without any discernible psychic disturbances.] . . . Now if the functioning of a single hemisphere can represent a person as soon as the other is destroyed, it is also possible that each hemisphere simultaneously possesses a distinct [einen anderen] complex of activity, from which each may represent a personality, when the functional connection of the two hemispheres is interrupted. (p. 366)

Although Bleuler had studied in Paris under Charcot and Valentin Magnan in the early 1880s, a time when interest in the role of the double brain in mental disorder was perhaps at its apex, he rather revealingly made no reference in his 1902 paper to any of the older French literature on this topic. The reputation of that literature had not fared well during the last years of the century, as Bleuler must have known all too well (see Chapter Nine; also Chapter Six, section 6-7). One finds the Swiss alienist stressing instead the extent to which his case of unilateral delirium might be profitably compared with the recent case of unilateral right-handed apraxia (loss of capacity to perform skilled movements) that had been exhaustively described by the German neurologist Hugo Liepmann and judged by the latter to have been caused by a lesion in the corpus callosum (Liepmann 1900; see section 5-8).[3]

[3] Bleuler's report also has at least a surface resemblance to a case of hysteria described in 1889 by William James. This case seemed too provocative not to be slipped into this study somewhere, even though James, rightly perhaps, did not explicitly interpret it in terms of the functional duality of the brain. The patient in question, a young American woman by the name of Anna Winsor, began at the age of nineteen to succumb to periodic spells of apparent unilateral delirium, tearing at her bedclothes and hair with her left hand, while the right hand tried to stop her from doing violence to herself by seizing the left hand and holding it still. It was not long before the patient—or, at any rate, the speaking consciousness—failed to recognize her right arm as her own but regarded it as a foreign object, "Old Stump" she called it, with its own intelligence. According to James, "When her delirium is at its height, as well as at all other times, her right hand is rational, asking and answering questions in writing; giving directions;

5-7. Objections to the Brain Duality Hypothesis and the Problem of the Corpus Callosum

Although, doubtless, all the reports reviewed in this chapter are vulnerable to criticism on various counts, at least they represent an honest attempt to correlate Wigan's old argument for duality of mind with a certain amount of physiological evidence and to cast the argument within a post-Broca conceptual framework. Again, though, the cases here are exceptional. Bruce was not wrong when he complained that the great majority of supposed examples of hemisphere functional independence were presented without an iota of physiological proof.

Ultimately, the tendency of advocates to exaggerate the explanatory power of the double brain began to tell against them. In Germany, Wilhelm Griesinger had been an early critic of Wigan and Holland's speculations, which he felt were "wanting in sufficient proof" (Griesinger 1861, 25–26), and his authority continued to be invoked through the end of the century. Théodule Ribot spoke disdainfully of those authors who indulged in "a kind of psychological Manichaeism," claiming that "cerebral dualism suffices to explain every discrepancy in the mind, from simple hesitation between two resolves, to the complete duplication of personality." Although he believed himself in the "*relative independence* of the two cerebral hemispheres," he pointed out that "contradictions in the personality . . . are not oppositions in space (from one hemisphere to the other), but oppositions *in time*. They are—to use a favourite expression of Lewes—successive 'attitudes' of the ego" (Ribot 1891, 109, 111). Alfred Binet, who developed a theory of hysteria as a form of double personality, regarded the double brain hypothesis as "an odd conception" that, he was glad to say, had been decisively refuted by such critics as Ribot (Binet 1891, 38n). Pierre Janet failed to find much explanatory value in the idea. "In fact," he said, "we have, all of us, two brains, and we are neither madmen, nor somnambulists, nor mediums" (1889, 414). Even Bourru and Burot (1888), who had reported on Louis V., recognized the limitations of the brain duality thesis in explaining cases of alternation like Félida, as well as the artificial provoking of personality in hypnosis, where the phenomena were seen equally on both sides of the body. And, in England, Charles Mercier put it quite unequivocably that "an

trying to prevent her tearing her clothes; when she pulls out her hair it seizes and holds her left hand. When she is asleep, it carries on conversation the same; writes poetry; never sleeps; acts the part of a nurse as far as it can; pulls the bedclothes over the patient if it can reach them, when uncovered; raps on the head-board to awaken her mother (who always sleeps in the same room) if anything occurs, as spasms, etc." (James 1889, 553).

hypothesis of so wild a character, and so utterly unsupported by evidence, does not merit serious discussion" (Mercier 1901, 509).

The duality of mind thesis was also criticized as it became increasingly clear that "we are not bound to the number two in considering the mass of conscious, subconscious, and unconscious states that may succeed one another in our body" (Rosse 1892, 187). At an 1885 Congress at Antwerp on Psychiatry and Neuropathology, Dr. Verriest presented a patient suffering from "double consciousness" similar to that seen in Azam's patient, Félida. He then revealed that this woman could also be thrown into a *third* state of consciousness through "hypnotic passes." "In such a case as this," he asked, "what becomes of Luys' hypothesis of the functional alternation of the two cerebral hemispheres?" ("Congress of Psychiatry and Neuro-Pathology at Antwerp" 1885–86, 621–22).

Finally, doubt was thrown upon the argument for hemisphere independent action, particularly in cases of personality pathology, because of the great uncertainty surrounding the functions of the corpus callosum, the great bundle of fibers joining the two sides of the brain. Most advocates of duality of mind believed that the healthy person experienced a sense of unified consciousness, in spite of the functional independence of his hemispheres, because the corpus callosum provided a bridge of conduction between the two psychic realms. In Germany, Theodor Meynert believed, on rather shaky anatomical grounds, that the main fibers of the corpus callosum united corresponding convolutions in the two hemispheres. William Broadbent in England endorsed the Meynert model but stressed that it particularly applied to those convolutions concerned with sensory activity. At this time, it was widely felt that sensory stimulation was an essential precondition for consciousness (see, e.g., Bastian 1880, 483–89). Broadbent's countryman, D. J. Hamilton (1884–85) also argued for the probable psychic functions of the corpus callosum, although he had challenged the Meynert-Broadbent anatomical schema. On the grounds of sheer size alone, it probably seemed to nineteenth-century neuroanatomists that this fibrous structure, which T. H. Huxley called "the greatest and most sudden modification exhibited by the brain in the whole vertebrate series," could not but play *some* sort of crucial role in higher nervous and mental functioning.

And yet cases were on record in which the corpus callosum had been "found to be entirely wanting, without any mental derangement or deficiency of intellect . . . and without any manifestation of a double personality." William Ireland spoke of the work of Hitzig, "who has studied the question carefully, [and] observes that no well-marked disorder of motion or sensation follows atrophy of the corpus callosum, nor is there any characteristic mental defect attendent upon this lesion" (Ireland 1886, 317–18; for varied cases, see A. Bruce 1889–90; Erb 1885; Ransom 1895). There *was* a school of thought that held that corpus callosum lesions were associated with a dulling of intelligence and even imbecility. D. N. Knox in Glasgow proclaimed himself in favor of this view; however, in virtually all of his fourteen cases, there was damage to other parts of the brain as well (Knox 1875). In the United States, Dr. Miles described a patient with a tumor that seemed almost exclusively restricted to the corpus callosum: "the only symptoms which the patient presented . . . was an exceeding slowness in reply to questions—a marked torpidity. His intellect was not so much disturbed, but it took him about four times as long to answer a question as it did the ordinary healthy person" (Wilder 1883, 510). What this meant about the functions of the corpus callosum was anyone's guess.

In France, Gabriel Descourtis tried, by appealing to common sense, to explain away the fact that loss of the corpus callosum rarely seemed to disturb the unified action of the personality. He pointed out that the brain's two hemispheres possessed the same hereditary dispositions and had lived through the same experiences; no mention was made of the data on functional asymmetry, which implies that in fact their activities and reactions might be consistently different. It was only to be expected, then, Descourtis concluded, that they should normally tend to feel and react in a unified fashion, regardless of whether they were physically joined by commissures or not. In this respect, they could be compared to two twin brothers who had long lived together, or even better, to the case of the Siamese brothers who, though able to manifest their independence, essentially shared a single personality (Descourtis 1890).

For William McDougall in England, the failure of corpus callosum lesions to result in any perceptible doubling of consciousness seemed cause to rejoice. In his view, the materialistic doctrine that the unity of personal consciousness is strictly dependent upon the unity of the brain's material connections was thereby refuted, chalking up a clear victory for "animism," or the belief that the workings of the mind were functionally distinct from those of the body (McDougall 1911, 117). Sir Cyril Burt, reminiscing about the early days of psychology at Oxford, recalled McDougall

> saying more than once that he had tried to bargain with Sherrington . . . that if ever he should be smitten with an incurable disease,

> Sherrington should cut through his corpus callosum. "If the physiologists are right"—and by physiologists, I suppose he meant Sherrington himself—"the result should be a split personality. If I am right," he said, "my consciousness will remain a unitary consciousness." And he seemed to regard that as the most convincing proof of the existence of something like a soul. (cited in Zangwill 1974, 265).

5-8. Unilateral Apraxia and the Problem of the Corpus Callosum: A Second Look

For better or worse, Sherrington never would be given the opportunity to plunge a scalpel into McDougall's brain, so the bet between the two men had to remain unresolved. Nevertheless, it would probably be imprudent to give McDougall a victory by default, because the situation actually was rather more complicated than yet has been indicated—more complicated, too, than McDougall likely realized. It is not strictly true that absolutely *everyone* was particularly confused about the corpus callosum. A school of thought was to take root early in the twentieth century with quite definite views on the functions of this structure. Focused in Germany around the person of Hugo Liepmann, the orientation and theoretical assumptions of this school stand in marked, almost jarring, contrast to those I have so far traced in this chapter. Not only that but, unlike those other traditions, the opinions of this school would prove themselves remarkably robust, surviving the social and cognitive upheavals of the early twentieth century, albeit usually in a derivative, predigested form, and going on to play a significant, if controversial, role in the continuing drama of the corpus callosum up through the 1960s (see Chapter Nine).

The story of the rise and consolidation of the Liepmann school begins in 1900, the year that Liepmann, a close colleague and admirer of Wernicke, completed a detailed report on Herr T., a forty-eight-year-old imperial councilor (Liepmann 1900). Hospitalized in Berlin after being diagnosed as having "mixed aphasia and dementia following apoplexy [stroke]," this man subsequently was transferred to the nearby mental asylum at Dalldorf, where it was decided that he was suffering from "aphasia and mental disorder" (Geistesstörung). In Dalldorf, he first came to Liepmann's attention. Liepmann saw that the patient indeed responded to questions and commands in a thoroughly senseless (tief blödsinnig) fashion and seemed to have lost the capacity both to understand language (Sprachtaubheit) and to interpret and make sense of objects in his environment (Seelenblindheit).

Liepmann noticed though that, almost invariably, the patient relied on his right hand in attempting, more or less vainly, to carry out a task. Since the left hand was not paralyzed, Liepmann saw no real reason why it should be so totally inactive and, virtually on a whim, he ordered that Herr T.'s right hand be impeded, so that he would be forced to carry out tasks with the left. A startling transformation occurred. Even though the patient's *right* hand had behaved in a deeply demented manner, the *left* hand proved itself quite competent at carrying out basic tasks on command. This proved, Liepmann felt, that actually the man had been comprehending the doctors' instructions all along and was neither aphasic nor demented but, in fact, was in possession of a fair level of intelligence. He was, however, strangely incapable of intelligently executing his "intentions" with the right side of his body (see Figures 4 and 5). This behavioral asymmetry was most strikingly revealed when the patient was given several simple tasks requiring coordinated use of both hands:

> He has to brush the doctor's coat. He takes hold of the bottom corner correctly with his left hand, and picks up the brush correctly in the right; but then he repeatedly executes rhythmical movements upwards and backwards, above his right ear. . . .
>
> He has to pour water from a carafe into a glass. The left hand picks up the carafe and begins to pour, while the right hand at the same time conveys the empty glass to his mouth. (Liepmann 1900, 34–35)

Now, incapacity to carry out tasks on command, *without* loss of verbal comprehension or any defects in general motor capability (paresis or paralysis), up to that point, had not been identified and carefully defined as a distinct clinical syndrome, although certain earlier authors such as Hughlings Jackson had called attention to the phenomenon in a general way. Liepmann proposed to call the disorder *apraxia* (from the Greek *prassō*, meaning "to do"), and in order to explain how such a syndrome could come to exist, he turned to Wernicke's "connectionist-associationist" model of brain functioning and aphasic disorder (Chapter Three, section 3-1).

Because apraxia was not an aphasic syndrome as such, Liepmann was forced to elaborate on Wernicke's model and locate the brain defect responsible for the syndrome somewhere other than in the speech centers and their connecting pathways. He pinpointed a spot between the patient's visual and auditory "memories" on the one hand, which, once awakened by the doctor's instructions, permitted him mentally to construct "the intention" in question, and his motor and kinesthetic "memories" on the other, which normally would allow

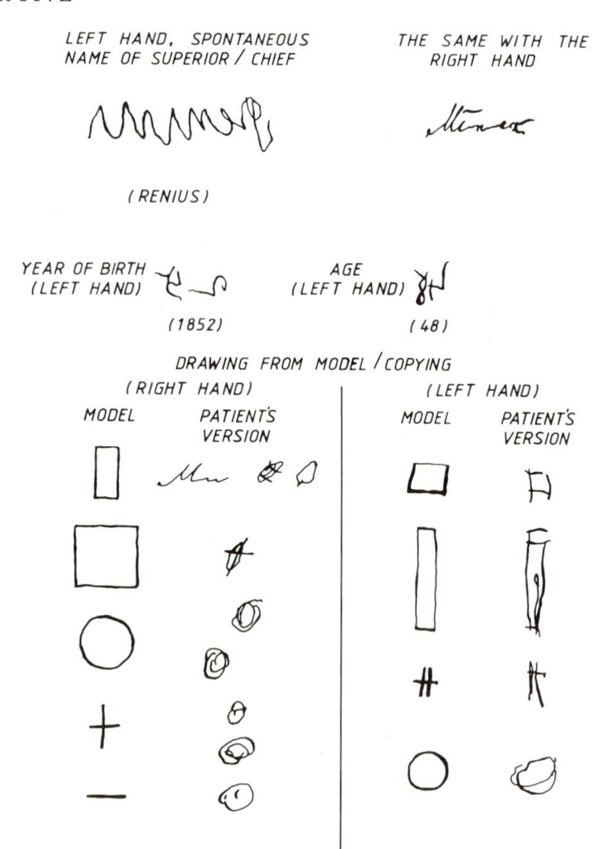

Figure 4. Liepmann's patient, Herr T., carries out writing and copying tasks, alternately using his apraxic, dominant right hand and his subordinate, intelligent left hand. (Source: Redrawn from a sample published in Liepmann 1900, 31.)

him to transform this mental "intention" into the act proper. Since visual and auditory memories were presumably located in the visual and auditory region of the cortex, while kinesthetic memories and motor memories were presumably located in the motor cortex and, perhaps, its surrounding areas (what Wernicke called the sensomotorium region), Liepmann concluded that apraxia resulted anatomically from a lesion of the association pathways between these two major cortical areas (Jeannerod 1985, 75–76). As he would later put it,

> Apraxia may be termed an *aphasia of the extremities.* . . . [The patient] is unable to awaken the motor associations of waving,

Figure 5. Liepmann's patient, Herr T., attempts in vain to carry out tasks calling for coordinated cooperation between his two hands. See Figure 4. (Source: Redrawn from a photograph in Liepmann 1900, 34.)

> clenching the fists, etc. . . . Thus an essential component is missing from his conception of fist-clenching, waving, etc; the words "make a fist" awaken in him something less than in us. . . . We see here the intermediary stages between noncomprehension and physical incapacity, and also how apraxic movements may lead to a misdiagnosis of circumscribed word-deafness. (Liepmann 1905, 25)

So far as the specific case of Herr T. was concerned, however, all this still was not quite a satisfactory explanation. One still had to explain how it was that the patient's *right* hemisphere, which had proved itself quite competent at comprehending instructions and constructing intentions, was unable to substitute for its impotent brother.

Why was it incapable of conveying an appropriate mental image to the patient's left motor region, which could take the place of the symmetrical image stranded in the patient's left sensory region? In reply, Liepmann suggested that, in addition to the left-sided intrahemispheric lesion, the association pathways between the two sides of Herr T.'s brain also must have been disconnected; i.e., there must have been a lesion of the corpus callosum. Only in this fashion, he believed, could one account for the striking unilaterality of the patient's defect. And indeed, autopsy of this patient *would* ultimately reveal a left parietal lesion along with a lesion of the corpus callosum (see Liepmann 1906).

Suddenly, the corpus callosum, that most baffling part of the human brain, seemed to be behaving as a pathway of communication, just as it was supposed to. Why it should have shaped up and fallen into line only under the sharp eye of the Germans is not immediately clear. What *is* reasonably certain, though, is that, while clinicians in other European countries, and the United States, continued to write confused and conflicting reports on the functions of the corpus callosum, Liepmann and his colleagues quietly went about creating a coherent predictive model of the role of this structure in higher brain functioning: a model that they backed up with a pool of remarkably unambiguous, internally consistent clinico-anatomical reports. One need not suspect any of these men of bad faith to wonder a bit about this phenomenon. Were they simply a group of exceptionally talented clinicians who knew, better than anyone else, what to look for; or could a certain amount of collective, and mutually reinforcing, theoretical bias have affected both what these men noticed and how they explained it?

In 1905, Liepmann gave a lecture before the Naturforscherversammlung in Meran, subsequently published as a major theoretical paper, "Die linke Hemisphäre und das Handeln" (Liepmann 1905). Building on the results of his continuing research into apraxia, Liepmann aimed to do nothing less than revise and enlarge the entire concept of left hemisphere predominance as it had been conceived up to that point. He began by laying out the relevant data, explaining how he had tested eighty-nine patients for evidence of apraxia (originally ninety were in his sample but one was eliminated for uncooperativeness). Forty-two of these suffered from left-paralysis caused by a right-brain lesion, forty-one from right-paralysis due to a left-brain lesion, five from aphasia without signs of paralysis, and one (the control subject) from apraxia without evidence of either aphasia or paralysis. Given a series of simple tasks (carrying out everyday movements, pointing to objects, etc.), a most surprising result was discovered. Among the forty-two left-sided hemiplegics, "it only happened occasionally that one or another of the tasks set could not be solved promptly, and it was never the case that a longer series of such could not be carried out." In contrast, among the forty-one right-sided hemiplegics, "no fewer than twenty showed clear disturbances in the execution of relevant exercises with the left, that is, the sound hand" (pp. 20–21). In other words, where Liepmann's *right*-brain damaged patients performed promptly and well with their right hands, their left hands being paralyzed, about half of his *left*-brain damaged patients performed in a significantly apraxic (or dyspraxic) fashion with their left hands, which they used since their right hands were paralyzed. Liepmann ruled out any possibility that this apparent left apraxia was simply due to some sort of failure, resulting from the left-brain damage, to understand instructions. The overwhelming majority of his patients not only revealed their capacity to understand language in ways outside of the immediate testing situation, but their distorted efforts showed that they understood the tasks that were expected of them and merely were incapable of properly bringing them to fruition. Not only that, their performance did not improve even after the doctor made the required movement himself for them to imitate, something that would have been expected if one really were just dealing with a language problem (p. 23).

For Liepmann, these results seemed to suggest that a lesion to the left side of the brain not only *paralyzed* the right arm but also made the nonparalyzed left arm *dyspraxic* to a greater or lesser extent. This pointed to the startling conclusion that an intact left hemisphere was not only necessary for proper functioning of right-sided movements, as everyone had long known, but was important as well for the execution of willed actions on the *left* side; in other words, it played an executive role in voluntary motor function for the *entire body*, both right and left side.

> Right-handedness implies that the right hand can do many things of which the left is incapable. Our results showed that even those things of which the left hand is capable are, to a large extent, not its own possession (that is, that of the right hemisphere), but are taken over from the right hand (that is, the left hemisphere). (p. 35)

Liepmann's expansion of the classical doctrine of left-brain domination to include voluntary or purposive movements ("Zweckbewegungen") not only raised some intriguing questions about the possible relationship between language and movement but opened the door to an important new conception of the purpose of the corpus callosum.

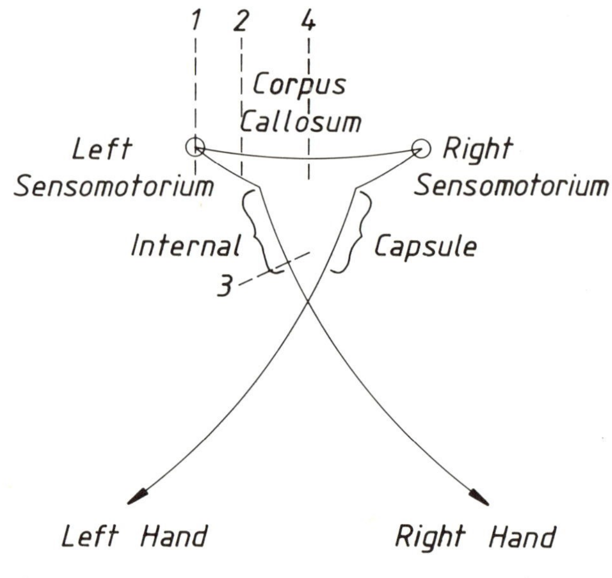

Figure 6. Liepmann's model predicting the different apraxic syndromes resulting from lesions in four critical areas of the brain. A cortical lesion of the left arm-center alone (1) paralyzes the right hand, and renders the left hand more or less dyspraxic. A supracapsulary lesion affecting the right hand's projection fibers in the left hemisphere as well as callosal fibers leading to the right hemisphere (2), paralyzes the right hand and renders the left hand dyspraxic. Damage limited to the internal capsule (3) paralyzes the right arm without causing dyspraxia to the left. A lesion strictly limited to the corpus callosum (4) leaves the right hand fully capable but renders the left hand dyspraxic. (Source: Redrawn after Liepmann 1907, 59.)

On the highest levels of voluntary motor functioning, it seemed to serve as a channel through which the superior *left* side of the brain, especially the frontal or motor region controlling the limbs, could work its influence—one is tempted to say its will—on the inferior *right* side (pp. 54–56). "Thus we arrive at the idea that the arm-center of the right cerebral hemisphere endures in a certain state of dependency upon that in the left hemisphere; i.e., that the latter rules the right by means of mediation [Vermittlung] through the fibers of the corpus callosum." From this conclusion, at least two important clinical predictions followed (but also see Figure 6 for a more comprehensive rendering of the clinical possibilities):

> Lesions that then affect the left-sided arm-center itself or [alternatively affect] its projection fibers as well as the callosal fibers, deprive the right cerebral arm-center of this dominance, and simultaneously paralyze the right upper extremity.
>
> . . . A lesion that only affected the corpus callosum to a suitable degree and in a suitable region would cause left-sided dyspraxia by depriving the right-sided hand-center of its dominance over that on the left-side, while the right hand would not necessarily be either paralyzed or apraxic. (p. 38)

In 1905, Liepmann's second postulated clinical syndrome, unilateral left-sided dyspraxia without right-sided paralysis through a lesion in the corpus callosum, was still just that, a postulated or theoretical entity. This unsatisfactory situation, however, was remedied remarkably quickly. A mere two years later, when Liepmann sat down to write an updated theoretical paper on matters concerning the corpus callosum (Liepmann 1907), he was able to point to no less than three recent case studies that essentially conformed to the clinical picture outlined in his 1905 paper (van Vleuten 1907; Hartmann 1907; Liepmann & Maas 1907). Two more clinical reports bearing favorably on the issue were to follow very close on the heels of these first three (Maas 1907; Goldstein 1908). One is struck, not only by the speed with which apparent confirmation followed in the wake of prediction, but also by how much of an "in-house" job the process of confirmation seems to have been. Not only were all the key actors German neurologists who had been trained in the tradition of the Wernicke school but all were thoroughly familiar with Liepmann's recent views on the functioning of the corpus callosum and all, except possibly Hartmann, knew Liepmann personally.

The case of Lorenz reported by van Vleuten, one of Liepmann's students, had led the way. It was a prime textbook study: unilateral left-sided disturbances of voluntary, skilled movements correlated with a lesion in the corpus callosum. Conclusion, "a simple interruption of communication between the left and right hemispheres [was sufficient to] destroy the left hand's capacity for coordinated movement [hat die Eupraxie der linken Hand aufgehoben]" (van Vleuten 1907, 237). In addition, van Vleuten stressed the extent to which the syndrome of callosal apraxia militated against the belief of the French neurologist Pierre Marie that apraxia was by definition an intellectual disorder, involving loss of the mind's capacity to generate "Bewegungsbegriffe." As van Vleuten saw it, all the thinking parts of Lorenz's brain were working fine; he just was unable to get them to communicate with one another.

Hartmann's 1907 paper was considerably more ambitious than that of van Vleuten, involving thoughtful discussion of three contrasting cases: a case of left frontal lobe tumor, a case of callosal tumor, and a case of right frontal lobe disease. His second case, involving destruction of the corpus callosum "from the splenium through to the plane of the forward commissures," is most relevant to our own immediate purposes. Briefly put, Hartmann's analysis of this case led him to conclude that isolated destruction of the corpus callosum produces the following clinical picture:

> 1. Amorphous motor performance, akinetic manifestations in the upper left extremity; that is to say, disruption of sequential movements whose execution initiates in the right brain, while capacity to perform independent actions [with the left hand] remains intact.
> 2. Disruption of sequential movements in the upper right extremity in that the maintenance of an action's continuity requires cooperation of at least the optical controlling [centers on both sides of the brain]; that is to say, damage to the capacity to combine those individual series of movements based purely upon performance memory [gedachtnismässigen Leistungen].
> 3. Retention of individual actions by the sensomotorium, on both sides, especially the left, loss of static [passive?] functions.
> 4. Retention of motor intention [i.e., patient retains a conception of the act to be performed]. (Hartmann 1907, 251)

Although Hartmann in no sense doubted Liepmann's views on the predominance of the left hemisphere in language and voluntary movements, his findings had led him to the conclusion, already suggested by Liepmann's 1900 case of Herr T., that communication along the corpus callosum perhaps was not always a one-way street. On more "elementary" levels of sensory communication, especially in the visual modality, a certain amount of mutual influence might operate and, to an extent, that could not be considered wholly negligible. Thus, in a case of callosal destruction, not only would the left hand be rendered more or less demented as a result of being deprived of contact with the left side of the brain, but the *right* hand might also suffer in its capacity to perform accurately (less dramatically certainly), from being suddenly cut off from sensory input, mostly visual, coming from the right hemisphere (see Hartmann's point 2). Bonhoeffer (1914, 124) would later call attention to this novel element of Hartmann's work, though he pointed out in the next breath that not all the later clinical evidence supported it.

Very soon after Hartmann's paper appeared, Liepmann and a colleague, Oscar Maas, reported on a patient they had been observing since the middle of 1905: a seventy-year-old cigarworker by the name of Ochs (Liepmann & Maas 1907). This patient was hemiplegic on the right side of his body, suggesting a lesion in the left hemisphere. His speech was rather indistinct and faint, but his language comprehension was said to be good, along with his general orientation for time and place. The *real* interest of the old man's case, however, was discovered only when the doctors attempted to test his writing capacity:

> Since the right hand was paralyzed and gave out after producing a legible *A*, the initial of his first name, the patient had to be allowed to use his left hand. . . . Our patient was incapable of accomplishing any writing act with his left hand, either spontaneously or under dictation. Nor could he be persuaded to copy; from this one may deduce that he was incapable of it, since he was normally happy to comply with tasks to which he was equal, only resisting, as is usual with such patients, when he felt incapable of performing what was asked of him. (p. 216)

The spirit was willing, but all the best "intentions" in the world were unable to coax the left hand into cooperating. Liepmann and Maas concluded that they were dealing with a case of so-called "isolated agraphia" (agraphia without loss of capacity to read and understand language), a rare—and disputed—syndrome to which Wernicke had drawn attention. More probing inquiry showed that in fact their patient suffered not only from agraphia but from dyspraxia as well. Although he demonstrated verbally that he understood what he was meant to do (stick out his tongue, smoke a cigar, brush his hair), the actions of his left hand bore only the crudest resemblance to the actual task at hand.

On strictly clinical grounds, Och's troubles seemed to conform rather well to Liepmann's clinical prediction that a left-sided lesion will paralyze the right side of the body and undermine the functioning of the left. The severity of his case, however, with its striking unilateral agraphia in the absence of aphasia, led the doctors to think that matters were more complex. The autopsy then permitted them to confirm their suspicions that Och's case bore importantly on the callosal question. Ochs was found to have been suffering from a lesion of the left frontal area, presumed to have caused the right hemiplegia and slight speech impediments. His right hemisphere seemed intact and healthy. Most significantly, there was marked damage to the "rear fourth to fifth" of the corpus callosum. "This lesion," concluded Liepmann and Maas, "must therefore be the cause of the severe apraxia and agraphia [suffered by] the nonparalyzed and nonataxic left upper extremity." The precise location of Och's callosal lesion, moreover, suggested to the authors that "the hintermost part of the corpus callosum is less crucial for coordinated movement [Eupraxie] of the left hand than the central parts" (p. 224).

Meanwhile, Oscar Maas independently had been investigating yet another likely candidate of callosal disorder and soon reported on a case of marked unilateral (left-sided) dyspraxia, in which there had also been a transient period of right-sided hemiplegia. Although anatomical confirmation had not yet been possible, he proposed to explain this combination of symptoms by assuming that a single lesion had damaged the descending pyramidal fibers in the left hemisphere, causing the paralysis, and also had broken up the neighboring callosal fibers where they separated from the white matter at the roof of the left lateral ventricle, causing the unilateral dyspraxia. Two separate lesions *could* have been responsible for the patient's problems, he conceded, but nothing particularly spoke in favor of this conceptually less neat alternative (Maas 1907, 791).

Kurt Goldstein's case in 1908, the final on our roster, was one of the most complex to emerge in the years immediately following Liepmann's paper. Ultimately, Goldstein would feel compelled to account for it, not only by assuming a lesion of the corpus callosum but destruction of a number of other connecting pathways as well. He had under his observation a fifty-seven-year-old woman who originally had suffered from left-sided paresis, while her right side moved freely. Gradually, she had begun to regain her strength and agility on that side; however, with one bizarre consequence. Her left hand started to act in a dyspraxic fashion, but more important, the patient complained nervously that it was acting *as if it had a mind of its own*. As Goldstein described the situation,

> On one occasion the hand grabbed her own neck and tried to throttle her, and could only be pulled off by force. Similarly, it tore off the bedcovers against the patient's will.—. . . She soon is complaining about her hand; that it is a law unto itself, an organ without will [willenloses Werkzeug]; when once it has got hold of something, it refuses to let go: "I myself can do nothing with it; if I'm having a drink and it gets hold of the glass, it won't let go and spills [the drink] out. Then I hit it and say: 'Behave yourself, hand' [literally, *mein Händchen*]" (Smiling) "I suppose there must be an evil spirit in it." (Goldstein 1908, 169–70)

The case was not wholly unlike Liepmann's 1900 case of Herr T., as Goldstein himself realized; it had even more points in common with Bleuler's 1901 case of "unilateral delirium" discussed in section 5-6, a case of which Goldstein seems to have been unaware. In contrast to Bleuler, though, Goldstein was wary of jumping too quickly onto the corpus callosum bandwagon in an attempt to explain his patient's apparent unilateral loss of volitional control and her strong subjective sense of being plagued by a "böser Geist" in her left hand. In conformity with Liepmann's model, Goldstein did think that the patient's unilateral dyspraxia could be attributed to a callosal lesion. However, he raised the possibility that this woman's more complex subjective symptoms could be attributed to some sort of rupture between the sensory and motor coordinating centers in her right hemisphere. Even this alternative explanation did not entirely satisfy him though, and he was forced to confess that, anatomically speaking, the subjective aspects of the case were distinctly puzzling.

Here we may let the matter of the corpus callosum rest for the moment, appropriately enough, on a note of uncertainty. It only remains to return briefly to the matters with which I began this section and ask plainly: in the final analysis, *should* McDougall at Oxford, fighting for the viability of the self-animating soul, have been worried by the work of Liepmann and his followers?

My cautious answer is that, in 1911 (the year he published his *Body and Mind: A . . . Defense of Animism*), McDougall probably would have been within his rights to conclude that the German work did not, on balance, outweigh the myriad of reports in which damage to the corpus callosum had *not* led to any of the syndromes reported by the Liepmann school. He also could have chosen a different tack and made the case that, even if callosal damage did produce the symptoms claimed by the Germans, it was not clear that these symptoms undermined belief in the essential inviolability of the human personality. The patients reported by these German doctors all perceived themselves as "one person"; true, they were having trouble performing as "they" wanted, but they were one person all the same. Even the difficult case reported by Goldstein, and leaving aside Goldstein's own doubts that his patient's subjective symptoms were a simple consequence of callosal damage, could hardly be compared to the phenomenon of true "double personality" as exemplified in such cases as Félida and Mary Reynolds. While this fact might have proved rather disappointing to psychiatrically oriented theorists of the double brain, McDougall could justifiably have drawn some comfort from it.

CHAPTER SIX

The "Experimental Evidence": Metalloscopy and Hemi-Hypnosis

I found to my astonishment that here were occurrences plain before one's eyes, which it was quite impossible to doubt, but which were nevertheless strange enough not to be believed unless they were experienced at first hand.
—Freud reporting on his visit to Charcot's Salpêtrière, 1886

IN CHAPTER FOUR, reference was made to certain striking studies in hysteria and hypnosis that were held up widely as definitive proof of the potential functional duality of the human brain. These experiments represent one very important reason why duality of mind theories seemed as convincing as they did in the late nineteenth century, particularly to workers in France. The aim of this chapter is to demonstrate just how this was so.

6-1. Hysteria and the Double Brain

I begin, then, with hysteria. Although this disorder has figured in discussions in earlier chapters, it is important that we look at it more carefully now, with particular attention to how it manifested itself in the hospitals of late nineteenth-century Paris, both clinically and under special experimental conditions. It will become clear that the role of hysteria in the history of thought on brain duality is far from incidental; indeed, in 1884, Edgar Bérillon, trained broadly in the tradition of the Charcot school, felt himself in a position to declare that "no illness presents an ensemble of characteristic symptoms more appropriate for furnishing, a priori, serious evidence of the functional independence of the cerebral hemispheres" (1884, 134).

The history of hysteria extends back twenty-five centuries, and it appears that much of it is a history of confusion. Although, up until the sixteenth century, most medical men still accepted the idea that hysterical disorders originated from the uterus, as the name implies, the protean parade of symptoms associated with the disorder seemed to defy all attempts at rational analysis. By the end of the seventeenth century, uterine theories were beginning to be rivalled by perspectives emphasizing the role of the nerves or the encephalon, but there was still no agreement and considerable incoherency in the literature. Ellenberger goes so far as to declare that the first "objective and truly systematic" study of hysteria had to wait for the publication of Paul Briquet's *Traité de l'hystérie* as late as 1859. Briquet denied that the disorder had anything whatsoever to do with the uterus; in fact, it could occur in men. He described it instead as a "neurosis of the brain," which chiefly manifested itself in disturbances of those vital or visceral acts concerned with the expression of emotions and passions (Ellenberger 1970, 142). Another influential French neurologist, Etienne Georget, also believed that hysteria was an encephalic disorder, but he focused on the so-called hysterical convulsion and stressed the disorder's close kinship to epilepsy. These two lines of thought would converge in the idea of "hystero-epilepsy," or *grande hystérie*, consecrated by the Charcot school in the late 1870s and celebrated by Charcot's disciple Paul Richer in his 1881 *Etudes cliniques sur l'hystéro-épilepsie ou grande hystérie* (Corraze 1983, 403).

Jean-Martin Charcot (1825–1893) came to the study of hysteria imbued with the doctrine of localization. Born in Paris, the son of a carriage builder, he had succeeded Vulpian in the chair of pathologic anatomy in 1872 and had gone on to achieve international fame with his penetrating analysis of various neurological disorders (amyotrophic lateral sclerosis, multiple sclerosis, tabetic arthropathies), all carried out according to his so-called clinico-anatomic method. This taught that, in order to understand the pathological anatomy underlying a disease, one must first study carefully all the symptoms presented by a large number of patients suffering from it, and then examine and compare the location and appearance of lesions discovered at each postmortem. The emphasis on *looking* owed a considerable amount to Charcot's personal temperament. He was, as he said himself, a "visuelle," a man who thought better in pictures than in words, and there was much of the artist in him.

Although Charcot realized that hysteria caused no observable structural changes in the brain, he nevertheless proposed to explain this disorder on analogy with the organic neurological disorders he had elucidated with such success. Hysterical ailments, he argued, were to be understood as "dynamic" or "functional" disorders of various cerebral centers, proper lesions of which would result in organic ailments of the same order. Underlying this initial theoretical formulation seems to have been an implicit hope, not to be realized, that once "present methods" of postmortem investigation were sufficiently refined, the putative differences between hysterical and organic disorders might well turn out to have been more apparent than real (see, e.g., Charcot 1878a, 38).

For the moment, however, hysteria was to be understood as a functional rather than organic, neurological disorder, or as Briquet had put it, a "neurosis." Nevertheless, in turning his attention to hysteria, Charcot the "visuelle" would remain convinced that the eye of the artist could be used to illuminate the world of neuroses no less than that of frank brain damage. He and Paul Richer would make extensive use of photography in an attempt to capture the elusive realities of brain and mind disorder (see Didi-Hubermann 1982). They would compile three major historical studies of illness, including cases of alleged hysteria, in art and pointedly compare such pictures with the phenomena they themselves observed daily in the hospital wards. There is reason to believe that this fascination with artistic renderings from the past had crucial, and ultimately fatal, consequences for the development of the Charcot school's thinking about disorders like hysteria. In their belief that one could interpret various time-bound representations of ecstasy and dionysian mania in terms of some historically transcendent medical truth, Charcot and his disciples became, as Gilman (1983) argues in a slightly different context, victims of their era's "ideology of realism." As victims, many historians agree that they probably made their patients into victims, too, at least partly conditioning them to produce a certain pattern of symptomatology compatible with a certain stylized, aesthetically dramatic preconception of the disease.

If all this is true, then the question as to what hysteria under Charcot "looked like" takes on a new interest. The following may serve as a rough thumbnail sketch (but see also the photographs in the Salpêtrière *Iconographie* [Bourneville & Regnard 1877–80] and those reproduced in Didi-Hubermann 1982): One finds, to begin, that although this disorder no longer had anything to do with the uterus, it nevertheless was felt primarily to strike women during the childbearing years. It was associated with hyperaesthesia (morbid sensitivity) in one or both ovarian regions (the testes in men), also with anaesthesia, paresis (disorders of movement), and painful trigger zones on various parts of the body. The eyes were often affected; many patients suffered from a peculiar form of achromatopsia (color-blindness) associated with narrowing of the visual field. Finally, and most dramatically, the disorder was distinguished by so-called crises, which presented phenomena in a definite order: epileptiform convulsions and "salutations" (stereotyped movements), followed by mental and emotional disturbances ("attitudes passionnelles" and "délire") during which the patient would act out some hallucination, often of a romantic or frankly erotic nature. These crises were usually preceded by an epileptiform "aura" commencing in the afflicted ovary, and it was said that one could often cut short an attack by applying firm pressure with the fingers or fist to the ovarian region (Gamgee 1878, 545; Richer 1881).

Charcot called all the permanent symptoms of hysteria, as opposed to the episodic crises, the *stigmata*. His choice of such a theologically loaded word in this context is worth lingering over a moment. Jan Goldstein (1982) has discussed the extent to which the Charcot school's early work on hysteria, under the patronage of an aggressively anticlerical Third Republic government, was directed towards a reinterpretation of various Catholic religious categories into "natural-pathological" ones. One important aim of Charcot and his workers, Goldstein argues, was to show how Catholicism's saints and inspired figures could be dismissed as simple cases of hysteria, mania, or ecstasy, such as one might find any day in the great municipal hospitals of Paris (Harris 1985). In pursuing this aim of demystification and denigration, Goldstein concludes, much like Gilman (1983) but for different reasons, that doctors may well have unconsciously elicited in their patients a certain complex of symptoms appropriate to their diagnostic preferences and preoccupations.

Now, of all the hysterical stigmata studied during the Charcot years, one of the most characteristic seemed to be hemianaesthesia, sensory disturbances affecting only one lateral half of the body. In both Catholic and Protestant doctrine, patches of anaesthesia on the body had been one important way of recognizing women who had been consorting with the devil. Given this, and in light of both Gilman and Goldstein's remarks just discussed, it is perhaps not irrelevant that a number of foreign observers would later suggest that there seemed to be something peculiarly *French* about hysterical anaesthesia in general; this symptom being not nearly so common in the hospitals of other countries (see, e.g., A. Myers 1889–90).

However, so far as *hemi*anaesthesia in particular is concerned, no immediate political or sociological explanation for its frequency in the French hysterical syndrome suggests itself. This, in itself, may be a significant fact, but I will eschew speculation and be content with simply noting the position of the doctors on this symptom. Paul Briquet's 1859 *Traité* had reported finding complete cases of hemianaesthesia (involving unilateral loss of vision, hearing, etc., as well as full cutaneous anaesthesia) in some 93 patients out of 400; mixed and incomplete cases were said to be still more frequent (Charcot 1875a). What qualified as a "mixed case" is not entirely clear. In the later literature, even examples of bilateral anaesthesia were sometimes interpreted as a variety of hemianaesthesia, according to one author, being nothing more than "a double hemilateral hysteria, of different degrees on the right and the left" (Barety 1887, 628).

In examining cases of "pure" hemianaesthesia, French medical men were profoundly impressed by the precision with which sensory capacity broke off at the median line of the body. Bérillon, for example, described one patient who, when plunged into a bath, only felt the water on half her body, and another who, when she drank, had the impression that the glass was broken at the center (Bérillon 1884, 135). Charcot, in an early paper on hemianaesthesia, also laid stress on the alleged fact that, as a rule hysterical hemianaesthesia occurred in conjunction with morbid sensitivity of the ovarian region on the same side of the body. He seemed to look upon this peculiar twinning of symptoms simply as an empirical reality that at the moment eluded any physiological explanation (see Charcot 1875b).

Attention was also drawn to the reputed finding that the various hallucinations seen by patients during their hysterical crises (rats, cats, fantastical beasts, gruesome heads) were also generally of a unilateral nature, no less than the permanent anaesthesias. That is to say, such hallucinations took shape to the right or the left side of the patient rather than directly in front, and it was significant that "the side where the hallucination takes shape is always that which corresponds to the hemianaesthesia and consequently to the ambylopia [impaired vision]" (Charcot 1878a, 38; Richer 1881, 9). For medical men who accepted Charcot's emphasis on the encephalic origins of hystero-epilepsy, the only possible conclusion from this correlation of symptoms was that both the hemianaesthesia and the periodic unilateral hallucinations were due to a malfunctioning of *one* hemisphere of the brain only. In Chapter Three, discussion was made of the fact that the hemisphere in question was often felt rather likely to be the *left*.

All this tended to encourage the conviction that hemilateral hysteria could cause a sort of mental dissociation resulting from the apparent independent functioning of the two brain-halves. After all, if one of the cerebral hemispheres in hysteria functioned normally while the other functioned abnormally, as one writer put it in 1887, "one realizes what dissension, what a struggle must exist between the left personality and the right personality, or in other, more familiar terms, between the good and the bad side. There then unfolds [in hysteria] a series of psychic disorders betraying this dissension and this struggle in the midst of all the most banal symptomatology" (Barety 1887, 628).

The frequent unilaterality of hysteria's physical and psychological symptomatology, however, was not the only, nor even the most compelling, reason why Bérillon felt no disease was in a better position to offer serious evidence of the functional independence of the two hemispheres. Toward the end of the 1870s, hysteria's role in the intellectual history of the double brain was to take on a striking new dimension.

6-2. *Metalloscopy and the Discovery of "Transfer"*

The story behind this development[1] begins in 1876 when an elderly doctor, Victor Burq, contacted the president of the prestigious Société de Biologie in Paris. Founded, according to official chroniclers, under the "impulsion of the positivist spirit," this was a society to which, as Paul Bert put it in 1884, "have been brought the foundations of all the discoveries made in this country about the history of living beings" (Gley 1899, 1023, 1077). In 1876, its president was none other than Claude Bernard, France's pre-eminent medical sage and physiologist.

In a personal letter to Bernard, Burq explained how early in his medical career, he had accidentally discovered that certain metals exerted a powerful physiological action on patients suffering from hysterical disorders, and soon after had found that he could cure hysterical hemianaesthesia by applying metallic discs to the afflicted side of the patient's body. The specific metal required varied from patient to patient (so-called metallic "idiosyncracy"): some found relief only from gold, others needed copper or zinc. For some twenty-five years, Burq had been making use of this treatment, which he called *metallotherapy*. The results spoke for themselves, but inexplicably the medical community had consistently disparaged or ignored his work. And this in spite of the fact that during the years 1848–1853 *alone*, he presented the Academies of Science and Medicine with no less than twenty-two reports of his findings (Burq, 1853)! Before he passed from this earth, Burq told Bernard, he hoped the Société would be willing formally to investigate and pass judgment on the scientific validity of his life's labor.

Bernard, with commendable open-mindedness, agreed. And he took Burq seriously enough to appoint a commission to the task that must have struck contemporaries as first-rate. The appointees consisted of

[1] I have discussed in detail the story of metalloscopy and its associated developments in my paper, "Metals and Magnets in Medicine: Hysteria, Hypnosis and Medical Culture in fin-de-siècle Paris" (Harrington in press). Portions of this chapter follow the arguments pursued in that study.

Jean-Martin Charcot, Jules Bernard Luys, and a certain Amédée Dumontpallier, about whom more later. Burq was invited to Charcot's Salpêtrière, where a series of experiments were carried out between 1876 and 1877. Although almost all the initial reports were said to have been positive, it seems that Charcot retained lingering reservations about the whole enterprise. Then, one day, in the course of making a point about anaesthesia in hysteria before an entourage of English doctors, who were skeptical about its occurrence, he suddenly pricked the arm of an habitually anaesthetic patient. She cried out with pain and the foreign visitors smiled smugly. Charcot was puzzled and embarrassed. It turned out that Burq had recently applied a gold plate to this particular patient's arm and had thus restored it to sensibility. According to de Courmelles, one of Luys' later students, after this episode, Charcot "seriously joined in the labours of the commission in metalotherapy [*sic*] and was soon convinced like the others of its truth" (de Courmelles 1891, 31–32).

The commission's report back to the Société de Biologie in 1877, then, was enthusiastic. True, there was doubt about the long-term *therapeutic* value of Burq's method; and, for this reason, the commission emphasized that it had focused most of its attention on "metalloscopy"; i.e., the production of phenomena without therapeutic intent (see Charcot 1883). At the same time, the genuineness of the metallic effects was unequivocably affirmed, and it was reported that other agents (especially magnets, static electricity, and electric current) had also been found effective; these various agents would all eventually be subsumed under the name of *aesthesiogens*.

More important, the Société de Biologie was informed that a *new* discovery had been made in the course of investigation. It seemed that when sensation was restored to a region on one half of the body using metals, symmetrical regions on the normally healthy side of the body *lost* their normal sensibility. As Charcot put it at the time, "It seems that with these hysterics, the nervous fluid, if one will pardon the expression, does not transport itself to one side until after it has in part abandoned the other" (cited in Féré 1902, 108). Although this phenomenon originally was discovered by one of the commission's assistants, Gellé, while experimenting with hysterical deafness, Dumontpallier, inspired by a visit to the bank in which he had moved funds from one account to another, christened it the *law of transfer*. The commission's report triumphantly concluded,

> Your Commission, Gentlemen, cannot affirm too strongly the existence of all these facts [concerning metalloscopy and transfer of

phenomena], and this affirmation is a consecration of the facts enunciated by the one who discovered them already more than twenty-five years ago. This affirmation is an homage rendered to Doctor Burq, who in spite of often severe criticism, never lost courage. (Dumontpallier, Charcot, & Luys 1877, 24)[2]

Within the framework of French thinking on hysteria, there could be no mistaking the implications of the commission's findings. In Bérillon's words, "there was evidently a cerebral action involved" (1884, 14). Somehow, the aesthesiogens were causing the "dynamic lesion" responsible for the patient's anaesthesia to transfer itself from its original site in one hemisphere to a symmetrical site in the other. It was a startling idea to consider, and in an early lecture on the new findings, Charcot quoted Claude Bernard as having said bemusedly that he personally could not comprehend the data at all, but that it did prove one thing: that science was not yet very advanced when it came to the physiology of the nervous system (1878a, 790).

Variations and elaborations on the original discovery followed rapidly. It was found, for example, hemiparalysis, as well as hemianaesthesia, was susceptible to transfer using aesthesiogenic agents. It was also becoming clear that transfer often, if not always, was followed by what Charcot called *consecutive oscillations*, in which sensibility shifted repeatedly back and forth between the two sides of the body. Properly speaking, this was not a mere extension of transfer, since no

[2] In the backlash that inevitably follows the consecration of any new hero, the question was raised as to how far Burq's ideas really could be said to be original to him. Medical librarian and physician L.-H. Petit (1880) found formal precedents for using metals and magnets as therapeutic agents that went back as far as Aristotle, Albertus Magnus, Galen, and of course, Paracelsus. The early mesmerists had also used magnetized metals in their work, and Charcot went back and personally consulted the mesmerically influenced 1779 report by the French physicians Michel-Augustin Thouret and d'Andry on the curative virtues of the magnet, published in the memoires of the newly founded Société de Médecine. The "metallic tractors" of the late eighteenth-century American physician Elisha Perkins also would be mentioned later as a clear forerunner of Burq's armatures and metallic plates, to such an extent that Burq felt compelled to belittle Perkins' importance in his 1882 work on the origins of metallotherapy. (For the story of Perkins and his metallic tractors, see Holmes 1842; also B. Perkins 1798). However, the most damaging challenge to Burq's originality was put forward in a lengthy 1880 article by a physician from Lyon, J. Monard. The article described the extensive investigations of the early nineteenth-century physician and mesmerist Antoine Despine, from Aix-en-Savoie, into the therapeutic properties of metals on magnetized patients. Although Monard admitted that Despine in his researches had failed to recognize such important phenomena as metallic "idiosyncracy" and transfer, he nonetheless felt justified in concluding that Burq's system was simply "the doctrine of Despine regenerated and perfected" (Monard 1880, 414). In the end, Burq was unable wholly to rebut this conclusion.

new application of the aesthesiogenic agent was required (Richer 1881, 539–40). Alfred Binet and Charles Féré, two of Charcot's more distinguished students, subjected a patient to these oscillations and recorded her complaints of a pain passing from one side of her head to the other: "It's as if there were doors banging between the two sides of my brain" (Binet & Féré 1885, 20).

Binet and Féré called the pain experienced by their subjects the *pain of transfer* ("le douleur de transfert") and considered it an important finding in its own right. In the majority of cases, they said, the pain-site, which always was quite discretely localized, was in absolute conformity with modern clinical and experimental findings on sensory-motor localization. "The subject consequently brings, without being aware of it, an interesting confirmation of the theory of cerebral localizations, sensory no less than motor" (Binet & Féré 1885, 24). It is unclear how Binet and Féré conceived the physiology underlying these alleged pain centers in the brain. Both animal experiments and clinical observation had long since established, by this time, that the living brain was insensible to pain (see, e.g., Ferrier 1876, 127).[3]

Matters were further complicated when doctors began to declare themselves able to transfer not only sensory-motor symptoms but voluntary behavior and intellectual activity as well. This was called *psychic transfer* (Binet & Féré 1887, 16–17). It was reported, for example, how one subject was made to write down figures in an ordinary way with her right hand while a magnet was concealed near her left. She was able to write only as far as the number 12 before she hesitated and felt *compelled* to switch the pen to her left hand. The figures she then set down were written in mirror-image, and she had become agraphic with her right hand.

Still another twist was added in 1886, when Joseph Babinski claimed that it was possible to induce "consecutive oscillations," not only between the two hemipsheres and two sides of the body but between two separate *individuals*. He described how he had seated two patients suffering from hemianaesthesia back to back: aping, one is tempted to think, the manner in which the two hemispheres stand in relation

[3] Belief in these cerebral pain centers in hysteria persisted through the end of the century and may indeed be seen as a new twist on Gall's belief that, by "reading" the bumps on the skull, one could make inferences about the localization of the cerebral centers below. For a further example of this "neo-craniology," see P. Sollier's (1900) description of how he had correlated hysterical, and hypnotically induced, anaesthesias with various "pain zones" in the brain, as revealed through cranial palpations. See also Gilles de la Tourette's (1900) crisp rejection of Sollier's pain zones in hysteria. "I carefully searched many times for the presence of these zones. I never came upon them" (p. 226).

to each other within the skull. By means of magnets, Babinski said, he had then caused subject A's nervous system to absorb the half-sensibility of subject B, making A *fully* sensible and B *fully* anaesthetic. The transfer had then reversed itself, B taking back not only her own sensibility but that of A, leaving A now anaesthetic on both sides of her body. This oscillation of sensibility continued back and forth, again and again, even after the magnet was removed (Babinski 1886).

Babinski would later repudiate this work (cf. Chapter Nine), but in the 1890s, Jules Bernard Luys at la Charité hospital went on to develop the idea of transfer between individuals into a dramatic therapeutic technique. A patient would be made to sit facing a trained hypnotized subject, usually a former hysteria patient, now paid for her services. The two would clasp hands. Using a magnet, Luys then would make passes over the patient, who was not hypnotized, and draw the current of his diseased "emanations" down his arms and into the arms and body of the hypnotized subject. The latter would respond to each new pass with a convulsive movement, as absorption supposedly was effected. At length, the hypnotized subject would have acquired the symptoms of whatever disease the patient had been suffering, while the patient would generally feel much better. The current would then be broken by breaking the handclasp, and the hypnotized subject would be rid of her acquired disease by being given the suggestion that she was now well (Robertson 1892, 523–24).

It was partly through these later exotic developments of transfer that some people came to the conclusion that the theory of biomagnetism first propounded by Mesmer must have some basis in reality after all (Dessoir 1887, 541). Thus, two French physicians, L.-T. Chazarain and C. Dècle, interpreted the metalloscopic findings in the context of certain sweeping views concerning the alleged "electrical polarity" of all living creatures and the "laws" of mesmeric healing that were derived therefrom:

> living beings, man, animals, and vegetables possess two types of radiating electricity; they are positive on certain points of their [body] surface . . . and negative on others. Animals, like man, are positive on the left side of their bust and on the external side of their limbs; they are negative on the right side of their bust and on the internal side of their limbs. There are no exceptions save for left-handers, among whom the poles are inverted. (Chazarain & Dècle 1887–88, 145)

While a handful of medical men, notably Jules Bernard Luys, would ultimately prove themselves quite open to various neo-mesmeric interpretations of metalloscopy, most of the circle of workers more immediately associated with Charcot worked very hard, especially in the early years, to keep theories sounding sober and scientific, which basically meant "physicalistic." All the phenomena of metalloscopy were to be explained in terms of known laws of biophysics and physiology and nothing mysterious was involved. Charcot declared (1883, 145) that "the problem is double, physical and physiological." The aim was therefore to explain: (1) how the aesthesiogenic agents acted on the nervous system; and (2) what physiological processes were involved in the shift of nervous "energy" between the two hemispheres of the brain.

On the biophysical side, there were many brave suggestions. Burq's original hypothesis was that the metals or magnets acted somewhat like a solenoid, producing a slight electrical current when placed in contact with the skin, which then acted on the nervous system. This electrical theory was painstakingly tested by one of the assistants to the commission, Paul Regnard, who eventually came down in favor of it. Another researcher, Ernest Nicholas Onimus, subsequently recast Regnard's theory in the context of Alfred Louis Becquerel's views on "electro-capillary" currents in living organisms. In contrast, French physician Antoine Rabuteau was inclined to favor a more chemical hypothesis; in his view, the initial catalyst was the oxidation produced by the contact of metal with the moist skin surface. Romaine Vigouroux took still another tack, finding an explanation for the varied findings in the changing state of "electrical tension" on the surface of the patient's body (Vigouroux 1878).

Meanwhile, in Germany, the medical community had responded to news of the French metalloscopy findings with considerable interest, and a number of notable contributions to the debate soon began to filter out of that country. Eduard Schiff, having reviewed all the French electrically oriented models and found them wanting, proposed, in an influential lecture before the Naturforscher-Versammlung in Baden-Baden, that the metalloscopic phenomena might be caused by molecular vibrations ("Molekularbewegungen") inherent to the aesthesiogenic agents, which then aroused the central nervous system through physical means. Schiff noted the broad similarities between this idea and Mesmer's belief that his "animal magnetism" was ultimately to be understood in terms of matter in motion; it was not impossible, Schiff concluded, that metalloscopy would eventually set a number of Mesmer's teachings on a solid scientific footing (Schiff 1879).

Focusing specifically on the transfer phenomenon, another German worker, A. Adamkiewicz from Berlin, rejected the various fancy electro-physiological theories circulating in the literature and argued that the results actually were just caused by a simple irritation produced mechanically and were simply an exaggeration of processes found in healthy persons (Adamkiewicz 1880). His American student, Albert Adler, agreed, having found, in a series of tests on nonhysterical persons, that unilateral irritation (using such items as mustard-plasters) increased the sensibility of the part of the body excited and decreased it in the corresponding part on the other side (Adler 1879; for an English-language summary of this study, see Adler 1880–81).

Back in France, Brown-Séquard had had a different idea. It seemed to him that one might be able to explain transfer in terms of "inhibition" and "augmentation" acting directly within the nervous system. He described how, in laboratory animals, one could heighten the sensibility on one side of the body and weaken it on the other side, by making a transverse cut at one lateral half of the base of the brain. A second cut at the dorsal cord on the anaesthetic side of the body *transferred* sensibility (Brown-Séquard 1882). In Germany, another alleged organic analogue to metalloscopic transfer was reported in 1884 by L. Hirtl. In five separate cases, this neurologist claimed to have transferred, using a "spanische Fliegenpflaster," the epileptic aura associated with the onset of Jacksonian (unilateral, "marching") epilepsy. The patients' fits now commencing from the "other" side of their bodies, the implicit conclusion was that the epileptic "discharging" lesion responsible for these attacks had shifted its seat from one to the other side of the brain. As in hysterical transfer, the effects were said to be only temporary (Hirtl 1884).

Focusing on the physiological side of the transfer problem, M. Rosenthal in Germany was one of the first to suggest that the aesthesiogenic agents acted unilaterally on the vaso-motor centers of the brain, thought to regulate dilation and contraction of blood vessels. Following this unilateral excitation, there might be a contraction of the blood vessels in the hemisphere serving the disordered side of the body. This caused a compensatory widening of the vessels in the other hemisphere—i.e., the hemisphere that served the normally healthy side of the body—bringing about the displacement of sensibility (Rosenthal 1882).

One finds from all these developments a slowly emerging idea that transfer in hysteria was simply one rather extreme example of a general physiological tendency for symmetrical parts of the body to act in an antagonistic relationship toward each other. This point of view received further strong, if indirect, support from Broca's doctrine of hemisphere substitution of function. Indeed, at the very end of the century, the concepts of substitution and transfer seem partly to have converged. Thus, one finds Marie de Manacéïne, a (woman!) psychologist from St. Petersburg, carrying out a series of experiments that suggested, she believed, that the nervous energy of the left hemisphere was largely transferred to the right during sleep. Working with fifty-two subjects, she found that tickling a sleeper's face on the right side of the median line always caused him to brush himself with his *left* hand, even when he was lying on his left side and the action of that hand was impeded. Significantly, left-handed people, of which she had a small sample, always brushed at their faces with the *right* hand, even when the left was hanging free. "These facts," Manacéïne believed, "may be explained on the hypothesis that the most active cerebral hemisphere is resting during the hours of deep sleep" (1897, 40–41; also 1894). She drew further evidence for the view that the brain's two hemispheres "substituted" for each other during states of unilateral nervous and psychological fatigue from her studies on the effects of enforced insomnia in puppies. It seemed to her that the puppies' reflexes disappeared and reappeared periodically in the limbs, first of one side of the body, then of the other. She thus concluded that the cerebral hemispheres were acting on a rotating basis, one sleeping while the other remained awake (1897, 41–42).

Manacéïne's work, first reported before the International Medical Congress of Rome in 1894, was hailed by French physician Camille Sabatier as definitive evidence in support of the idea that human mental life and personality were fundamentally dual (Sabatier 1908, 1918; see also Chapter Four, section 4-4). Her studies also made a strong impression on J. H. Sarda, a physician from Toulouse, who tried the tickling experiment on young children (Sarda 1908). He found that very young infants (three subjects, aged two weeks to one month) responded to stimulation in an undirected and purely reflexive fashion, agitating their limbs bilaterally. A somewhat older group of five children, eleven to twenty months, gave evidence of purposive movements, which Sarda took to be of cerebral origin. In this second group, "the impression produced on the left provoked movement of the left arm; the impression produced on the right provoked movement of the right arm" (p. 190). The implication was that these children, being preverbal, had not yet developed the functional asymmetries that caused right-handed adults to become left-handed in their sleep, and vice versa.

In 1902, Charles Féré, who now ran his own laboratory at the Bicêtre hospital in Paris, argued in a paper, "L'alternance de l'activité des deux hémisphères cérébraux," that substitution or transfer of energy between the hemispheres, exhibited spontaneously in cases of brain disease and fatigue and produced artificially in experiments with hysterical subjects, was an artifact of an inborn tendency towards symmetry in all our bilateral movements. "It is known that, during the first months [of life], virtually all the movements children make are symmetrical. . . . Muscular asymmetry grows with age." Even after the acquisition of asymmetrical skills, he went on, there always remained traces of the individual's original symmetrical inclinations. He found, for example, that, when writing with the right hand, "one's left hand did not remain immobile; first off, the phalanx [joint] of the thumb became animated by small movements, then the index [finger] began to join in." Because of this innate sympathy between bilateral organs, for example, raising one arm involved, in the healthy person, an equal act of motor inhibition in the other arm. However, in cases of unilateral weakness, whether brought on by brain-damage, hysteria, or fatigue, the influences that normally inhibited the "wrong" side of the brain from participating in a unilateral action failed to function. This led to the perceived "substitution" or "transfer" (Féré 1902; cf. Féré 1901a, 1901b).

Although the general paradigm suggested by transfer lingered on until the end of the century, starting about 1885, metalloscopy and transfer proper came under increasingly effective attack by Charcot's chief rival, Hippolyte Bernheim of Nancy. Speaking before the Société de Biologie and later in published articles, Bernheim accused the Salpêtrière investigators of having unwittingly trained their subjects to transfer symptoms and to expect that such and such an effect was supposed to occur. At Nancy, where his patients knew nothing of metalloscopy, he consistently had been unable to produce the sorts of dramatic results reported by the Paris workers—that is, when he attempted to do so simply by silently manipulating metals and magnets. He only had to drop a bare *hint*, however, about what was expected of the subject and the effects would soon appear, with appropriate variations. Although he would not go so far as to deny categorically that the magnet could have an action on the organism, he nevertheless concluded that the sorts of findings being reported by the Parisian metalloscopy workers could effectively all be accounted for in terms of "suggestion" (Bernheim 1885, 1886).

Bernheim's criticisms found an echo in England, where early Hack Tuke had pointed out the likely role of "expectant attention" in the production of the metalloscopic effects. He took pains to stress that he did not deny the reality of the phenomena observed by the Paris investigators, but "facts are one thing and their *rationale* another, [as] Charcot would be the first to admit" (Tuke 1878–79, 601).

At first, the Charcot school fought back against the charges that the mental state of their subjects somehow had a hand in the creation of the metallic effects. They seem to have taken the view that to admit even a partially subjective origin for the phenomena they were observing would be tantamount to saying that the phenomena did not "really" exist. Some researchers stressed the extensive experimental precautions taken against any possibility of deception or unconscious suggestion. Others pointed out that a number of the metalloscopy findings had been reproduced on laboratory animals, who after all could hardly be accused of fraud or over-excitable imaginations (Petit 1880, 24–25).

Back in the 1870s, in reply to his English critics, Charcot had also made the point that transfer could not be explicable in terms of "expectant attention" because it was a phenomenon "that was discovered through chance alone, and that the patients would not know how either to invent nor simulate" (1878d, 219). How, for example, could an ignorant girl know Landolt's physiological law concerning the order in which colors were meant to appear and disappear as hysterical hemiachromatopsia was transferred from one eye to the other? This line of argument was now resurrected in the battle with Bernheim's suggestion. Janet would later raise the possibility, however, without committing himself to it as a wholly satisfactory explanation, that the Parisian doctors were underestimating their subjects. Someone who lives in a hospital where people are always discussing medical issues is likely to be at least as familiar with them as many medical students (Janet 1925, II:826–27). Indeed, one of the metalloscopy researchers would later make the telling admission that, in the Charité hospital run by Luys, some of the patients involved in experiments could be said practically to have had a formal medical education. It appears that the medical students often made their subjects who wrote well "copy their notes of lectures or their observations" (de Courmelles 1891, 161–62).

The Paris workers also argued that suggestion could not explain metalloscopy because sometimes the effects of the metals and magnets were contrary to what suggestive circumstances should reasonably have led the patient to expect. De Courmelles recorded how Vulpian's early skepticism toward metalloscopy was overcome after he had witnessed an experiment in which two metal plates, both wrapped in linen to make them indistinguishable, were applied to a subject. The first produced no effect; the second did. The first plate was then shown to have been of gold, and the second of copper; and it was pointed out that previous studies had established that gold did not affect this patient. *If* the metalloscopy effects were simply due to the imagination of the patient, then sensation should have been affected in both cases. "It was therefore evident that a real action had taken place," declared this worker of Luys' school. "All the reports of the Biological Society unanimously agreed on this point . . . [and] show what searching investigations must be made in establishing scientific data" (de Courmelles 1891, 32).

In the end, though, these and similar sorts of arguments failed to convince. The long-term effect of Bernheim's criticisms, as Janet would put it, was "that of a thunderbolt," leading to dissent in Charcot's ranks and a widespread feeling among outside observers that the whole enterprise had somehow been a cheat. Janet felt that the Paris investigators should have persevered in the fight, that in fact transfer *was* "a real and very interesting phenomenon . . . in close relationship with the psychological exhaustion and the restriction of the field of consciousness." Writing in 1925, he opined that Burq's investigations were "still worthy of attention" (Janet 1925, II:796–98).

Nevertheless, Janet felt strongly that the Paris workers, in refusing to recognize the role of *any* psychological factors acting in metalloscopy, essentially doomed their investigation from the start. They should have realized, he said,

> that all kinds of moral precautions were essential to successful research in this field; they thought only of the physical aspects of their experiments. . . . They correctly weighed the metallic balances, and they registered the most trifling muscular tremors with the aid of Marey's tambour; but they saw nothing wrong in carrying on experiments in public, amid the chatter of casual spectators, and they would themselves discuss the meaning of the experiments in their patients' presence." (Janet 1925, II:794–95)

Although Bernheim's suggestion ultimately triumphed over Charcot's physiology and called attention to the psychological element in the metalloscopy studies, it is highly unlikely that its success can be understood purely in terms of its greater cognitive coherency.[4] Later

[4] Taking this point further and considering how far the general battle between Charcot and Bernheim over metalloscopy can be simply understood as a clash between opposing intellectual perspectives, it is worth at least calling attention to the following: Very shortly before turning himself into the chief critic of the Charcot school's metalloscopy enterprise, Bernheim *himself* had carried out extensive investigations at Nancy into the clinical potential of so-called magnetotherapy; he was an enthusiastic believer in the

commentators would point out that it was an unanalyzed term, as vague and all-encompassing as Mesmer's animal magnetism or Reichenbach's odylic fluid. On its own, for example, it could offer no insight into *how* the mental state of the hysterics could bring about the wide range of physiological alterations and cures to which the Paris workers were witness.

Even if one did not insist on that touchy psycho-physiological point, at the very least one could complain that the term glossed over a dynamic interplay of psychological factors that were not to be understood simply. Owen, for example, suggested that the technological mystique of the magnets and electrical currents contributed to their efficacy; patients who took no stock in religious miracles might yet find the gadgets of science numinous and possessing mana (Owen 1971, 164). Significance might also be found in the nature of the relationships in these hospitals between the mostly young, mostly lower-class female patients and their authoritative male doctors; doctors "both divine and satanic" (Foucault 1973, 275), whose approval and attention they at once coveted and feared. Early it had been noticed by investigators that, using the same subject and the same aesthesiogen, one physician would invariably obtain positive results, where another would not. One may readily believe, then, that the experimenter's personality was no less important to the outcome of an experiment than the metals, magnets, and other technical equipment. Overawed by a doctor, who in their eyes possessed mana, it might be supposed that, on some level, these young women were afraid *not* to respond positively to the treatments and experiments.

All the same, Janet argued the symptoms of these patients were valuable to them. He thus proposed at one point that transfer represented, at least in part, a compromise between the patient's desire to please the doctor and her desire to hang on to her symptoms. The bilateral symmetry of the human body allowed her at once to be cured of her original symptom (as the doctor wanted), *and* to replace that symptom with one that was psychologically and physiologically almost identical (as she wanted). Janet called this the *disposition to equivalences* (Owen 1971, 156–57).

"aesthesiogenic virtue of the magnet" as late as 1882 (Bernheim 1881a, 1881b, 1882). His subsequent, rapid eschewal of all physiological approaches to the metalloscopic phenomena is at least suspicious, especially since his first theoretical formulation of the principle of suggestion the following year (Bernheim 1883, 1884) had *not* led him to reject the work of Charcot and his disciples (see Bernheim 1883, 1884). For other perspectives on the Charcot/Bernheim debate, see Harris (1985); Owen (1971); Ellenberger (1970); and Hillman (1965).

6-3. Hypnotizing the Double Brain (Phase One)

Whatever the final verdict on transfer, it was at any rate clear to Charcot and his colleagues, experimenting in the late 1870s and 1880s, that metalloscopy and transfer represented a striking proof of the potential functional independence of the brain's two hemispheres. But matters were still more complex than has been intimated. It is well known that, during the period about which we have been speaking, roughly 1877 through the mid-1880s, Charcot began experimenting with hypnosis, finally reading a report of his first findings before the Académie des Sciences in 1882. In this paper, he described hypnosis as an artificially induced modification of the nervous system that could be produced only in hysterical patients and that manifested itself in three distinct phases: catalepsy, lethargy, and somnambulism. Each phase was effected by strictly physical means and could be identified by special physiological signs: waxy rigidity, neuro-muscular hyper-excitability, somnambulistic contractures, etc. (Charcot 1882).

As Janet would later put it, it was a true tour de force for Charcot to succeed in having a report on hypnosis listened to, seriously and respectfully, by the same Académie that had roundly condemned this phenomenon three times within the past century under the name of animal magnetism. (After the last inquiry, the Académie had resolved that it would treat all further papers on animal magnetism in the same manner as it treated propositions on perpetual motion and squaring the circle). With one short report, Charcot was able to use the weight of his international prestige to give an aura of scientific respectability to a previously shunned and suspect subject.

There is little doubt about the correctness of these basic details. The historical record is less clear, however, about *why* Charcot became involved in hypnosis at this time and why his perception of it took the precise form that it did. Veith (1965, 238) was of the view that it is "astounding" that Charcot, "the most scientific and productive neurologist of his day," should have gotten actively involved with hypnosis. "It can only be explained," she declared, "as a result of his broad interest in all aspects of brain functions, including the new field of psychology." Sarbin and Coe suggested (1979, 511) that Charcot began to study hypnosis because it appeared to have a bearing on certain mental processes associated with diseases of the nervous system, especially hysteria. Indeed, it is quite true that Charcot looked upon *grand hypnotisme* as a type of artificial pathology, which could only be induced in the "suitable soil" provided by the hysterical nervous system. Other historians suggest that Charcot was influenced by the physiologist Charles Richet, who had become interested in hypnosis in the early 1870s and wrote articles insisting that hypnosis was a genuine phenomenon and that the subjects were not simulating. Because Charcot was eager to demonstrate that hysterical patients were victims of a genuine disorder and also were not simulating, as quite a few cynical neurologists believed, it perhaps occurred to him that a better understanding of hypnosis might be useful (Ellenberger 1970, 90; Owen 1971, 184).

Owen (1971) is one of the few recent historians to at least consider the possibility that Burq's metalloscopy had something to do with Charcot's growing involvement with hypnosis in the late 1870s. He limits himself, however, to the simple suggestion that metalloscopy might have led to a willingness on Charcot's part to consider certain subjects like hypnosis, which more conventional colleagues would have shunned as too bizarre (p. 184). A more recent biographical study of Amédée Victor Dumontpallier makes a stronger case for the importance of metalloscopy in the rise of hysteria and hypnosis research in Paris, but in very general and speculative terms. The author merely observes that "starting in 1878, the studies on hysteria begin to take on extraordinary proportions, and it is logical to think that the two reports [on metalloscopy] are not extraneous to the sudden interest that is now attached to hysteria" (Miloche 1982, 41).

Logical, yes, but more to the point, the primary sources speak quite clearly about the extent to which Charcot's involvement with hypnosis evolved directly out of his involvement with metalloscopy. This is an important story that unfortunately cannot be developed in any detail here.[5] Very briefly, however, it can be said that the treatment of hysteria using metals was tangled up with mesmerism from its earliest beginnings under Burq. Burq had proposed an intimate relationship between the action of metals in the mesmerized subject and the action of metals in the hysterical patient and believed that the precise bond between the two conditions was the presence of insensibility, or anaesthesia, in both. The metals were able to break a mesmeric trance, as he believed, *or* cure a patient suffering from hysteria because of their influence on this anaesthesia (see Burq 1853).

The commission's later research into Burq's metalloscopy in the 1870s led to the further conclusion that certain aesthesiogenic agents, especially perhaps the magnet, could be used both to inhibit and to

[5] For the more comprehensive treatment of the claim of a link between metalloscopy and the renaissance of hypnosis in late-century France, see Harrington, "Metals and Magnets in Medicine" [in press].

provoke certain trance states in the hysterical nervous system. What next happened, it seems, is that this idea became gradually generalized, increasingly dissociated from its metalloscopic roots. Finally, by 1882 or so, when Charcot presented his hypnosis paper to the Académie des Sciences, it had been more or less transformed into the well-known Salpêtrière doctrine, that hypnosis could be induced and broken using a wide range of physical agents that worked by irritating the perculiarly sensitive nervous system of the hysterical patient through action on the sensibility, including the special senses.

The words of Dumontpallier, speaking at the first International Congress of Experimental and Therapeutic Hypnosis in 1899, could hardly be more clear on the subject:

> It gave us keen satisfaction to have been able to render justice to a researcher whose merit had been all too long unrecognized. . . . Thus, it was magnetism that revealed to Dr. Burq the action of metals on hypnotizable hysterics, and twenty-five years later, it was the researches into metalloscopy that drove the members of this commission to study the action of electricity, of electro-magnets, of magnetised iron, and of the different procedures [employed by] the magnetizers to determine somnambulism, catalepsy, and lethargy. . . .
>
> One could say, then, that Burq was the catalyst, unconscious perhaps, for the renaissance of hypnotism. (Dumontpallier 1889, 80)

In terms of the immediate purposes of this study, the chief importance of the link between metalloscopy and the rise of hypnosis in France, lies in the fact that the interest in hemisphere functional independence aroused by metalloscopic transfer was now carried over into this new hypnosis research. Quite simply, it was proposed that one might hypnotize the brain's two hemispheres separately.

The idea was not wholly without precedent. In the 1840s, Scottish-born surgeon James Braid in Manchester, who, before Charcot, was one of the first to attempt to gain proper medical and scientific recognition for hypnosis, had produced a type of hemi-hypnosis within a phrenological context. For a while, it was Braid's view that, in the hypnotized subject, all the different phrenological faculties could be made to manifest themselves by rubbing the appropriate regions of the skull, stimulating the cortical centers below. In the course of developing this "phreno-hypnosis," he had found that "by exciting *antagonistic* points in the *opposite hemispheres* of the brain, the patient may be made to exhibit correspondingly opposite feelings in the different sides of the body." One young lady hypnotized by Braid had her "friendship" and "adhesiveness" faculties excited on one side of her brain and her "destructiveness" faculty excited on the other. Braid recounted how "she protected me [with one arm], and struck her own mother [with the other]." Phrenologist and mesmerist John Elliotson, who would savagely attack Wigan for his unwarranted claims of originality (Chapter One), felt that Braid's studies of "opposite influences on the two sides" were "the most astonishing and beautiful experiments that all physiology affords" of the truth of Mayo's aphorism that "each lateral half of a vertebral animal is separately vitalized," with "preservation of consciousness in one half . . . independent of its preservation in the other" (Braid 1843, 131–32, 138, see also p. 64; for some examples of Elliotson's own striking experiments with hemi-hypnosis, see Elliotson 1845–46a, 469; 1845–46b, 53, 74).

In the United States, the Universalist minister Joseph Rodes Buchanan (d. 1899) more or less confirmed Braid's findings on the effects of stimulating antagonistic faculties in the two hemispheres—and did the British researcher one better. Well known later in his life for his theory of *Nervaura* (a subtle force midway between the physical and the spiritual, which was supposed to emanate from the cerebral centers of the brain), Buchanan, in the 1840s and 1850s, also was a staunch advocate for the use of mesmeric techniques to advance neurological understanding. In an 1850 paper, "Duality and Decussation" (published in his own *Buchanan's Journal of Man*), he described how he had produced artificial dissociation between the two sides of the brain by making mesmeric passes with his hand along the median line of his subjects' skulls. Operating in this fashion upon one young man "of the highest susceptibility, a complete discord of the hemispheres was produced. The eyes, directed by dissociated influences, lost their focal tendency and looked apart in different directions; the expression of the countenance became wild, and the movements confused." In another case, that of a "lady of delicate structure, a few movements of the hand forward along the median line would produce immediate prostration of the whole system. She described the sensation as being somewhat like that of a knife passing through the head, though unaccompanied by the pain" ("Duality and Decussation" 1850, 515–16). The overall effect of Buchanan's research in this area was to convince him, notwithstanding the contrary opinion of most authorities of his time, that decussation of function in the human brain was "complete and unqualified" (p. 521).

In France, the first experiment in hemi-hypnosis seems to have been carried out sometime between 1877 and 1878, and was the brain-child of Gabriel Descourtis (whose work was discussed in Chapter Four), then an "élève du service" under Charcot at the Salpêtrière. The procedure employed was very simple. A patient was plunged into the first stage of hypnosis (catalepsy) by the normal physical means: a bright light, a sudden noise, etc. Since it was held that shutting the eyes stimulated the second phase of the trance (lethargy),[6] here the procedure was varied and only *one* eye was shut, presumably affecting only the hemisphere on the contralateral side; a modern-day anatomist, however, would realize at once that the anatomical reasoning here was faulty. The experimenters declared that the patient immediately became lethargic, but *only* on the side of the body that corresponded to the shut eye (see Figure 7). She remained in a state of catalepsy on the other side. She thus was hemi-cataleptic and hemi-lethargic at the same time (Charcot & Richer 1878, 970).

This constituted the basic formula from which all manner of variations were attempted. Hemi-catalepsy and hemi-lethargy were induced in a patient and then transferred from one side of the body to the other. Patients were hypnotized, and then only *half* their brain was awakened; Jules Bernard Luys, late in the century, commented on the "cold-blooded" manner with which patients then noticed that they had been divided in two (Luys 1890a). In one early study reported by Richer, a number of patients in the cataleptic state were made to wash their hands, do a bit of needlework, or perform some other task, which was always carried out in a mechanical fashion held to be devoid of all consciousness and will. One of their eyes would then be closed and the corresponding side of their body would at once become lethargic and incapable of further action. The other side of the body would continue to perform the same movements as before, even though all action had now been rendered useless. Richer noted, though, that patients often seemed unconsciously to try to make amends for their unilateral deficiency, by supporting their still functioning hand against their knee or breast (Richer 1881, 371–72, 391–93).

One early experiment, which attracted a good deal of attention both in France and abroad, involved an attempt experimentally to demonstrate, in vivo, the truth of Broca's left-sided localization of language. Conceived by a Professor R. Lépine in 1878 while working under Charcot at the Salpêtrière, this study exploited the fact that the faculty of language was allegedly abolished in the cataleptic state while it was preserved in somnambulism. Lépine proposed that one might

[6] At this point in the evolution of Charcot's thinking, hypnosis only consisted of two phases, lethargy not being clearly distinguished from somnambulism.

Figure 7. Hemi-hypnosis at Charcot's Salpêtrière. The right hemisphere (left body-half) is in a state of catalepsy and the left hemisphere (right body-half) is in a state of lethargy. (Source: Bourneville & Regnard 1877–80, Vol. 3, Planche XVI. Reprinted with the permission of the Bodleian Library, Oxford University.)

induce catalepsy unilaterally in first one than the other side of the brain and compare the effects on language. Gilbert Ballet, soon to become distinguished for his own work in aphasiology, psychopathology, and the history of medicine, described how the first test was carried out; in the later literature, one finds countless variations:

> A patient is hypnotized [somnambulistic stage]. . . . If we then invite the patient to speak, to write, to gesture, she obeys us. Now we open the left eye, meaning that we plunge the right cerebral

> hemisphere into catalepsy. Nothing is changed in the picture, at least from the point of view of language. Speech remains possible, the patient continues to write, to draw, to gesticulate if she is so ordered.
>
> But let us modify things. We close the left eye that corresponds to the right hemisphere, [while] we open instead the right eye, meaning that we put the left hemisphere in catalepsy; immediately then all communication with the exterior world is abolished. Vainly, we question the patient; she responds no longer. Vainly, we place a pen in her hand, ordering her to write her name; the pen does not move. In vain we call for a gesture or a play of the physiognomy; the hand remains immobile, the facial features remain dumb. . . .
>
> The conclusion speaks for itself. (Ballet 1880, 740; cf. Richer, 1881, 391–93; Charcot 1883, 158–59; Bérillon 1884, 98; Luys 1891)

Richer felt that this study in artificially induced aphasia strongly militated against the possibility of any simulation going on in hemi-hypnosis; for how, he asked, could an ignorant girl know that the nervous system was contralaterally organized and that language was localized in the left hemisphere only (Richer 1881, 579–80)? In fact, one *does* find a number of later studies in which it was reported that subjects had actually been rendered aphasic following manipulations, not only to their left hemisphere but to their right as well. This does not seem to have led people to reconsider the possibility of simulation, however; instead it was suggested that perhaps the right hemisphere played a certain role in the exercise of language after all (Luys 1890a; Bérillon 1884, 98).

Meanwhile, in Germany, Breslau professor of physiology Rudolf Heidenhain (1835–1897) was making his own study of Anton Mesmer's animal magnetism. In apparent ignorance of the French work, he and several colleagues, notably O. Berger, also had begun to experiment with a form of hemi-hypnosis (Berger would later repudiate this research). In their version, unilateral sensory-motor disturbances were produced by means of slight and prolonged friction on one side of the scalp. Stroking the forehead produced crossed catalepsy; stroking the occipito-temporal region, catalepsy on the same side. In this state of unilateral hypnosis, the half of the face that was awake would smile, while the other remained in an immobile, waxy cataleptic state; one arm and leg would be moved at will while the other was helpless; one eye would see distinctly, the other very imperfectly or not at all. When the right side was affected, aphasia would be produced, but only exceptionally when the left side was affected (Heidenhain 1880; Berger 1880; Hall 1881).

Although all the hypnosis effects appeared to be strictly lateralized, the German researchers felt that they nevertheless interfered with the normal working of the nonhypnotized side of the body. Persons hypnotized on the left side, for example, often experienced a certain difficulty in writing fluently with the right hand, even though it was not paralyzed; letters would be written very close together and sometimes in mirror-image. Thus, Heidenhain concluded that "the movements of each arm are influenced not merely from one cerebral hemisphere, but to a certain extent from the other" (Heidenhain 1880, 102; cf. the remarks on this same issue by the French worker, E. Chambard 1881). Heidenhain believed that all the unilateral phenomena were probably caused by expansion of blood vessels on the affected side and laid stress on the fact that, in general, anemic persons were most easily affected (Hall 1881, 100–101).

In Britain, after the early work of Braid and Elliotson, no systematic research on hemi-hypnosis seems to have been carried out. There is evidence, though, of at least one brief, informal investigation having been made. In the spring of 1881, an eleven-year-old girl was admitted to Queen Square Hospital in London, suffering from what was diagnosed as hysterical catalepsy; the cause was attributed to her having been shut up in a coal cellar all one morning and very much frightened. At Queen Square, she came under the supervision of John Hughlings Jackson and David Ferrier, neither of them obviously an expert in hysteria but both doubtless intrigued by the chance to investigate first-hand the dramatic disorder that had been receiving so much attention in France.

From certain laconic notes in Jackson's casebook, it seems that the two British neurologists had also read about hemi-hypnosis, most likely the German version as recently described in Hall's 1881 article, and they could not resist the temptation to try it out on their young patient. On April 9, they sat her down and applied friction to the temporal region on one side of her scalp, hoping to induce (ipsilateral?) unilateral catalepsy. They found that, unless she were made to count the number of strokes, she was indeed apt to "go off as usual," but "no unilateral effects [were] produced." A second experiment made four days later, which may have been an attempt to invoke aphasia by rubbing the left side of the head, also failed to produce any exciting results. Efforts then seem to have been abandoned. Neither Jackson nor Ferrier ever appear to have made any public allusion to this little study and may have been a bit embarrassed by it. The case report in which Jackson refers to the experiments, however, was discovered in the 1950s by historians of neurology, Riese and Gooddy. They published it with an "interpretation," but without seeming to have understood the significance of all the unilateral scalp-rubbing exercises (Riese & Gooddy 1955).

6-4. *Hypnotizing the Double Brain (Phase Two)*

Back in France, hemi-hypnosis entered a new phase with a series of studies that went beyond the relatively mundane trick of suppressing the activity of one or the other hemisphere. Instead, people began to explore the possibility of inducing a type of hemisphere disconnection or duality of mind, analogous to that presumed to occur spontaneously in certain types of insanity. The man who pioneered this line of research, and whose name consequently became more intimately tied up with hemi-hypnosis than any other, was Amédée Dumontpallier (1827–1899). Before his involvement with metalloscopy, which profoundly changed the direction of his career, Dumontpallier's main research interests had been neuropathology and especially obstetrics and gynecology; he obtained his medical degree on the strength of a thesis entitled "l'infection purulente et l'infection putride à la suite de l'accouchement." He was part of the bevy of Claude Bernard's favored students, known as "la Pléiade" (he appears in the famous painting by L'Hermite of Bernard surrounded by disciples at the Collège de France), and he later served as Armand Trousseau's chef de clinique at the Hôtel-Dieu. In 1870, he was named chef de service at the Pitié Hospital, where he was to remain, working there that first year in war conditions and subsequently under siege. In 1879, he became secrétaire perpétuel of the Société de Biologie, a post of which he was very proud. Colleagues would later recall "his olympian head, his lofty stature, the fullness of his movement, his frank and open gaze, the warmth and authority of his speech, his captivating familiarity" (Vallin 1899, 601). These admirable qualities would eventually be immortalized in a statue of an aging, bearded Charlemagne on horseback, which still stands on the Notre-Dame parvis in Paris, for which Dumontpallier served as the model.[7]

[7] The story of Dumontpallier's connection with this statue of Charlemagne has been told by Miloche (1982, 16). The statue was the work of the Rochet brothers, who had long searched in vain for a model to represent the features of the emperor. Knowing Dumontpallier, it finally occurred to them to ask him if he would do the job. He accepted with joy, and the statue was completed and put on public display.

At the end of the century, however, the Paris city council seems to have become disinclined to continue to honor the memory of the old emperor and was on the point

Like Charcot and Luys, Dumontpallier's interest in hypnosis was first awakened by the 1877 investigation of Burq's metallotherapy. Assisted by his student Paul Magnin, he began a series of studies on the role of physical agents in the production of hypnosis and their relationship to agents acting in metalloscopy (see, e.g., Dumontpallier & Magnin 1882a, also Miloche 1982, 53–60). Before the middle of 1882, his interest in the phenomenon of hemi-hypnosis was relatively minimal, though he had dutifully reproduced a number of the results reported by Descourtis, Charcot, Lépine, and others. The real turning point came on May 16, when he and his colleagues arrived at the Pitié one morning to find that a certain hysterical patient, Maria C., had fallen into a profound lethargic sleep. Gazing upon her closed eyelids sufficed to awaken her. Although awake, she remained incapable of speech but made doctors understand by means of gestures that she wished to write. She was then able to explain that, during the night, a delirious neighbor had terrified her by approaching her bed, so that she had felt herself become paralyzed all over her body. What was remarkable about her case, in the doctors' eyes, was that although to all external appearances she had been completely unconscious, her hearing was conserved and her memory was clear for all the events that had passed (Dumontpallier 1882b).

This bout with lethargy apparently shook up young Maria C.'s nervous system, for shortly thereafter she began to act in such a way during hypnosis as to suggest that her two hemispheres were functioning unequally. When Dumontpallier fixed his gaze upon her eyes to induce trance, he noticed that she began to turn her head repeatedly from right to left, as if she were trying to follow an object with her right eye. When hypnosis was produced, the characteristic physical signs could only be provoked on the right superior member and left inferior member. A light shown in the left eye and pressure on the left vertex were without effect. When she was awakened, it was discovered that she had lost sensibility in her left superior member and sight in her left eye. It thus seemed that, during the hypnotic state, "only the left brain had conserved its functional activity."

On the following day, using metallic plaques, Dumontpallier was able to cause only the *right* hemisphere to respond to the hypnotic manipulations and later found that, by applying plaques to the two

of selling the statue to a certain town in Germany. Dumontpallier's indignation knew no bounds—an ardent patriot such as he, to see a part of himself exiled to Germany! He kicked up a noisy fuss, and the city council, fearing political fallout, backed down. It is thanks to Dumontpallier, then, not only that the Charlemagne statue was completed but also that it still stands on the Notre-Dame parvis today.

sides of the skull simultaneously, some sensibility returned to both sides of the body and hypnosis could be bilaterally induced but much more feebly. It seemed that Maria C. had only a very limited supply of nervous energy, which was weakened through diffusion. As Dumontpallier put it, the "sum of the nervous system's activity being equally distributed in the two hemispheres of the encephalon, that activity appeared notably less for each hemisphere than in the instances when one hemisphere alone was the seat of nervous activity" (Dumontpallier 1882a, 394–96).

Intrigued by these results, Dumontpallier resolved to carry out further research to see what light hypnosis could throw on the functional independence of the two hemispheres. He claimed to be particularly interested in (1) whether certain functions depended primarily on one or the other hemisphere; and (2) whether substitution of function between the hemispheres existed and, if so, under what conditions and in what proportions (p. 397).

Several weaks later, he returned to the Société de Biologie with the first of his reports. The experiments he recounted, in the words of Edgar Bérillon, "had not yet been described anywhere, either in France or abroad, and to the interest that they offer in themselves is adjoined the priority of their description" (Bérillon 1884, 169). Dumontpallier had begun by returning sensation to both sides of his subjects' bodies and inducing bilateral hypnosis. Then, having gotten both hemispheres into the same phase of trance (catalepsy or somnambulism), he implanted within his subjects' brains *two* distinct suggestions or hallucinations, each allegedly mediated by a different hemisphere. In one study, for example, focusing on the visual and auditory modalities, a subject in the somnambulistic phase of hypnosis was made to hallucinate a pleasant country scene in her right hemisphere, in response to suggestion spoken into her left ear. Then, that ear being blocked up so as to isolate the right hemisphere from further sensory input, the left hemisphere was equally made to believe itself witness to a bloody wild-boar hunt. Declared Dumontpallier, "*her face expresses fear on the right side*, whereas *the left side of her face expresses contentment* at the sight of the pastoral scene." On being questioned alternately in each ear, the subject was able to "retrace each of the episodes of this double visual hallucination, whose origin had been in the organ of hearing" (Dumontpallier 1882b).

Over the next several years, Dumontpallier would continue to develop this novel line of research, assisted now by Edgar Bérillon, whom Paul Magnin would later affectionately recall as "a young student with a skull already bald, a highly developed forehead, [and] piercing eyes sheltered under a pince-nez" ("Banquet en l'Honneur du Docteur Bérillon" 1906–07, 44). Bérillon's much-acclaimed medical dissertation on the functional independence of the cerebral hemispheres was the result of this collaboration with Dumontpallier. In addition to recounting a series of new innovations in hemi-hypnosis, the work defended the case for hemisphere independent action with evidence drawn from anatomy, physiology, embryology, cerebral thermometry, comparative pathology, human pathology, mental pathology, and introspective psychology (Bérillon 1884).

The hypnosis experiments, however, were regarded by Bérillon as the most original and significant part of his work. For example,

> A patient, the girl by the name of Pauline G. . . . aged twenty years, is put into catalepsy. It was made certain beforehand that her sensibility was equally distributed in the two sides of her body. The experimenter places the fingertips of the patient's left hand on her lips . . . and transmits to the arm on the same side those movements used in the act of blowing kisses with the hand. Immediately, the patient repeats the movement, at the same time as the left side of her face takes on a smile and brightens up. While the movements of the left arm continue and the left side of the face retains its smiling expression, one sets the arm and hand on the right side in the posture that a person would take to push away an object, an image that filled one with horror; the patient retains her right arm in this posture, and the right side of her face takes on an expression of terror, her eyebrow contracts, her right labial commissure [lip] droops. [See also Figures 8 and 9.]
>
> In this state, the patient's face displays a *double* expression that agrees with the attitudes [adopted by] her limbs. (Bérillon 1884, 173; for more hypnosis work in this vein, see also Dumontpallier & Bérillon 1884)

Word of Dumontpallier's studies began to excite interest in the Paris hospitals. At Ste. Anne's, Valentin Magnan invited Dumontpallier to give an experimental demonstration of these bilateral but differing hallucinations (Dumontpallier & Magnan 1883). At the time, he was in the midst of studying the same sort of phenomenon, but in a clinical context (see Magnan 1883; also Chapter Three). A great deal was made of the claim that the double expression seen on these subjects' faces was so bizarre that it could not possibly be simulated. "There is no mime," Dumontpallier told his audience at Ste. Anne, "however able he may be, that can simultaneously convey on each side of his face such opposing expressions" (Dumontpallier & Magnan 1883, 874). Why this was felt to bear so directly on Magnan's clinical work is far from clear; none of Magnan's patients contorted their faces along the median line to express joy and despair simultaneously.

At the Salpêtrière, Charcot seems to have decided to see if he could duplicate Dumontpallier's results and indeed found that, by putting the two arms of a cataleptic subject into two attitudes, each suggestive of a contrary emotion, he too could cause a smile to appear on one half of the subject's face and an angry puckering of the brow on the other half (Charcot 1883). An English visitor to the Salpêtrière, George Robertson, was shown this particular trick and declared, one imagines, with some disquiet, "I had before me a person angry with one side of the face and laughing with the other" (Robertson 1892, 505–506).

Sensational though Dumontpallier's experiments might have been, later in the century at the Charité hospital, Jules Bernard Luys was to do him one better. This hypnosis worker believed himself to have produced not only a doubling of the intellectual faculties, like the Pitié researchers, but a doubling of *emotional* experience as well. Using magnets, he said, he could cause joyful feelings to migrate to one side of the brain and feelings of despair to the other, while in the exact middle there would appear a line of experimentally induced indifference (Luys 1890b, 126–27). He had also managed to produce emotional polarization, he said, by exploiting his patients' presumed sensitivity to medicines sealed in glass tubes. One young woman, for example, was put in a hypnotic trance, and a tube containing a gram of pilocarpine was pressed against the left side of her neck. She felt a sensation "of heat, of discomfort, and of suffocation" and then "began to express a series of tender sentiments . . . to an imaginary being. 'Come here, come here,' she said, 'draw close to me, you know very well that I love you.' " The tube was then transferred to the right side of her neck, and her emotional state dramatically reversed itself. " 'I'm cold,' " she said, " 'I'm drenched in cold sweats.' " She then addressed herself again to the same imaginary being. " 'Go away,' "she said, " 'I detest you,' " and the gesture she made with her hands "expressed aversion and repulsion" (Luys 1890b, 129–30).

6-5. *Contemporary Reaction and the Decline of a Research Program*

Although Luys had his share of loyal followers, many of his contemporaries were suspicious of his hypnosis findings; and in the end, his

Figure 8. Hemi-hypnosis at Dumontpallier's ward in la Pitié hospital: Maria C. in a state of catalepsy. The two sides of her body have been put into attitudes suggestive of different acts and moods. Waving farewell with her right hand, she smiles on the right side of her face; gesturing "menacingly" with her left hand, she scowls on the left side of her face. (Source: Bérillon 1914, 40–41. From the private collection of Dr. P. Morel, Hôpital Psychiatrique du Bon Sauveur, Caen.)

increasingly fantastic work in this area robbed him of a good part of the scientific reputation it had taken him some forty years to acquire (see, e.g., the scathing remarks in Hart 1893). None of his critics ever doubted his sincerity but, as the *British Medical Journal* tartly remarked on the occasion of his death in 1897, "in dealing with the mysteries of hypnosis the investigator has more need of the wisdom

of the serpent than the simplicity of the dove" ("[Obituary] Luys" 1897, 619).

In contrast, the rather less extravagant hemi-hypnosis experiments of Charcot and Dumontpallier, at least for a while, were received quite well, attracting the admiring notice of such eminent French scientists as Pasteur, Milne-Edwards, Chevreul, Henry Bouley, Brown-Séquard, and Paul Bert. This last, who had become president of the Société de Biologie after the death of Claude Bernard, declared before a session of the Société on June 24, 1884, that all the new researches in hypnosis had uncovered nothing that the old magnetizers had not already known—with *one* exception. Here, he made reference to Dumontpallier's work:

Figure 9. Hemi-hypnosis at Dumontpallier's ward in la Pitié hospital: Maria C. in a state of somnambulism. She is under the influence of two simultaneous but opposing visual hallucinations that have been separately given to her two brain-halves. Her face expresses "gaiety" on the right side and "gravity" on the left. (Source: Bérillon 1914, 42–43. From the private collection of Dr. P. Morel, Hôpital Psychiatrique du Bon Sauveur, Caen.)

> The only really new fact, [the one] that has impressed me most and that the old magnetizers had never realized, is that of dividing the hypnotised man in two and making of him a double individual. I consider therefore that these studies must be pursued on account of the exceptional interest that they offer. (cited in Bérillon 1899, 13; and 1884, 185–86; but not recorded in the Société's proceedings)

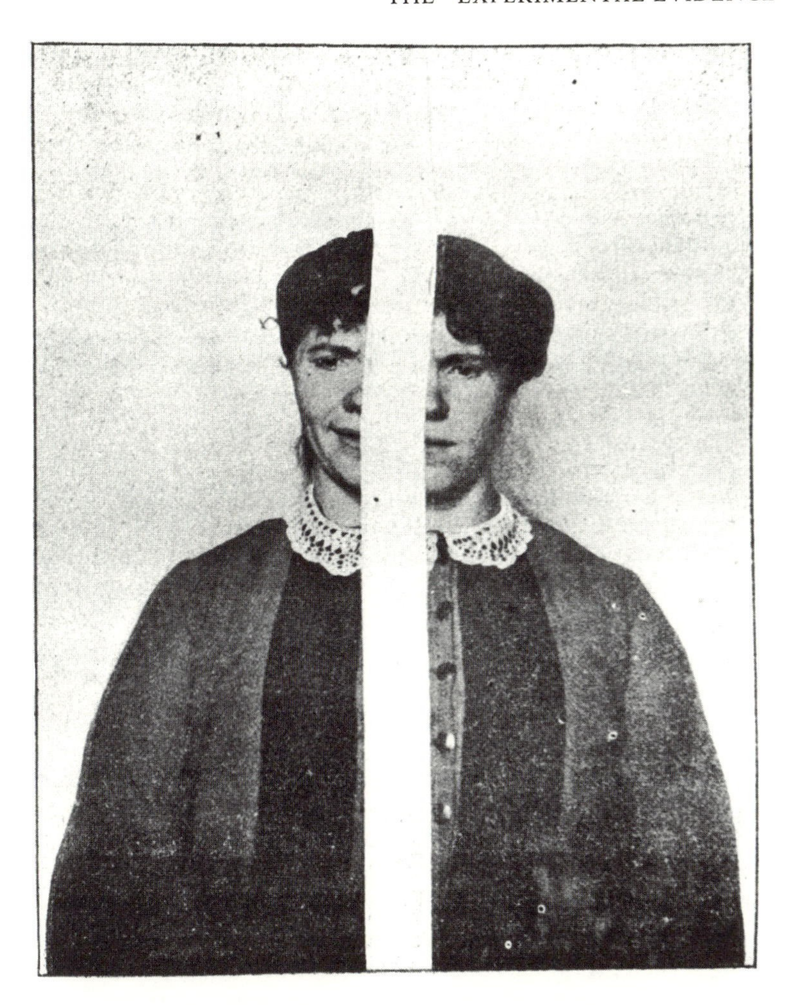

In light of this ringing endorsement by a leading representative of the scientific elite, it should come as no surprise to learn that professional popularizers were not long in jumping onto the hemi-hypnosis bandwagon. In 1884, a rather lurid advertisement for a "grande Séance de Magnétisme" given by Carl Hansen of Copenhagen began making the rounds in Paris. One of the many attractions the public was promised was a "scientific experiment" entitled "the two-faced man" (advertisement reproduced in Bottey 1884, 260). In 1886, sophisticated Parisians ("gens du monde") were given the opportunity to hear all about the new work on cerebral dualism and hypnosis by the popular "spirituel conférencier" Achille Poincelot. Speaking before a presumably rapt audience at the salle des Capucines, Poincelot described both the hemi-hypnosis experiments and the evidence for hemisphere independent action from other quarters; all of which demonstrated "definitively that we have *two functional brains.*" He was happy, he said, to be in a position to affirm this important truth in the lecture hall that evening (Poincelot 1886).

The great stir of interest in France over the hemi-hypnosis work also had inevitable ripple effects abroad. In the United States, the *Journal of Nervous and Mental Disease* was thoroughly enthusiastic about the studies, announcing unqualifiedly to its readership that "Dumontpallier has demonstrated the functional independence of each hemisphere." It did not hesitate to grant precedence to certain of Dumontpallier's results over some conflicting evidence from anatomical research on the question of the decussation of the olfactory nerves. On anatomical grounds, there was reason to believe that these nerves passed ipsilaterally to the cortex, but in Dumontpallier's subjects, sense impressions in all the modalities seemed to have been processed in an equally crossed fashion. "Hence experimental physiology demonstrates a total or partial decussation of the olfactory, auditory, and optic nerves" (Ott 1883, 324; cf. Dumontpallier 1882b, 794n on this same issue).

Max Dessoir in Germany, one of that country's best-read authorities in the field of hypnosis (see Dessoir 1888), also warmly saluted Dumontpallier as "the first one to show in a decisive manner that the duality of the cerebral system was proved by these hypnotic phenomena." He went on to make the intriguing remark that "the whole doctrine of hemi-hypnotism" had important implications for another problem currently of interest to workers in the field, so-called thought transference (Dessoir 1887). It is not entirely clear what Dessoir was driving at here, although one can see how the idea of two "minds" simultaneously holding forth inside a single cranium might be construed in such a way as to seem relevant to the problem of alleged telepathic communication (indeed, cf. Word 1888).

Not all foreign commentators were so charitable though. In a detailed critique of those early Salpêtrière experiments that claimed to be a "new proof" of Broca's left localization of language, Frederick Bateman in England pointed out that, because Charcot's hemi-hypnosis depended upon a method of presumed cerebral manipulation via the retina, the assumption had to be that all the fibers of one eye's optic tracts passed from the retina to the opposite hemisphere. Physiology in the 1880s was still not clear on this point. Many physiologists believed, correctly according to modern teaching, that the fibers crossed partially at the chiasmas and that was all, as Newton had suggested as early as 1704. If they were right, this meant that stimulation of one retina should affect *both* hemispheres, "and the Salpêtrière experiments would lose all their import" (Bateman 1890a, 339–40). Charcot accepted that decussation at the optic chiasma was only partial; however, in order to account for the phenomenon of one-eyed blindness in hysteria, he had proposed in the 1870s that there was an additional crossing of the fibers in the corpora quadrigemina, which made in the end for full decussation (Charcot 1875a). As the century progressed, this schema came under increasing attack by such writers as Michael Foster and David Ferrier in England (the latter of whom had originally praised it), and Seguin and Jules Soury in France (Soury 1896; Bateman 1890b, 292–93). The debate was further complicated by a third school of thought, largely centered in Germany, which argued that *complete* decussation of the human optic nerve actually took place at the chiasma, in the same manner as had been shown for a variety of other, lower animal species (Bateman 1890a, 339–40; 1890b, 292–93).

But optic tract decussation was not the only anatomical consideration raised by Charcot's work. Bateman pointed out that, before any of the hypnosis experiments could be accepted as bearing on the question of Broca's speech localization, it would be necessary to prove that at least some of the optic tract fibers found their way into the anterior lobe of the brain, where the motor speech centers were supposed to be located. This was by no means univerally admitted. True, Gratiolet in France and Hamilton in England believed that the optic fibers connected with every major area of the cortex in man, but Ross and Gowers denied this, arguing that the connection was exclusively or chiefly occipital. Bateman made reference to some private communication with Gowers, who "speaks most emphatically upon this point, and says that his views of the physiology of the cortex would exclude altogether the idea of any passage of fibers from the optic tract to the motor speech region" (Bateman 1890b, 294; 1890a, 342).

In England, the French hemi-hypnosis work was further criticized, *not* on the grounds that the work itself was defective but on the grounds that one was never justified in drawing conclusions about human beings in their normal state from observations made on human beings in an abnormal state. A reviewer of William Ireland's (1886) *The Blot Upon the Brain*, which discussed Bérillon's hypnosis work, declared, "It is the same difficulty which exists in drawing conclusions from post-mortem examination of the brain, when it is impossible to show that the brain exists under the same conditions as it did during life." The reviewer argued that, in the normal state, there exists "a psychical *unity* underlying the duality of the hemispheres, of which the facts of hypnotism do *not* take cognizance." For this reason, "the only conclusion that can truthfully be drawn from the experiments seems to us to be that in the hypnotic state, so far as there is automatic production of induced hallucinations, there is duality of function." This duality disappeared, he said, the moment "psychic unity"—an entity that sounds suspiciously like a soul—reasserted control over the brain ("[Review] *The Blot Upon the Brain*" 1885–86, 548–49).

Although Hippolyte Bernheim in France explicitly criticized the Salpêtrière hemi-hypnosis studies on several occasions (Frederick Bateman, for one, wrote both to him and to Hack Tuke asking for their opinion on the language localization experiments), it would probably be more accurate to take a broader view of the Nancy school's role in undermining hemi-hypnosis. This line of research was just one among many to fall victim to Bernheim's campaign to discredit the Paris school and to redefine hypnosis as a strictly psychological phenomenon, rather than a physiological one, that could be induced by means of suggestion in virtually any normal, healthy individual.

There is no shortage of evidence to show that Charcot was also aware of the importance of suggestion and other psychological factors both in hysteria and hypnosis, particularly during the last years of his life. Yet he continued to stress the *physiological* side of these states at least in part, and perhaps a bit ironically in retrospect, because he did not wish to be led astray in his studies. According to Janet,

> he was never weary of repeating that these psychological phenomena were of a very complicated kind; that their study was a delicate matter; and that all the mistakes of the magnetizers had been due to faulty methods, and their unfortunate way of beginning the study of a problem at the most complicated end. For his part, he would be guided by Descartes' rule, and would begin by the study of simple facts, of those in which scientific investigation is easiest. Before studying psychological intricacies, the involved happenings in the mind of a person in an abnormal state, we must first of all ascertain the precise characteristics of this abnormal state and must learn how to recognize it by definite signs which cannot be counterfeited. (Janet 1925, I:166)

Even though Bernheim and other critics of the Paris school greatly dampened enthusiasm for phenomena like hemi-hypnosis, it is important to realize that interest in this subject did not die completely. What one sees instead is a partial migration of hemi-hypnosis out of orthodox medicine and into a mesmerically oriented, occult fringe of French hypnosis research. This same fringe, significantly, was no less appreciative of Burq's metalloscopy after it had ceased to be respectable within mainstream circles. For example, the neuroanatomist-turned-mesmerist Jules Bernard Luys at the Charité hospital kept the hemi-hypnosis candle burning into the 1890s. Another physician, A. Baréty, one of the pioneers in the effort to resurrect and rehabilitate the biomagnetic fluid of Mesmer, also still was actively involved in hemi-hypnosis research and theory in the late 1880s. He even went so far as to suggest that that phenomenon could be used as a diagnostic aid in hysteria. The doctor, by provoking a "doubling of the personality," became able to "*interrogate* each cerebral hemisphere separately." This meant that he could readily determine which of the patient's two hemispheres was disordered and from what it suffered; or, where both sides of the brain were afflicted, how the trouble differed from one side to the other (Baréty 1887, 572).

At the turn of the century, Hector Durville, founder of the Société Magnétique de France and the École pratique de Magnétisme et de Massage, introduced a new twist in the old hemi-hypnosis story. Experimenting on a number of entranced subjects, he found that touching various points on their skulls determined, in accordance with the laws of "human polarity," either the excitation or sedation of the corresponding cerebral faculty below. There was nothing new about this discovery in itself; the wonders of phreno-hypnosis had been known to people like Braid and Elliotson since the 1840s. Durville, however, stretched beyond these early workers by using the procedure to investigate the differential functioning of the two sides of the brain. Where all sensory and motor centers had proved, under mesmeric stimulation, to be bilaterally symmetrical, Durville had unearthed a distinct pattern of asymmetry in the localization of the "moral and intellectual" faculties. In general—predictably—the left side of the brain was found to preside over positive, high-minded, and joyous emotions and instincts; the right side of the brain proved to be the seat of melancholy, hypochondria, and discontentment (de Rochas 1904–05, 49).

Meanwhile, back in more orthodox medical circles, matters were taking a rather different turn. Beginning about 1887, Dumontpallier gave up hemi-hypnosis and began to focus increasingly on the therapeutic possibilities of hypnosis and less on its physiological aspects. He was greatly impressed with the power of Bernheim's suggestion but still did not think it could explain all that he had observed. Presiding over the first International Congress of Experimental and Therapeutic Hypnosis, held in 1889, an event that had been organized by Bérillon, he declared that "the truth lies in the schools of Paris and of Nancy" (Bérillon 1899, 14–16).

Edgar Bérillon also effectively gave up hemi-hypnosis after completing his doctorate, but he was to remain preoccupied with hypnosis for the rest of his career, particularly its pedagogical potential. Although increasingly he admitted the importance of psychological factors in hypnosis, he also invested considerable energy into an intense, even vitriolic, campaign to undermine the Bernheim school and its vacuous, monolithic concept of suggestion. It seems that everything that was new about the Bernheim approach was wrong; everything that was right had been said before. In 1906, a banquet was held in his honor at the Palais d'Orsay to celebrate his nomination into the Legion of Honor and to commemorate his many scientific contributions. Here, it was ringingly declared over the cheese, fruits, and assorted sweets (gâteau Henry Bouley, bombe Dumontpallier) that, just as Claude Bernard had come to personify physiology and Paul Broca had incarnated anthropology, so Bérillon, "whose name is universally known, more so even abroad than in France, may be identified with hypnotism itself" ("Banquet" 1906–07, 74).

In spite of his later, wide-ranging interests, Bérillon always retained a soft spot in his heart for his early doctoral work, whose validity he seems never to have questioned. This is shown plainly in a short report he wrote in 1901 with his old friend Paul Magnin, describing a thirteen-year-old hysterical girl who was hypnotized by means of eye fixation and suggestions of deep sleep. When the trance state was achieved, the authors discovered, to their surprise, that one half of her body had fallen into a state of catalepsy and the other half into lethrgy. That this could happen, they said, demonstrated the fallacy of assuming, as so many of the Nancy school did, that simply because one admitted the importance of suggestion in producing hypnosis, the trance state itself was not a genuine, physiological distinctive phenomenon. "The verification of this fact demonstrates, with evidence, the distinction that one ought to draw between suggestion [as] *cause* and hypnotism [as] *effect*; the same cause [in this case] being able in a simultaneous fashion, without any design on the part of the experimenter, to provoke the appearance of a state of hypnotism of a different degree for each side of the body" (Bérillon & Magnin 1901–02, 283). Referring to this case in his own 1903 book on hypnosis, J. Grasset thus would affirm that hemi-hypnosis demonstrated the functional independence of the two hemispheres, and that it demonstrated it *even if* such hemi-hypnosis was produced through suggestion (Grasset 1903, 170).

The number of those who walked a middle course between psychologism and biologism, however, would diminish as the century progressed. Oscillations in the history of psychology seem to occur with the same inexorability as those that guide the fates of governments, and the onset of a swing in a new direction is in some sense symbolized by the triumph of the Nancy school over that of Charcot in the 1890s—a triumph abetted in part by the master's death in 1892. Where the last decades of the nineteenth century were marked by a desire to reduce mental pathology to brain pathology, the first decades of the twentieth century were no less marked by an interest in studying "mind" and its disorders on their own terms. In this new environment, not only would metalloscopy and hemi-hypnosis be made to seem ridiculous, but equally the French duality of mind literature that was inspired so greatly by those unhappy experiments would fall into neglect and disrepute (see Chapter Nine).

CHAPTER SEVEN

The Hughlings Jackson Perspective

> *[The doubleness of the nervous system] . . . is a very striking fact, but one so well known that we are in danger of ceasing to think of its significance—of ceasing to wonder at it.*
> *—J. Hughlings Jackson, 1874*

NINETEENTH-CENTURY thought on brain duality and functional asymmetry reached its philosophical climax in the person of John Hughlings Jackson (1835–1911). The maverick of this study, Jackson's writings on hemisphere specialization and the duality of the mental operations collectively constitute a dense forest of physiology and mental philosophy that so far has been very inadequately traversed by historians and other modern commentators. This chapter does not pretend to be the last word on the subject, but I hope at least to cut through enough brambles to bring into focus the essential roots and branches of Jackson's thought in this area.

7-1. Dissent from the French Faculty School

A modest man of reclusive habits and few diversions, Jackson was born in Yorkshire in 1835. At fifteen, he became apprenticed to Dr. Anderson, under whom he began his medical education at York Hospital Medical School. He continued his studies at St. Bartholomew's Hospital Medical School in London and qualified as a surgeon in 1856.

An older friend and fellow medical man, Jonathan Hutchinson, helped Jackson secure his first appointment, as a lecturer in pathology at London Hospital Medical College in 1859. At this time, Jackson was suffering doubts about his chosen career and considered abandoning medicine altogether to study philosophy. Hutchinson, however, managed to dissuade him from making such a move and, a bit reluctantly perhaps, Jackson went on to take an M.D. degree in 1860 (the positions of surgeon and of physician are distinguished from each other in England). He decided to specialize in neurology after being firmly told by Brown-Séquard, who then was teaching in London, that "it was foolish to waste his efforts in wide observation of disease in general and that if he wished to attain anything he must keep to the nervous system" (Hutchinson, cited in Greenblatt 1965, 363). In 1862, Jackson joined the staff of the newly created National Hospital for the Paralysed and Epileptic at Queen Square, and in 1863, he became attached as a physician to the London Hospital as well. He would remain at these two posts for the rest of his career, slowly gaining respect for his work on epilepsy and speech disorders, yet partly alienating and confusing his colleagues with his complicated, unconventional notions, his pronounced philosophical bent, and his opaque, painfully precise style of writing.

Jackson made his first published reference to Broca's work on aphasia in a May 1864 letter to *The British Medical Journal*, in which he asked readers to search their records for cases of right hemiplegia coinciding with loss of speech. "M. Broca believes that disease of the brain *on the left side only* produces loss of speech; and if I were to judge from the cases under my care, I should think so too" (Jackson 1864a, 572).

It is a small historical mystery precisely how Jackson could have become acquainted with Broca's views on language localization at that date, since the details of the French aphasia debate did not become generally known in England until several years later. Greenblatt (1970, 561) wrote that "despite many discussions of this early period in the history of aphasiology, the exact nature of the intellectual relationship between Jackson and Broca has never been clarified." Greenblatt began by suggesting that Brown-Séquard may have known of the language localization debates going on in France and alerted Jackson to them; the two men served together on the staff at Queen Square in 1862 and presumably remained in contact afterwards. He admitted, however, that there is "no concrete evidence" for this idea. Alternatively, Greenblatt said, Jackson may have seen the February 1864 report in *The British Medical Journal* describing the letter Broca sent Trousseau on the relative merits of the terms *aphemia* and *aphasia* to denote loss of speech (see Chapter Two, note 3). This second suggestion, though, fails to explain how Jackson could have known of Broca's views on language asymmetry, since the letter to Trousseau did not address that question. Finally, Greenblatt argued, Jackson may have come across some of Broca's original articles in the French literature by sheer chance, for he read French (Greenblatt 1970, 562–63). This is possible but still fails to explain why Jackson would have come to feel that Broca believed only left-sided damage produced loss of speech. Before 1865, all of Broca's formal, published pronouncements on the problem of language localization either disregarded the unilateral issue or else affirmed a great reluctance to concede that the brain could be functionally asymmetrical (see Chapter Two).

I would reject all of Greenblatt's suggestions, then, and offer one of my own: that the actual original source of Jackson's knowledge about the French debate on aphasia, including the debate over the problem of language asymmetry, was a July 1863 report by the Parisian physician Jules Parrot, published in the Paris *Bulletins de la Société Anatomique*. This described a patient who had suffered atrophy of the right third frontal convolution without speech defect (see Chapter Two). In the lengthy discussion that followed this report, a discussion dominated by Broca, virtually all the important issues involved in the French controversy over language localization to date were aired and perused. Most significantly, Broca discussed in some detail the "strange predilection" that lesions causing speech loss seemed to have for the left hemisphere. Unless some counterevidence could be found, Broca said—some solid cases correlating speech disorder with *right* brain damage—"it will be necessary indeed to recognize that the faculty of articulate language is localized in the left hemisphere" ("Atrophie" 1863, 380–81).

What reason is there for supposing that Jackson had studied the Parrot article by May 1864? The answer lies in a passage buried in an 1866 article on speech defects, where Jackson quotes Broca from the 1863 Parrot article and remarks in passing that Broca *himself* had been "good enough" to send him that piece "a year or two ago" (Jackson 1866c, 661). A year or two ago could well have been sometime in 1864, the year Jackson first alluded to Broca's views on language asymmetry. Jackson's laconic comment here also casts a bit of light on the more personal side of the relationship between the two men, suggesting that they were interacting on a cordial professional level from the earliest years of aphasiology.

Though he never ceased to speak of Broca with respect, Jackson soon, and almost alone among the important early aphasiologists, began to suffer serious doubts about the reality of Broca's "speech faculty." Historians (e.g., Young 1970; Greenblatt 1970, 1977; Engelhardt 1975) have shown in various ways how the "faculty" approach to mind and brain was essentially foreign to Jackson, who had been reared in the British philosophical tradition of associationism. As a student at York, he had been strongly impressed by the teachings of Thomas Laycock (1812–1876), who had pioneered the idea that one could account for all nervous functioning, including that of the cortex, in terms of a reflex principle operating according to associative laws (Chapter One). Jackson had also imbibed the writings of such associationist psychologists as Alexander Bain, George Henry Lewes, and especially Herbert Spencer. We may conclude that he ultimately found it impossible to accept the essentially dynamic view of mental processes implicit in associationist theory and, at the same time, go along with an approach that broke up aspects of mind and localized them as static entities. Thus one finds him, in the late 1860s, warning that "while we may localise the *damage* which makes a man speechless, we do not localise language. It will reside in the whole brain (or whole body)" (Jackson 1867, 669).

Jackson's dissent from the French faculty approach to language, however, can be even more decisively linked to his growing puzzlement over the "strangeness of this association of 'the loss of a purely mental faculty' [i.e., speech] with a decidedly physical defect such as hemiplegia" (Greenblatt 1977, 424). It soon became clear to him that he was dealing with an issue that struck straight at the heart of the mind-body problem. Was it possible that aphasia and hemiplegia (a "mental" and a "physical" defect, respectively) could be caused by a *common* lesion? For a short while, Jackson denied it, proposing instead that the association of symptoms was brought about by an obstruction of the left middle cerebral artery that simultaneously damaged the left corpus striatum (causing the paralysis) and destroyed the speech center of the left frontal lobe (see e.g., Jackson 1864c). When this view became untenable, however, largely due to the paucity of postmortem evidence in favor of it, Jackson was forced to re-examine some fundamental premises. On what grounds could one justify the practice of differentially localizing mental functions on the one hand and physical ones on the other? How, indeed, could one speak of spatially localizing something *mental* at all?

The answer, quite simply, was that one could not. Some seventy years earlier, the philosopher Immanuel Kant had denounced the doctrine of the seat of the soul, pointing out (in Riese and Hoff's summary) that the soul "perceives itself only by the internal sense; thus it cannot assign to itself any place in the body." *Motion and sensation alone*, Kant went on to say, admitted "of a spatial relationship with the cerebral organs" (Reise & Hoff 1950, 59, 61).

This was essentially Jackson's conclusion as well. The only logically possible field of inquiry for physiology was that of sensory-motor relations. There could be no "physiology of the mind," he declared, any more than there could be a "psychology of the nervous system." Thus, the explanation for the "strange association" of a "mental" defect, like speech-loss, with a "physical" motor defect, like hemiplegia, lay in realizing that, at least so far as the physiologist was concerned, hemiplegia and aphasia were *both* motor disorders, to be understood within a common framework. Combining his understanding of reflex theory with the evolutionary ideas of the time, particularly as interpreted by Herbert Spencer, Jackson now took the crucial step of rejecting the traditional view of the cortex as the terra incognita of the mind. As Laycock had done before him, albeit more speculatively, Jackson instead proposed to extend the principle of sensory-motor reflex action right up into the highest cerebral centers. Young (1970, 204–210) concluded that this was his most important contribution to the history of cerebral localization.

7-2. Basic Principles: Concomitance, Evolution, Localization, Compensation

Jackson's solution to the problem of mind/brain relations was motivated by both heuristic and philosophical considerations. To begin, as Engelhardt (1975) has pointed out, commitment to psycho-physical parallelism (or "concomitance") meant that neurology and medicine would no longer be able to get away with pretending that neurophysiological functions could be accounted for by vague references to mental events. These sciences were concerned with the world of physics and had little hope of progressing so long as they persisted in talking about the brain "as if it were a kind of mind made up of cells and fibers" (Jackson 1879, 240). More fundamentally, because the doctrine of the conservation of energy had shown that the physical world was a closed system, the idea that mental events could cause physical ones implied, in effect, a denial of "natural law" (Engelhardt, 1975, 144–45). For Jackson, in short, "the universality of natural causation," as Spencer had put it, made it impossible to grant mental states any independent causal status and, thus, led logically to belief in some form of psycho-physical parallelism (see Jackson 1879; but see also Morton Prince's 1891 criticism of psycho-physical parallelism, made with special reference to the views of Jackson).

In an 1857 article, "Progress: Its Law and Cause," Spencer had described the law of development in evolution as "an advance from homogeneity of structure to heterogeneity of structure" (cited in Mandelbaum 1971, 90). Jackson's integration of reflex doctrine with Spencer's evolutionary philosophy thus led him to take a view of nervous functioning that, in its barest outline, can be summarized as follows: the nervous system had evolved over time into a pyramid of sensory-motor ganglia of ever-increasing complexity, heterogeneity, and flexibility. The lower, more rigid, and less specialized (automatic) centers represented comparatively early evolutionary acquisitions, whereas the higher, more flexible, and specialized (voluntary) centers had evolved at a later stage. The very highest and most differentiated levels of cortical activity, those associated in human beings with conscious experience, operated according to the same physical laws governing activities in the more basic, homogeneous levels of the spinal cord. Indeed, in a certain sense, they could simply be considered more complex "re-representations" of those lower levels.

> I am supposing the nervous system to be a sensori-motor mechanism from bottom to top; that every part of the nervous system represents impressions or movements, or both. . . . The periphery is the real lowest level; but we shall speak of three levels of central evolution. (1) The lowest level consists of the anterior and posterior horns of the spinal cord, and of Clarke's (visceral) column and Stilling's nucleus and of the homologues of these parts higher up. It represents all parts of the body most nearly directly. . . . (2) The middle level consists of Ferrier's motor region, with the ganglia of the corpus striatum, and also of his sensory region. It represents all parts of the body doubly indirectly. (3) The highest level consists of the highest motor centres (prae-frontal lobes), and of the highest sensory centres (occipital lobes). They represent all parts of the body triply indirectly. These highest sensori-motor centres make up the "organ of the mind" or physical basis of consciousness; they are evolved out of the middle, as the middle are evolved out of the lowest, and as the lowest are evolved out of the periphery; thus the highest centres re-re-represent the body—that is, represent it triply indirectly. (Jackson 1887–88, 29–30)

In addition to performing their own functions, Jackson held that "the higher nervous arrangements inhibit (or control) the lower" (1882, 412). Their dominance, however, was a fragile affair. Being younger (phylogenetically) and thus less perfectly organized in structural terms, the higher, more voluntary centers were more vulnerable to damage and disease than the older, more automatic ("more perfectly organised") centers. "That parts suffer more as they serve in voluntary, and less as they serve in automatic operations is, I believe, the law of destroying lesions of the cerebral nervous centres" (Jackson 1873, 84). By voluntary, it should be stressed, Jackson intended no reference to some metaphysical will. He used the word, as he put it, physiologically to refer to "those movements which have at once great speciality and also great independence of other movements" (1871, 642).

Because brain damage rendered the higher nervous centers functionless and removed their control over the lower centers, Jackson proposed that it should be considered a reversal of the normal evolutionary process: a dissolution. The contrast of the term *dissolution* with *evolution* had also been borrowed from Herbert Spencer. Spencer had hoped to demonstrate that all phenomena (mental and physical, living and nonliving) were governed by the same laws of development and decline, and he had attempted to describe these laws in terms of the universal characteristics of force discovered by physics: "Evolution under its most general aspect is the integration of matter and concomitant dissipation of motion; while Dissolution is the absorption of motion and concomitant disintegration of matter" (Spencer, 1862, 261).

If, then, the highest cortical centers of the human nervous system represented the highest and latest stage of development that nature had achieved *and* if it were true that, within these centers, all lower forms of functioning were somehow combined and integrated, then it could be believed that disease involved a reversion to what was normal at lower, earlier levels of existence. Jackson believed that when a person's higher centers underwent dissolution, his or her lower centers *rose* in activity. For this reason, in every nervous malady, two elements had to be considered: the "negative" symptoms, or the loss of function produced by the damage; and the "positive" symptoms, or the functioning left over in the remaining intact structures.

Jackson's physiological rendering of Spencer's evolutionary principles also led him to adopt an innovative perspective on the nature of sensory-motor localization. All the traditional localizers of his time, he noted, believed that "there is a centre for, let us instance, the movements of the face only, one for those of the arm only, and one for those of the leg only"; in other words, they assigned different centers in the brain to individual limbs and muscles. "This seems to me," Jackson said, "to ignore that there is in healthy operations cooperation of movements of different regions" (Jackson 1882, 414).

In contrast, Jackson proposed that in fact "nervous centres represent movements, not muscles" (1882, 411). That is to say, a group of muscles that cooperated in, say, ten different movements would be represented in a center consisting of ten different parts. Each of these parts would represent the entire group of muscles but each would represent it differently, in accordance with the movement in question. For Jackson, then, a man with a paralyzed hand and partially paralyzed arm had not damaged his hand and arm centers but had suffered a dissolution of certain special movements, such as those employed in writing, which involved those two parts of the body. Because the "law of destroying lesions" taught that the most voluntary of movements, the ones represented highest in the functional hierarchy, were always the first to disappear, the man who was incapable of writing nevertheless might be able to employ his paralyzed limb in certain other, cruder motions.

This conception of localization led Jackson to an interpretation of the phenomenon of "recovery" that stands in marked contrast to the French idea of cerebral substitution. In many cases, he taught, the brain was able to compensate for a loss of nervous functioning, *not* because other nervous arrangements took on duties they had never had before, as the French held, but because the original damaged center retained patterns of sensory-motor innervation closely related to those that had been lost. "When the first term is destroyed," he wrote, "there *appears* . . . to be decided paralysis of X only, but slowly the other terms come to serve for moving the whole region more efficiently, and what is called the paralysis of X diminishes. There is *some* compensation" (Jackson 1882, 434).

7-3. *Unilateral Disorders and "Sparing"*

What, then, were the implications for Jackson of the fact that the nervous system, a hierarchy of sensory-motor representations, was at once *double* in structure and, at the highest level of cerebral representation, functionally *asymmetrical*? My reply begins obliquely with focus on an 1874 lecture by Jackson, in which he declared to an audience of medical students that hemiplegia, "the commonest kind of paralysis," was "the most important of all nervous symptoms. . . . Without a thorough knowledge of this symptom, the student is not fit to begin the *scientific* investigation of the more difficult nervous diseases, epilepsy, chorea, affections of speech, etc." (Jackson 1874b, 69).

One of hemiplegia's most revealing features, Jackson taught, was its *selectivity*. Regardless of the extent of unilateral brain damage, some muscles always were more severely paralyzed than others. Limbs, especially the arms and hands, generally suffered the most, while the muscles of the trunk and upper face often escaped with little or no weakness. Why should this be? Jackson first became interested in this problem around 1863, before the development of his mature views on the physiology of nervous functioning. In a pamphlet circulated privately among his colleagues, *Suggestions for Studying Diseases of the Nervous System on Professor Owen's Vertebral Theory*, he wrote of hemiplegia:

> I think we find only paralysis of the muscles of the limbs. . . . Is this another way of saying that those muscles less under the control of the will escape, and hence that the corpus striatum and thalamus

> opticus are able to direct (voluntarily) the limbs only? (cited in Greenblatt 1977, 417)

Jackson had not yet come to the view that movements, not muscles, are paralyzed in hemiplegia nor had he advanced to the point of looking upon the cortex as a sensory-motor organ capable of directing the limbs directly, rather than via the mediation of the (motor) corpus striata and the (sensory) thalamus optici. Moreover, at this point he was unable to offer any physiological reason why those parts of the body "most under the control of the will" should suffer most in hemiplegia. He had merely noticed the association, and that was all.

Then, in 1866, William Broadbent, a fellow British neurologist, also addressed himself to this problem of selective paralysis in hemiplegia, and he noticed something else. He pointed out that the parts of the body that suffered most in hemiplegia were those the individual moved "independently of the corresponding parts of the opposite side"; that is, they tended to be muscles that generally or exclusively were *unilaterally* innervated by the contralateral side of the brain. The parts of the body "spared" in hemiplegia, on the other hand, were those generally capable of being moved only "in concert with the corresponding muscles of the opposite side"; that is, parts of the body innervated only *bilaterally*, or by the two sides of the brain together. The reason, according to Broadbent, that these latter muscles escaped damage in hemiplegia was that, though they were indeed normally innervated by the two sides of the brain in tandem, they were "*capable of being excited by either singly*" (italics in the original). Thus, when an individual suffered unilateral brain damage, the muscles of bilateral use on the paralyzed side could continue to function, because the undamaged ipsilateral brain-half could continue to call them into service (Broadbent 1866).

The influence of the Broadbent theory of bilateral innervation on Jackson's thought was profound, and always profusely acknowledged, for it gave him the key to making physiological sense of that intriguing relationship he had observed between paralysis in hemiplegia and body parts "under the control of the will." In his hands, the Broadbent hypothesis was to undergo a significant mutation; for, of course, he was soon to reject both the concept of will as a metaphysical entity and the common view that sensory-motor centers in the brain represented muscles. Broadbent's observations on hemiplegia, he felt, showed that the hierarchy of the nervous system was complicated on the level of the cerebral centers: movements on this level tended to be represented *unilaterally* in proportion as they were voluntary; they tended to be represented *bilaterally* in proportion as they were involuntary, or automatic (cf. Gowers 1885, 51–53). The apparent explanation for this was that, because the nervous system must accommodate itself to a limited space within the body, it was physically incapable of evolving indefinitely upwards. On the very highest (voluntary) levels, then, it was forced to begin differentiating along the vertical axis.

Now, it could be understood why, in a unilateral disorder such as hemiplegia, only the most voluntary movements—what Jackson had previously been calling the movements most under the control of the will—were lost. Furthermore, having by this time come to the conclusion the entire nervous system must be conceived in sensory-motor terms, that one could not distinguish between physical and mental functions in the brain, Jackson saw no reason why the insights gleaned from hemiplegia should not be valid for another unilateral motor disorder: aphasia.

7-4. Propositional (Voluntary) versus Emotional (Automatic) Speech

Clinical observation, Jackson felt, strongly suggested that the aphasic, like the hemiplegic, lost most voluntary (speech) movements, but was spared the loss of his more automatic movements. That is to say, he generally was severely deficient in the highly voluntary and specialized capacity to form "propositions," or string together units of speech to express ideas, but might often manage to burst forth with samples of involuntary "emotional" speech: swear words, exclamations, etc.

Jackson credited the French alienist Jules Baillarger (1809–1890) with having been the first to notice this striking characteristic of the aphasic condition (Jackson 1866c), although there is reason to believe that he had in fact partially misunderstood Baillarger. The Baillarger passage Jackson had in mind (cited without reference) does indeed refer to a perversion of language characterized by a "substitution of automatic speech for voluntary verbal incitation." When one reinserts the passage in its original context (the 1865 Académie de Médecine debates), however, it becomes clear that Baillarger was not speaking of "emotional" language in the sense that Jackson thought but was pointing out that aphasia seemed to consist of two types: simple, Broca-style loss of speech, and a disorder of the language faculty "that consists of pronouncing incoherent words without connection to the ideas that one wants to express" ("Discussion sur la faculté du langage articulé" 1864–65, 828, 832). It seems, then, that rather than distinguishing between propositional and emotional speech, Baillarger was instead anticipating the distinction between Broca's and Wernicke's aphasia. Jackson more appropriately might have credited *Broca* with having been the first to point out that aphasics may retain certain "emotional" and stereotyped aspects of speech. In 1861, Broca had noted how certain patients "have, in a certain way, two degrees of articulation. Under ordinary circumstances, they invariably pronounce their favorite word ('mot de prédilection'): but when they experience a fit of anger, they become capable of articulating a second word, most often a crude oath" (Broca 1861c, 333).

The question of priority is in any event probably a moot one, for a case can be made that Jackson's distinction between higher propositional speech, on the one hand, and lower emotional speech, on the other, was not so much a clinical discovery (although obviously finding some support in the clinic), as a philosophical discrimination. In his 1772 prize essay, *Abhandlung ueber den Ursprung der Sprache*, German philosopher Johann Gottfried Herder (1744–1803) had declared that the various exclamations human beings used to express their feelings were remnants of an inferior natural language that they shared with the animals. Such articulations were not to be confused with language properly speaking, which was a strictly intellectual capacity, inseparable from human reason and, therefore, unique to human beings (Marx 1966, 328–29, 349).

There is no doubt that Jackson, no less than Herder, meant his propositional speech to be identified with abstract, rational thought, although, as will be seen, he also believed one could equally "propositionise" in visual images. This is why he would argue that the essential character of speaking was not the capacity to utter words but the capacity to refer words to one another in a particular manner. By themselves, the components of speech were nonsignificant, they took on meaning only by being organized into logical relationships (Jackson 1868d, 527). In fact, whether the individual actually spoke the words aloud was almost beside the point. "Internal" speech (thinking in words) and "external" speech (articulating) were physiologically and psychologically differentiated from each other only by their degree of relative faintness and vividness.

Because there has been considerable misunderstanding in the historical literature on this issue, it may be worth stressing that Jackson's distinction between internal and external speech was *not* that of Jean Baptiste Bouillaud, even though he made use of Bouillaud's nomenclature. For Bouillaud, internal speech was the symbolic, intellectual side of language and fundamentally different from external speech, which was a strictly mechanical capacity. The two occupied different parts of the brain, and it was possible to lose one without losing the other (see Chapter Three, sec. 3-3). In contrast, Jackson was quite clear that, in his view, internal and external speech involved the same centers of the brain and were essentially of a kind. When one was lost, both were lost:

> That the speechless patient cannot propositionise *aloud* is obvious—he never does. But this is only the superficial part of the truth. He cannot propositionise internally. The proof that he does not speak internally is that he cannot express himself in writing. . . . He can say nothing to himself, and therefore has nothing to write. (Jackson 1874d, 130–31)

As we have said, for Jackson propositional speech, whether internal or external, by definition was a highly voluntary activity, whereas emotional speech, swearing and exclaiming, was highly involuntary; in Gower's words, "the will is needed not to cause, but to restrain [it]" (Gowers 1885, 53). Following Jackson's interpretation of Broadbent's "bilateral innervation" theory, this implied that automatic or involuntary speech was essentially a bilateral activity and consequently, in the left-brain damaged aphasic, it could be set into motion by the undamaged right side of the brain. In contrast, the highly "special" and educated" movements of propositional speech were represented only unilaterally in the brain, like the special voluntary movements of the hand and arm. These movements therefore were lost when half the brain was damaged. As Jackson put it in a lecture before the British Association for the Advancement of Science in 1868:

> The movements of speech are educated movements, and thus differ widely from those movements which may be said to be nearly perfect at birth, such as those for respiration, smiling, swallowing, etc. All the movements represented in the corpus striatium unilaterally require a long education, and the most special of these are those engaged in the movements of speech and next those of the arm [later in his career, Jackson might have preferred to speak of the cortex here]. The muscles always acting bilaterally, and chiefly represented bilaterally in the corpus striata, are born with their centres for movements nearly perfect. Thus, then, the term "Intellectual Language" merges in the larger term "Special movements acquired by the Individual," and the term "Emotional Language" in the term "Inherited Movements" (common to the Race). [These in turn, Jackson went on to say, merge into the still more fundamental terms,

voluntary and *involuntary*.] (Jackson 1868a, 275; see also, 1868c, 358, 1871, 642; 1873, 84; Gowers 1885, 126)

Aphasia, of course, differed from hemiplegia in one crucial respect: the latter could result indifferently from damage to either side of the brain, whereas the former almost always resulted from left-sided damage only. Jackson accepted part of the Broca-school explanation for this curious fact and spoke of Gratiolet's" statement to the effect that the frontal convolutions on the left side are in advance of those on the right in their development. Hence, if this be so, the left side of the brain is sooner ready for learning. It is the elder brother." Jackson denied, however, that *only* the left hemisphere learned language, as he felt Broca and Moxon had argued. Both brain-halves, he said, were educated to the extent that both laid down representations for words. Nevertheless, because the left was more precocious, it "began to act" first, and therefore it became proficient in the voluntary movements that constituted propositional speech, both internal and external (1866c, 661; 1868b, 209).

The use of the term *voluntary movements* here in quite precise, for Jackson seems never to have wavered in his view that "the anatomical substrata of words were motor (articulatory) processes" (1874a, 391). He declared his debt to the associationist psychologist Alexander Bain for this insight; Bain having declared, in his 1855 *The Senses and the Intellect*, that "a *suppressed articulation* is, in fact, the material of our recollection, the intellectual manifestation, the *idea* of speech" (cited in Bastian, 1869, 210).

Now, the general idea that aspects of thought and consciousness might be fairly correlated with dynamic *motor* nerve processes, represented one of Bain's more innovative and enduring corrections of traditional associationist psychology, a system that since its inception had tended to see cognition in passive sensory terms (Smith 1973, 95). Bain's more specific proposal, though, that the global cognitive processes of language were mediated by cerebral motor impulses, had received a much more mixed reception from his countrymen. British neurologist Charlton Bastian, who in 1869 was the first to describe and name *word deafness*, declared his opposition to this view about the same time that Jackson seems to have adopted it. Bastian proposed instead that words "are revived in the cerebral hemispheres as remembered Sounds, which when revived in an unconscious and automatic manner, call into play . . . the various muscles necessary for the articulation of the sound" (Bastian 1869, 210). For most nineteenth-century aphasiologists, the work of Carl Wernicke in 1874 would seem an emphatic confirmation of the essential correctness of Bastian's position over that of Bain (see, e.g., James 1890, I:54). Jackson, though, seems to have been curiously oblivious to the implications of Wernicke's work, for reasons that may become at least partly clear as we proceed. I found one, very late reference to Wernicke's word deafness in a footnote (Jackson 1893b, 205n), where Jackson denied it was a speech disorder and instead rather cursorily explained it as a special sort of "imperception" (see section 7-5).

Without an awareness of Jackson's attitude to Wernicke and the whole question of "sensory" aphasia, one might have a hard time understanding how he could have included among the automatic (bilateral or either-sided) functions of language not just emotional outbursts but the very important process by which we "receive the propositions" of another, by which we understand language. This, Jackson proposed, perhaps was the most involuntary language function of all. Every healthy English-speaking person was "*compelled*" to understand the meaning of, say, the word *horse* every time it was spoken, he was helpless to prevent some image of the animal from springing to mind (Jackson 1868d). Thus, both sides of the brain, having the capacity for automatic functions "in which words serve," could "understand" language. This was why, Jackson said, the left-brain damaged aphasic patient "cannot speak at all, but understands all we say" (Jackson 1873, 85).

For Jackson though, and this point is crucial, even while both sides of the brain could automatically "understand" speech, only the left could go one step further and become *conscious* of this understanding; only the left could ever become "conscious in words." This aspect of Jackson's thought will be explained more fully when we come to look at his views on "the duality of mental operations." For the moment, suffice it to say that the justification for this claim is bound up with Jackson's belief that *only* the left side of the brain was capable of voluntary speech or propositionizing. One will recall that Jackson's use of the term *voluntary* had nothing to do with some metaphysical free will but simply described the functioning of the highest, but still lawfully determined, levels of the nervous system. From a *psychological* perspective, then, the difference between a voluntary and an involuntary act was simply that the former was always initially "preconceived" or "*represented in consciousness*" (Spencer, quoted in Jackson 1868c, 359; italics in the original). By definition, in other words, actions could be deemed voluntary if they were accompanied by consciousness. Conversely, "the more operations are automatic, the less we are conscious of them" (1874b, 141). This explains why

Jackson could argue that the speechless man whose intact right-brain could still (*automatically*) understand speech, and who therefore could instantly hand him a brick on command, had no "memory" of the word *brick*, was not "conscious of the *word itself*. He has no consciousness of it, but of the thing it is a symbol of—a very different thing" (1874b, 140–41).

7-5. *"Imperception" and the Voluntary Functions of the Right Hemisphere*

If the right side of the brain lacked the capacity to be "conscious in words," since it lacked voluntary speech, it was not unconscious altogether. The speechless man had lost his capacity for verbal consciousness but, Jackson argued, he remained fully conscious of *things*. For Jackson, this suggested that, alongside the sensory motor processes underlying speech and verbal thought, there was a second series of "sensori-motor processes concerned in the *recognition* of objects (not in *seeing* objects), and in putting images of things in 'propositional order,' so to speak" (1872, 514). These two functions, object recognition (perception) and visual thought (putting images in propositional order) were seen by Jackson as the external and internal sides, respectively, of this proposed "second" series, relying on the same cerebral centers. The distinction, then, between recognition of objects in the outside world and thinking rationally in pictures was merely one of degree, directly comparable to Jackson's distinction between speech and verbal thought.[1]

[1] The basic assumption in Jackson's writing, that thoughts and experiences engage the same nervous centers and thus are fundamentally the same from a physiological perspective, is worth lingering over a moment. If nothing else, it highlights the extent to which Jackson's neurological thinking was soaked in the native philosophy and psychology of his time. In effect, his external/internal distinction can be seen as a "neurologizing" of the British tradition of epistemological scepticism, with roots going back to the sensationalist psychology of David Hume. Starting from the empiricist premise that "all ideas are derived from impressions, and are nothing but copies and representations of them," Hume had concluded that consequently there was no way of proving the existence of an independent, outside world. The only difference between ideas and those perceptions one *believed* were caused by objects outside of one's self was priority (in time) and vividness.

Jackson demonstrated his personal sensitivity to the Humean dilemma even at the basic clinical level: "A girl under my care for intracranial tumour became quite blind (quite deaf too), and later insane. One of her illusions (or perhaps I should here say hallucinations) was that there was a man in the corner of her room who mocked her. I submit that this girl's 'illusion' was her perception; it was what should have been her morbid ideation, but because certain arrangements of her visual centres were strongly discharged, the image this blind girl had rivalled in vividness, etc., the perceptions she

Jackson's earliest suggestion that this "other" series was mediated in its most voluntary aspect by the *right* side of the brain appeared in the *Lancet* in November 1864. After suggesting that cases of optic neuritis (inflammation of the optic nerve leading to loss of function) tended to occur more frequently in conjunction with left hemiplegia than with right, he declared,

> Now amaurosis [blindness] is due to disease of a *highly specialised nerve of sensation*, while there are good reasons for believing that defect of articulate language, which M. Broca calls aphemia, is a kind of ataxy of articulation, and therefore a *disorder of motion*. Again, the contrast may be made more general. Sight is a department of general perception, whilst articulate language is a department of general expression. If, then, it should be proved by wider evidence that the faculty of expression resides in one hemisphere, there is no absurdity in raising the question as to whether perception—its corresponding opposite—may not be seated in the other. (Jackson 1864b, 604)

Several years later, he slightly modified his position. His 1864 comparison of the two sides of the brain had been *illegitimate*, he said. He no longer accepted Broca's view that the capacity for expression resided in only one hemisphere. Both sides of the brain were "educated in expression," but the left was the "leading" side; that is, the side concerned with the voluntary (conscious) functions of speech. In light of this, he concluded, "I would still advocate the view I brought forward in the *Lancet* with the important qualification that the right side may be the [conscious] side for perception—educated sensation. So, then, we should have 'crossed action' betwixt the two brains, and thus we might suppose a more perfect mental co-ordinating" (Jackson 1868b, 208).

One may assume that for Jackson, with self-confessed weakness for philosophical systems, there must have been something seductive about

had when well. It may be said that this doctrine confuses reality and unreality. But what reality and whose reality? The mocking man was the poor girl's reality; she took a poker to strike him" (Jackson 1893a, 208).

Though the "representation" theory of knowledge may have convinced Jackson, the claim that ideas and sensations differ only in *degree* was certainly not without its critics. Consider, for example, William James' stern remarks on Hume: "There is no reason to suppose that when different states of mind know different things about the same toothache, they do so by virtue of their all *containing* faintly or vividly the original pain. Quite the reverse. The by-gone sensation of my gout was painful, as Reid somewhere says; the *thought* of the same gout as bygone is pleasant, and in no respect resembles the earlier mental state" (James 1890, II:6).

the idea that the two halves of the brain cooperated in a precise dance of corresponding opposites. What evidence, however, did he use to back up his views? The early claim for a link between amaurosis and left hemiplegia was quietly dropped, after it was bluntly denied by a colleague in 1868 (Albutt 1868). He then began to stress that his speechless patients seemed to retain a capacity to recognize objects, even though they could not name them. At best, though, this is only a form of negative evidence, hardly constituting a convincing case for the conclusion that object recognition and visual ideation therefore "must" be mediated by the right hemisphere. Jackson was too good a clinician not to have been aware of this. One wonders if he was being totally honest with himself in not at least entertaining the possibility that nondamaged parts of his patients' left hemispheres, parts other than those involved in voluntary speech, could have been playing a "leading" role in the series of operations required for making propositions of objects.

Jackson's *positive* clinical evidence for a right-sided localization of voluntary visual perception was chiefly drawn from only two cases. The first of these, described in 1872 and not, incidentally, included in the *Selected Writings*, involved a middle-aged man suffering from a relatively rare form of left-hemiplegia, in which the leg was more affected than the arm. Jackson believed that this form of paralysis was particularly apt to be complicated by psychological disturbances, other than loss of speech. "Trousseau has spoken of the evil omen of cases of left hemiplegia in which the leg suffers more than the arm. . . . He appears to refer—in part, at least, to mental affections." The most striking thing about the case was that the man seemed to have lost his ability "to recognise places and persons. At one time he did not know his wife, he gave his watch away, and having wandered from home was unable to find his way back" (1872, 513).

By the time Jackson described his second, similar case, that of Eliza T., in 1876, he had christened this inability to recognize objects, persons, and places *imperception* and argued that it was defect "as special as aphasia." It resulted, he believed, from an insult to the right posterior lobe. "I think . . . that the right posterior lobe is the 'leading' side [for visual ideation], the left more automatic. This is analogous to the difference I make as regards use of words, the right is the automatic side for words, and the left side for that use of words which is speech" (Jackson 1876b, 148).

Just as the aphasic patient trying to speak often found that all he could do was swear or endlessly repeat some stereotyped phrase, so the patient suffering from imperception tended to make errors indicating that he retained certain automatic, stereotypical perceptual skills; skills that presumably were mediated by the left side of the brain. Jackson described a London Hospital patient who "supposed that she was at the place in Holborn where she had worked for some years—a place the image of which was in her mind more automatic. It was a case of misrecognition, analogous to a mistake in uttering words more automatic than intended" (Jackson 1874c, 174).

Jackson himself "confessed" that he actually had little clinical evidence, beyond the two cases just mentioned, that all cases of imperception really did result from morbid changes in the right posterior lobe. Indeed, he went so far as to admit that, taken together, the evidence on the clinical side was "slight and doubtful" (1874c, 143). It was *anatomy*, rather, that he saw as providing the truly "strong" support for his views (1872, 514). What was the anatomical evidence then? The answer is an important one, for it seems to have escaped notice in the secondary literature to date. It will be recalled that Jackson had accepted Gratiolet's claim that the left frontal lobe grew in advance of the right, and he had used this claim to make the case that the left hemisphere was the "elder brother" and took the "lead" in voluntary speech processes. His writings on the functions of the right hemisphere show that he equally accepted Gratiolet's belief that the right posterior lobe grew in advance of the left; "the 'important' part of the right hemisphere is the posterior lobe," he affirmed, "whilst the 'important' part of the left is the anterior lobe" (1872, 514).

Accepting for the moment that the right posterior lobe really does grow in advance of the left (the highly uncertain merits of the Broca school's views here were discussed in Chapter Three), why should Jackson have come to the conclusion that the leading function served by that lobe was visual perception and imagistic thought? The answer has nothing to do with the localization of vision in the occipital regions of the brain. Munk would not place his visual center in the occipital lobes until 1877, and Jackson began arguing for a link between posterior lobe functioning and visual perception at least as early as 1868, the year he seems to have become aware of the French anatomical argument for asymmetrical, complementary growth between the hemispheres.

Jackson's views on the complementary activities of the two hemispheres seem, rather, to have been significantly shaped by certain a priori convictions about how the hierarchy of sensory-motor functions in the nervous system "should" be organized. It will be recalled that Jackson, following Bain, believed that the anatomical substrata of words were motor processes and that voluntary speech represented the pinnacle—the highest and most voluntary—of all possible movements. In a similar way—it is not absolutely clear why—he felt that visual perception represented the highest and most voluntary of all possible sensations. Since physiological research had shown that sensory functions were localized in the posterior regions of the spinal cord, and since, functionally speaking, the brain was nothing more than a complex "re-representation" of the spinal cord, it followed that the highest and most discriminating sensory functions must be located in the posterior lobes of the cortex. Jackson's discovery of the French anatomical studies led to the revelation that the left anterior and right posterior lobe each were slightly more advanced evolutionarily than their twins. For him, this meant that where the left lobe took the lead in speech, in a complementary fashion the right very likely took the lead in visual perception. The whole scheme is marvelously rational and neat. What, though, if Jackson were forced to accept Wernicke's interpretation of "sensory" aphasia? Would this not have required him to introduce a messy new variable into the left/right, anterior/posterior symmetry of his own system? Is it possible, therefore, that semi-aesthetic, semi-philosophical sentiments contributed to his apparent reluctance to look closely at the work of his German colleague?

On further reflection, Jackson decided that, although his original conception of complementary activity between the hemispheres was broadly valid, the anatomical relationship between the two functions (speech and visual perception) was not always *perfectly* symmetrical. Visual perception and ideation were not exclusively sensory processes, he affirmed, but often had a significant motor component, involving the ocular muscles. This component was more or less active depending on what aspect of the object was being noticed or imagined:

> When we think of the brick as coloured, the *leading* activity is of nervous arrangements in the right posterior lobe. This is intended to mean that it is of a part where retinal impressions are most represented in complex combinations and definite associations. When we think chiefly of extension (the shape and size) of the brick), the *leading* activity will be in the left anterior lobe—that is, in a part where certain ocular movements have the leading representation. (1876a, 131)

It is worth noticing that Jackson's idea, that the left hemisphere served voluntary movements (including speech) and the right voluntary sensation (including visual perception), does have certain points in common with the Continental argument that the left hemisphere "predominated" in motor processes and the right in sensory ones (Chapter

Three). It will be recalled that the men who put forward this view of lateralized functioning not only pointed to the evidence for asymmetrical, complementary development between the hemispheres but also to certain clinical evidence suggesting a statistical tendency for left and right brain damage to result differentially in motor and sensory disabilities. (Jackson seems to have been unaware of these data.) Nevertheless, the differences between the Jackson perspective and that of such men as de Fleury, Delaunay, and Exner far outweigh the similarities. Not least of these differences is the tendency of Continental writers to see the alleged asymmetrical development of the two hemispheres as a sign of the left brain's superiority and the right's inferiority. Coming from a different tradition, Jackson had managed to avoid absorbing the dubious doctrine that taught that all the intellectual, uniquely human faculties were housed in the frontal area of the brain, while all the instinctual, bestial faculties skulked about behind. His views on the relative importance of the two hemispheres were based instead on his perception of the importance of the two functions he believed each chiefly served, with the result that the right hemisphere came out looking good indeed.

"I think, as Bastian does," he said, "that the posterior lobes are the seat of the most intellectual processes.[2] This is in effect saying that they are the seat of visual ideation, for most of our mental operations are carried out in visual ideas" (1876b; 148). Once he had come to this conclusion, he never budged, affirming as late as 1897 that "if all Visual ideas were cleared out of a man's mind, he would be practically

[2] The British generally seem to have taken the lead in defending the posterior lobes of the brain and opposing the old prejudice that everything that counted in a person was to be found up front. In 1881, the British neurologist C. Clapham presented what seems to have been a quite well-known paper, "The Noble Forehead." Here he gave five reasons why he believed the occipital, rather than the frontal lobes, represented the chief seat of intellectual activity.

"1. The occipital lobes occur only in primates, being absent even in the lowest of monkeys, whereas the frontal lobes are present in all the mammalia."

"2. The occipital lobes, where present, are the latest developed, whereas the convolutions first make their appearance in the brain of the human embryo in the frontal lobes."

"3. The occipital lobes are not occupied, as are the frontal lobes, by extensive motor areas; indeed, they have no motor cells at all in their cortical substance."

"4. The occipital lobes are small and ill-developed in idiots (a straight back to the head being a common feature of idiocy), whilst the frontal lobes are usually large, relatively speaking."

"5. Wasting of the occipital lobes is always accompanied by dementia; not so wasting of the frontal lobes" (Clapham 1881–82, 623; cf. also "The posterior lobes of the brain" 1883).

mindless" (1897, 85). Certainly, he admitted that words were required for "conceptual thought" (1897, 86) and for "thinking on novel and complex subjects" (1878–79, 167). But of the two functions—visual ideation and verbal thought—he seems to have been quite certain that the former was more fundamentally important. One cannot help but wonder why he apparently never considered that congenitally blind individuals generally were far from "demented," a fate he held must afflict the individual who had experienced a total loss of the capacity to propositionize in images.

7-6. *The "Duality of the Mental Operations"*

So far, we have determined that, for Jackson, the left anterior and right posterior lobes served the conscious, voluntary sides of speech and visual perception, respectively. The right anterior and left posterior lobes, standing at a functionally lower level of evolution than their elder brothers, also laid down representations of words and images but were incapable of forming propositions with them. Rather, they only were able to associate on an unconscious, automatic level, making links between representations on the basis of crude resemblances, and lacking the finer discriminatory capacity to appreciate differences. In cases of pathological dissolution, these more automatic levels of speech and perceptual activity revealed themselves in the automatic speech of the aphasic and the recognition errors of the imperception patient.

It should not be thought, however, that during healthy, normal functioning, the lower nervous arrangements in the right anterior and left posterior lobes were idle. Jackson believed, in fact, that they played a vital role in the sensory-motor activities underlying propositional speech and perception. The energizing of the lower arrangements, he argued, "although unattended by any sort of consciousness," was essential for and led to the energizing of those higher, more voluntary arrangements in the left anterior and right posterior parts of the brains, arrangements whose activities *were* attended by consciousness (1878–79, 167).

In short, *mentation was a dual process*, played out between the two hemispheres of the brain. During the first automatic stage, words and images were revived according to the associative law of resemblances. An unconscious physiological struggle for the "survival of the fittest" words and images in the fittest relation then ensued (1880, 328–29). The end of this automatic, preparatory phase represented the beginning of thought in its voluntary aspect, as the words and images that had been "selected," in the Spencerian-Darwinian sense, arranged themselves in "propositional order" and then welled up, as if from nowhere, into the conscious corners of the brain (Jackson 1868c; 1868d; 1874a; 1876a, 173–76; 1878–79; 1880). "It is, I think," Jackson mused, "because speech and perception are preceded by an unconscious or subconscious reproduction of words and images that we seem to have 'faculties' of speech and perception, as it were, above and independent from the rest of ourselves. . . . We seem to ourselves to Perceive, as also to Will and to Remember, without prior stages, because these prior stages are unconscious or subconscious" (1878–79, 168).

For Jackson, then, there were *two* distinct series of operations (verbal and imagistic) involved in mentation, and each of these consisted in turn of *two* distinct stages: an unconscious, automatic stage, and a conscious, voluntary one. Since most human cognition involved rapid interplay between words and images, the two series of operations normally engaged in a sort of rhythmic give-and-take, making for four-way "crossed action" between the two sides of the brain. As Jackson wrote in an early description of this model of brain functioning,

> I wish to speak of the links *between* the revivals [he is referring to the two revivals (unconscious and conscious) of words]. The links are, I imagine, the perceptions [or images] of which the words are the arbitrary signs. I wish also to speak of the links between the revived perceptions. They are the aforesaid revived movements of words. In fact there is, I conceive, an alternate play between the higher "movements" and "sensations" of mind, as there is in the commoner movements and sensations of the body. . . . It is convenient to speak first of the two revivals of perceptions, and next of revivals of words, although . . . I conceive the revivals to be alternate—involuntary motor, involuntary perceptive, voluntary motor, voluntary perceptive. (1868d, 527)

It has been seen that, for Jackson, the speechless man continued to understand language "automatically" using his right hemisphere. It can now be understood that this was because he was supposed to have retained the first, automatic and unconscious half of his verbalizing series that, following Broadbent, was bilaterally represented in his brain. He had lost, however, the second (unilaterally represented) voluntary/conscious half of that series, the half that constituted propositional speech. The duality of his perceptive/ideational series, however, remained intact, and he therefore could "be conscious in objects." The patient suffering from imperception was supposed to be in the converse situation, able to speak but incapable of dealing properly with images and objects in the outside world.

Jackson believed that his discovery of the duality of mental operations meant that one could finally begin to make sense of a puzzling but very basic feature of human cognition: the fact that, in thinking, the individual was aware on some deep intuitive level that his thoughts belonged to him and him alone; the fact that his consciousness consisted not just of knowledge but of a knower, not just of an object but of a subject. This was simply because, Jackson argued, every thought was actually activated twice, first unconsciously and then consciously. The first, unconscious activation gave rise to that tacit awareness "deeper than knowledge" (1887, 96) that one was experiencing processes of thinking and experiencing, which collectively constituted the foundation of one's existence as a thinking self. The second, conscious activation corresponded to awareness of the content of one's thoughts and experiences in their own right. Jackson proposed to call the first, automatic half of thought *subject consciousness* (even though it was not actually conscious at all!) and the second, voluntary half of thought *object consciousness*. He tried to clarify matters with the following illustration:

> [When seeing or thinking of a rose] we seem to have two mental states together in reversible succession; we seem as if we had not only the idea of the rose, but *we* seem to know that *we* have it. We seem to remember the rose as if we had a general consciousness in addition to and yet distinct from another more special consciousness, a sort of "faculty" of remembering the rose. . . . Let us state an hypothesis explanatory of this feeling. It is that the substrata of consciousness are double, as we might infer from the physical duality and separateness of the highest nervous centres. The more correct expression is that there are two extremes. At the one extreme the substrata serve in Subject-consciousness, at the other extreme in Object-consciousness. (1876a, 73)

In calling the two "halves" of thought *subject* and *object* consciousness, Jackson was making use of terms with a long, respectable history in British mental philosophy. It generally was supposed that subject consciousness corresponded to introspection, or contemplation of the self and its thoughts, while object consciousness corresponded to contemplation of objects recognized as distinct from the self; i.e., consciousness directed outward. This emphatically is *not* what Jackson meant by subject and object consciousness, however; and his unfortunate choice of such familiar nineteenth-century expressions practically invited misunderstanding of his views. For Jackson, *only* states of object consciousness could be contemplated or known; the knower could not make himself the object of his knowledge, since that implied the absurdity that a self could at once think and observe itself thinking. As Mercier, one of Jackson's students, would later put it, subject consciousness involved "a colouring or modification of the mental self directly, and without the intermediation of any processes of thought." What most philosophers called *self-awareness*, then, was possible only through a kind of "fiction," since the moment one tried to observe one's present subjective state, that state changed its character and became an object to be contemplated. The self proper was left still elusive—a stubborn "hole in being at the heart of being," as Sartre would later put it—contemplating "an objectified image of itself which is not itself" (Mercier, 1901, 493–94, 503, 494; cf. Smith 1982). In Jackson's own words:

> What is called the introspection of consciousness is of states of object-consciousness already "come out of" subject-consciousness. Subject-consciousness, *that which has these states* (popularly, "contemplates," "reflects on," "arranges," "attends to," etc.) *is not to be known in any such way*. . . . [W]e only know in the ordinary sense of the word know, subject consciousness on its ceasing to be—on its becoming object consciousness. . . . [W]e may say that what is commonly called consciousness (object consciousness), is "a revealing of self." (Jackson 1887b, 96, Italics added)

Individuals, in short, could become conscious of changes in their "selves" (changes in the highest levels of their nervous systems) only indirectly, only through the mediation of "symbol-images" (Jackson 1878–79, 167–69). The symbols exploited by the brain consisted of two main types: verbal and visual.[3] This is a vital point to grasp in order to understand the final significance of the double brain for Jackson. He held that the words and images served up by the left and right sides of the brain were the symbolic "stuff" that cloaked, or "objectified," otherwise invisible thoughts and sensations and let them be consciously "perceived," or known. In the final analysis, then, we see that Jackson's propositionizing double brain was charged with a task of fundamental existential proportions, doing nothing less than pro-

[3] Late in life Jackson also began to speak of the importance of what he called *tactual ideas*. In an 1897 lecture, for example, he noted that "Herbert Spencer has pointed out that intelligence in animals is proportionate to the development of the tactual organs; . . . again he writes 'that the most far-reaching cognitions, and inferences the most remote from perception, have their roots in the definitely combined impressions which the human hands can receive" (1897, 439). It does not seem, however, that Jackson ever fully integrated these later thoughts with his earlier psycho-anatomical theories concerning the role of words and images in consciousness.

viding the self with a continuous, lawfully ordered, if "fictionalised," commentary on its own personal reality. One could hardly hope to grant brain duality and hemisphere specialization a more profound role in human mental life.

7-7. Applications of the Thesis

Jackson applied his theory of subject/object consciousness or mental duality to a great many, apparently, differing disorders. He argued, for example, that certain speech errors made by healthy people were due to "slightly imperfect co-ordination in the action of the two sides of the brain" (Jackson 1870, 460). Everyone knew of cases, Jackson said, in which a person said something along the lines of "mukes from Boodies" when he meant "books from Mudie's" (a London lending library), or "get a cash chequed" when he meant "get a cheque cashed."

> I believe that these troubles of speech are owing to hurry on the right side of the brain, to hurried reproduction of the words of the subject-proposition. Believing that images and words are subjectively revived in an order the reverse of that in which images and words are finally arranged, I suppose, speaking roughly, that the words of the subject-proposition "come over" to the left side prematurely. . . . What is the cause of this hurry? I believe it to be strong emotion. (Jackson 1880, 340)

Jackson had noticed quite early in his studies of speech disorders that a certain proportion of his patients were capable of articulating not only exclamatory and highly organized or stereotyped expressions ("hello," "thank you") but what seemed to be "rags and tatters" of complex propositional speech. Their repeated use of these complex utterances, Jackson took pains to stress, in no sense violated the general principle of dissolution. Like paraphasias in healthy persons, such "recurrent utterances" resulted from a disruption of the normal processing from the subjective (right-brain) to the objective (left-brain) stage of verbalizing. Although seemingly complex, recurrent utterances in fact were highly organized, inflexible units of speech used simply to express emotion and without propositional significance. Like oaths and exclamations, they were activated by the right side of the brain. They had been retained in the patients' limited repertoire of words because "*they were being said, or about to be said, when the patient was taken ill*" (Jackson 1880).

In short, recurrent utterances simply were subject propositions in the act of being turned into object propositions when trauma struck.

Jackson supposed that, under the impact of that trauma, the usually transitory nervous innervations involved in verbal subject consciousness (in the right anterior lobe of the brain) became "stuck," as it were, frozen in the act of discharging and compelled to retain their high level of energy in a permanently activated circuit (Forrester 1980, 18–19). In Jackson's words,

> The hypothesis . . . unfolded . . . is that the words of the recurring utterance had been revived during activity of the right half of the brain, when destruction of that part of the left half occurred which caused loss of speech, that they constituted the last proposition, or rather that stage of verbal revival (what we have called the subject-proposition), prior to the last proposition (the object-proposition). Ever after, they were meaningless or, figuratively speaking, dead propositions. (Jackson 1880, 330)

Jackson did not limit application of his views on hemisphere differences and mental duality to the realm of speech disorders. The duality of the nervous system was also important, he felt, to understanding what constituted his other chief area of interest: epilepsy. His discussions of epilepsy and the double brain, however, might seem puzzling to readers who had dilligently followed the arguments laid out in his writings on aphasia. Somewhere in the transition from one disease to the other, a subtle theoretical shift seems to have taken place, without explanation and apparently without any conscious awareness of inconsistency. Where Jackson carefully explained in his aphasia essays how both verbal and imagistic thought (the motor and sensory sides of consciousness) were equally a double process of subjective and objective states, in his writings on epilepsy, he began to declare that subject consciousness was the "chiefly sensory" half of the organism's double relationship to the environment and object consciousness "the chiefly movement half" (1876a 173–74). During subject consciousness, Jackson argued now, impressions from the environment were passively received, the organism was acted *upon*, and this was a largely sensory process. During object consciousness, in contrast, the organism *reacted*, and this was a largely motor process. Such theoretical assumptions may well have been adopted from Alexander Bain, who had proposed a parallelism between the passive and active sides of consciousness and the sensory-motor model of nervous processes (Smith 1973, 97).

Be that as it may, Jackson's advocacy of this view meant that the right hemisphere, with its presumably more highly evolved sensory area, must now be conceived as the more passive, "subjective" side of the brain, while the left hemisphere, more immediately concerned with voluntary motor activities, must be considered the more active, "objective" side of the brain. "Since most people are right-handed, the left cerebral hemisphere in most people represents the most objective movements, those movements for most specially operating on the environment. Moreover, speech is a process by which are symbolised relations of things in, or as if in, the environment or things considered objectively" (1880–81, 199).

Having introduced his readers to this somewhat new way of looking at things, Jackson then went on to propose that various epileptic "warnings," the characteristic symptoms that many patients experience just prior to an attack, could be understood as "excessive and crude developments of objective or of subjective sensations" and could be correlated with disease in the left and right hemisphere, respectively. Careful clinical observation had convinced him that epileptic fits that commenced with a spasm on the right side of the body (pointing to an initial discharge in the left hemisphere) tended to be associated with such "objective" warnings as vertigo (a disorder of the organism's basic adjustment to the environment), anger (an outwardly directed emotion), visual and auditory hallucinations (these senses being more "objective" than taste and smell), memories of recent experiences, and automatisms of a purposive nature. Fits that commenced with a spasm on the left side of the body (or an initial discharge in the right hemisphere) tended, Jackson believed, to be associated with such "subjective" experiences as fear, epigastric and other "systematic" (chiefly visceral) sensations, gustatory and olfactory hallucinations, vague memories of long-past experiences, and repetitive, apparently pointless actions.

There also was a correlation, Jackson believed, between left-sided (right-brain) epilepsy and that classic symptom of temporal lobe epilepsy known as the "dreamy" state: a complex, elusive syndrome in which a patient may experience a sense of *déjà vu* or *jamais vu*, a vivid, dreamlike revival of distant memories, a panoramic rush of ideas, or a terrifying sense of dissolving personal identity (cf. Dewey 1896). Jackson defined the dreamy state as a pathological heightening of subject consciousness accompanied by a diminishing of object consciousness; a process that had the effect, as one of Jackson's colleagues would later put it, of disrupting "those intuitions that yield the consciousness of continued existence" (Crichton-Browne 1895, 8). The tendency, Jackson felt, for patients to describe the dreamy state as a feeling of double consciousness was quite proper, since the syndrome represented a "revelation of the normal duality of all healthy mental action," the person being dimly aware of both the objective *and* the subjective stages of his thought processes (Jackson 1876c, 702).

Even though he admitted that the dreamy state often seemed to be associated with left-sided epilepsy, Jackson was reluctant to take the view that the same right-brain discharge presumably responsible for such "subjective" symptoms as "a stench in the nose," crude gustatory movements, and epigastric sensations could also be "the physical condition for an infinitely more elaborate psychical state." He therefore suggested that the latter might actually be caused by "slightly raised activity" in the left hemisphere as it reacted sympathetically to the loss of control in its twin (1880–81, 200). Crichton-Browne, however, having found repeated evidence that the right-brain tended to be damaged in cases of epilepsy involving dreamy state, declared unequivocably that these states generally corresponded "with involvement of the right or more subjective of the two hemispheres of the brain" (Crichton-Browne 1895, 21). Arnold Pick, an overseas colleague of Jackson, also felt the syndrome pointed to "a particular participation of the right cerebral hemisphere." He had studied several epileptics in which the dreamy state had established itself as a permanent condition; and he had observed, among other things, a marked tendency for instances of paraesthesia (abnormal cutaneous sensation) to manifest themselves on the left side of the body and for convulsions to be experienced more severely on the left side than the right (Pick 1903, 247).

Rather unexpectedly, Jackson's paper on subjective and objective warnings in epilepsy (1880–81) has more broad affinities with continental writings on hemisphere differences (Chapter Three) than any of his many works on speech disorders. The Jacksonian identification here of the left brain with active, objective (conscious, voluntary, intelligent) processes, and the right brain with more passive, subjective (unconscious, involuntary, visceral/emotional) ones, by now should have a familiar ring.

It would be nice to be able to explain away the apparent contradiction between Jackson's views on hemisphere differences in speech disorders and those in epilepsy. I am not convinced, though, that such an exercise in mental gymnastics would be useful or appropriate. Certainly, historical studies are more satisfying if they can show a connected pattern underlying a particular flux of thought and events, but they should not seek to do this if it means misrepresenting the flux in its own right. Jackson had an extremely wide-ranging, restless mind, and he worked out his ideas in the course of responding actively to different sorts of clinical problems and writing scores of different sorts of reports and essays. It would be astonishing therefore if he were wholly consistent in his thinking.

Inconsistency, indeed, is a chronic failing of any great systematizer and, like Spencer, this is what Jackson was. He aimed to do nothing less than accommodate all varieties of normal and pathological nervous and mental functioning within a single comprehensive theoretical edifice, and he built this edifice by creatively bringing together a broad range of clinical and anatomical data with some of the most characteristic intellectual currents running through educated English debate and discussion at this time: associationist psychology, psycho-physical parallelism, scientific naturalism, determinism, and Spencerian (rather than strict Darwinian) evolution.

If the time-bound, place-bound nature of his thought has not always seemed clear, if the neurological holists of the interwar years (Chapter Nine) could hail him as a spiritual comrade-in-arms and the neoclassical localizers of the present day claim him equally as one of their own, for the historian that must suggest something interesting about his role in the history of science. Is it possible that his wide-ranging difficult concepts, often expressed in oblique and tortured prose, have resulted historically in his becoming a kind of Rorschach inkblot for the generations of neurologists that successively have rediscovered them? If so, one is tempted to speculate that, at any one time, history's judgment of the nature and extent of Jackson's contribution to neurology might say almost as much about the theoretical preoccupations of the generation in question as about Jackson himself.

CHAPTER EIGHT

Freud and Jackson's Double Brain: The Case for a Psychoanalytic Debt

THIS BRIEF CHAPTER attempts to examine the possibility that certain aspects of Freud's metapsychological teachings were influenced by Jackson's views on dual mental functioning and hemisphere differences. Differing somewhat both in its aims and methodology from the rest of the book, this chapter might be most appropriately regarded as an extended coda to the previous discussion on Jackson. Reading Jackson led me gradually to the conviction that it might be necessary to reconsider Freud's place in the history of the double brain. Freud's role as the founder of a psychological system that did much to undermine double-brain theories of personally pathology had never been in doubt (Chapter Nine). However, I was now forced to entertain the rather paradoxical idea that, in a certain way, psychoanalysis might also be indebted to the double brain.

That Freud was on *some* level influenced by Jackson is generally conceded by scholars (Dewhurst 1982); and there is something close to a consensus that the Freudian concept of "temporal regression," the idea, roughly speaking, that neurosis involves a reversion of the libido to a more primitive condition, was largely inspired by Jackson's view of dissolution in the nervous system. Indeed, the evidence for an intellectual link between dissolution and regression is fairly straightforward (see, e.g., Stengel 1953, 1954, 1963; Riese 1958; S. W. Jackson 1969; Sulloway 1979, 270–72). Whether or not one is justified in arguing for influence on other fronts, however, is a subject of more dispute. The question of regression is exceptional; in most cases, Freud so consistently neglected to credit the true sources of his ideas that authors wishing to make claims for various sorts of influence on his thinking are forced to rely heavily upon inference and analogical reasoning. The present exercise is no exception.

8-1. Jackson and the Freudian Monograph "On Aphasia"

Although it would seem that no one to date has made the case for a psychoanalytic indebtedness to Jackson's double brain, I have nevertheless found that a certain group of Freudian commentators (Stengel 1963; Fliess 1953; Katan 1969; Forrester 1980) have adopted a particular perspective on the nature of Freud's debt to Jackson that is broadly compatible with my own. Therefore, I would like to briefly review both this perspective and the evidence upon which it is grounded and then use both as a springboard for developing my own argument.

To this end, I turn to what has been the most important source of evidence for *all* current arguments claiming one or another sort of psychoanalytic indebtedness to Jackson: Freud's 1891 *Zur Auffassung der Aphasien* (in the English translation used here, *On Aphasia*). Although written during the last years of his so-called pre-psychoanalytic period, this monograph has been dubbed the first Freudian book because of its largely dynamic, functional, and developmental approach to the problem of speech disorders. Having undertaken an extensive review of the literature, Freud came down hard on the Wernicke-Lichtheim anatomic theory of the aphasias, which dominated clinical neurology at that time, especially in the German-speaking countries (for a detailed exposition, see Lichtheim 1886, also Chapter Three, section 3-1). Wernicke had taught that all varieties of speech disorder could be accounted for by assuming the presence of a lesion in one or more of the three main linguistic centers of the brain or, alternatively, in one or more of the paths of connection between these centers. "The contrast between central aphasias and conduction aphasias was assumed to be the key for the understanding of the speech disorders" (Freud 1891, 100). Freud warmly acknowledged the extent to which Hughlings Jackson had helped him to recognize the errors inherent in such a perspective and declared Jackson the chief inspiration behind his proposed alternative approach to speech disorders:

> In assessing the functions of the speech apparatus under pathological conditions we are adopting as a guiding principle Hughlings Jackson's doctrine that all of these modes of reaction represent instances of functional retrogression (dis-involution) of a highly organized apparatus and therefore correspond to earlier states of its functional development. This means that under all circumstances, an arrangement of associations which, having been acquired later, belongs to a higher level of functioning, will be lost, while an earlier and simpler one will be preserved. (Freud 1891, 87)

Freud was particularly indebted to Jackson for offering him a means of making sense of the aphasic recurrent utterance, that persistent complex unit of speech which Jackson had taken pains to stress did not contradict the general principle of dissolution (see Chapter Seven). Within the anatomic model of Wernicke and Lichtheim, the phenomenon was very difficult, if not impossible, to explain. After all, if the Germans were right and aphasia did indeed result from damage to one or another linguistic center in the brain, then language should be uniformly disordered; it was extraordinary that a tiny cluster of nervous arrangements serving one particular sentence could persist in splendid isolation. Yet, there was no doubt that recurrent utterances were a genuine fact of the aphasic syndrome, which had to be accounted for. Freud cited two of Jackson's examples of the phenomenon (taken from Jackson's 1879–80 article on speech disorders): a clerk who had completed a catalogue just before suffering a stroke, and who was left with the phrase "List complete"; and a man who had been rendered aphasic after being struck on the head in a brawl and whose only remaining words following injury were "I want protection" (Freud 1891, 61). Following Jackson, Freud affirmed,

> Such examples suggest that these speech remainders are the last words produced by the speech apparatus before the onset of its morbid condition, perhaps already under the influence of its realization. . . . If this last modification of speech through the aphasic process occurs at a moment of great inner excitement its very intensity should serve as an explanation for its persistence.

Freud made clear that he recognized the wider implications of Jackson's interpretation of recurrent utterances. He stressed that the production of such speech-remnants need not be exclusively restricted to cases of trauma involving a physical lesion of the brain. Any situation, rather, that had the effect of "lowering . . . associative activity" in the brain could cause a fixation of thought, provided that such a lowering took place during a moment of "great inner excitement." As evidence, Freud offered an analogy to the aphasic recurrent utterance in an experience of his own where, suddenly finding himself in a life-threatening position, his thinking narrowed and endlessly circled around the phrase "This is the end." "I heard these words," declared Freud, "as if somebody was shouting them into my ears, and at the same time I saw them as if they were printed on a piece of paper floating in the air" (189, 62).

Stengel (1963) declared that Freud's early interest in aphasic speech-remnants foreshadowed his later preoccupation with the relationship between emotional trauma and fixation of thought content. Robert Fliess (Wilhelm Fliess' son) believed that Freud looked upon the production of speech in dreams as a type of recurrent utterance consisting of samples of emotionally charged language uttered during the course of the day (Fliess 1953). Maurits Katan has argued that there is a continuity between Freud's early views on recurrent utterances and thought-fixation in a situation of danger and his mature psychoanalytic views on the significance of fetishism and construction in analysis, this last being a reference to the peculiar defensive response seen in certain patients when confronted with a plausible construction (see Freud 1937, 266). In all cases, "the ego reacts by bringing into the foreground the situation when mastery still seemed possible, for the trauma had not yet struck" (Katan 1969, 552).

John Forrester (1980) has put forward one of the most detailed arguments for intellectual continuity between Freud's early interpretation of recurrent utterances and his later psychoanalytic teachings. Focusing on Freud's researches into hysteria with Josef Breuer in the mid-1890s (see Breuer & Freud 1893–95), Forrester lays stress on Freud's discovery that most hysterical symptoms expressed traumatic ideas that the patient was incapable of expressing verbally and that, consequently, had got "caught up in his body."[1] For Freud, the hysterical symptom was a chronic somatic expression of some unarticulated memory that had been repressed from the patient's consciousness. As such, according to Forrester, Freud came to look upon hysteria as a type of recurrent utterance, analogous to the complex "rags and tatters" of speech retained by some aphasics. "It was a piece of language that once had meaning but which, by becoming cut off from the dynamic structure of contingent elements . . . was doomed to repetition" (Forrester 1980, 20).

Some commentators also paid attention to the interest that Freud showed, in *On Aphasia*, in Jackson's views on speech errors in healthy persons. Stengel declares (1963, 349) that we see in Freud's aphasia monograph an anticipation of his later interest in "slips of the tongue" and other verbal errors in everyday life (Freud 1905). Just as aphasiclike recurrent utterances could occur in stressful situations that did not involve structural damage to the brain, so was it equally plausible to assume, as Jackson had, that there was an underlying physiological identity between aphasic speech errors and instances of reversion or misnaming in normal speech.

> Paraphasias, i.e. mistakes in the use of words by aphasic patients, does not differ from the incorrect use and the distortions of words which the healthy person can observe in himself in states of fatigue,

[1] Freud excluded the classical stigmata described by Charcot from his semantic interpretation of hysterical symptoms (Breuer & Freud 1893–95, 265; cf. Breuer's remarks, 244–45); indeed, in a letter to Breuer in 1892, he described the origins of this aspect of the hysterical syndrome as "highly obscure" (cited Freud 1896, 192–93n). This disclaimer seems pertinent in light of the nineteenth-century idea that the presumably

or of divided attention, or under the influence of disturbing affects—the kind of thing that frequently happens to our lecturers and causes the listener painful embarrassment. (Freud 1891)

Now, as mentioned in Chapter Seven, Jackson's views on recurrent utterances and speech errors in healthy people were conceived within the framework of his views on the duality of mental operations (subject/object consciousness) and hemisphere differences. If, then, the commentators just cited are right and Freud's theory of neurosis was influenced by Jackson's views on paraphasia and especially recurrent utterances, it also is possible at least that the "metapsychology" that served as the framework for Freud's theory of neurosis was influenced by Jackson's "metaneurological" (if one will pardon the expression) theory of mental duality in the double brain. The two articles cited by Freud in *On Aphasia* (Jackson, 1878–79 and 1879–80), after all, contain a detailed exposition of that theory; and there is no reason why Freud might not have been even more familiar with Jackson's work than the references given in *On Aphasia* imply. According to Dewhurst (1982), Freud subscribed to *Brain*, which in 1915 republished a significant selection of Jackson's papers, including the 1874 essay on the double brain.

8-2. *Freud and Jackson on Language and the Possibility of Consciousness*

What is the evidence for a Jacksonian influence of this sort? In Freud's next important work after *On Aphasia*, the ultimately aborted and posthumously published *Project for a Scientific Psychology* (Freud 1895[2]), his goal was nothing less than to construct a model of the human mind using building blocks derived solely from the world of late nineteenth century physics. The "neurone," which had been formally introduced into neurology by Waldeyer in 1891, served as the material scaffolding for this model and a nervous energy Freud called quantity (or Q) was meant to provide the necessary dynamics.

Modeled on the reflex arc, Freud's neurones were divided into motor and sensory components. Sensory neurones in turn were of two sorts, serving two mutually exclusive functions: perception and memory. In contrast to Jackson, who had imagined that the recollection of a rose involved a weak revival of the same nervous centers that had served in the rose's initial perception, Freud, following Breuer, did not believe

nonverbal right hemisphere plays a particular role in the production of such stigmata (see Chapter Three).

[2] Large parts of the following exegesis are also indebted to Stewart (1969).

that perception and memory could be served by the same nervous arrangements. In Breuer's words: "The mirror of a reflecting telescope cannot at the same time be a photographic plate" (Breuer & Freud 1893–95, 189).

In the *Project*, then, the sensory neurones serving perception were "permeable." That is to say, they received impressions from the outside world but retained no record of them, and therefore were constantly fresh to transmit new impressions. Freud called these phi or φ neurones. The neurones serving memory were "nonpermeable." When quantity flowed through these neurones, resistance was set up, and a permanent modification in the nervous matter occurred. This second set of neurones, which Freud named psi or ψ neurones, cooperated with φ neurones, storing the perceptions received by the latter for later recall or modification by future experience. Ψ neurones were identified with psychical processes in general.

Up to this point, Freud was able to cast his model in strictly quantitative terms, using matter and energy. So long as perception and memory were independent of consciousness, the quantitative approach was adequate. However, human beings were capable of introspection on their own mental operations, which forced Freud to introduce something new into his scheme: a third set of neurones capable of detecting the quality or message encoded within the flow of psychic energy running through the system. Freud called these consciousness neurones, omega (ω) neurones. Although they were described as a subdivision of the ψ (memory-processing) system, they were permeable like the φ (perceptual) neurones, because their function was to perceive the quality contained in psychic energy, Q. Consciousness, as Freud would state clearly a few years later, was to be conceived as "*a sense-organ for the perception of psychical qualities*" (Freud 1900, 615).

This idea was found in Jackson, but since there is no evidence that it was particularly original with him, it cannot be assumed that Freud was necessarily influenced by Jackson here. Far more compelling evidence for influence emerges, however, when Freud moves on to consider the problem of *how* the psychic energy (Q) manages to make the information it carries, its qualities, perceptible to the ω neurones. He concludes that psychical processes achieve quality by becoming associated with "*indications of speech-discharge.*" Coupling thoughts to "motor speech-images" and "sound-presentations" (nascent verbal hallucinations) "help . . . put thought-processes on a level with perceptual processes, lend them reality and *make memory* [i.e., conscious recollection] *of them possible*" (Freud 1895, 365–66).

This view that psychic processes must be "cloaked" in words in order to be perceived by consciousness, a view that closely parallels Jackson's view on the necessity of "objectifying" thoughts, would be retained by Freud in essentially unaltered form long after the *Project* had officially been abandoned. In *The Ego and the Id* (1923), he writes,

> The part played by word-presentations now becomes perfectly clear. By their interposition internal thought-processes are made into perceptions. It is like a demonstration of the theorem that all knowledge has its origin in external perception. When a hypercathexis of the process of thinking takes place, thoughts are *actually* perceived—as if they came from without—and are consequently held to be true. (Freud 1923, 13; cf. 1900, 617; 1911, 221; 1915, 200–203)

Jackson had not felt that words ("indications of speech-discharge") were the only medium that human beings could use to "objectify" their thoughts and, thus, become conscious. In his system, visual images also played an essential role in most forms of conscious mentation, to such an extent that he felt a man without visual images would be "practically mindless." In *The Psychopathology of Everyday Life* (1901), Freud similarly acknowledged that thought need not invariably represent itself to consciousness in verbal form. Most people, it was true, made more frequent use of (verbal) "audito-motor" images, but a few individuals were more visual, relying to a larger extent on what Jackson had called the "retino-ocular" series of thought.

> [Conscious] remembering in adults, as is well known, makes use of a variety of psychical material. Some people remember in visual images; their memories have a visual character. Other people can scarcely reproduce in their memory even the scantiest outline of what they have experienced. Following Charcot's proposal, such people are called *auditifs* and *moteurs* in contrast to the *visuels*. In dreams these distinctions disappear; we all dream predominently in visual images.[3] [Similarly] . . . in the case of childhood memories: they are plastically visual even in people whose later function of memory has to do without any visual element. Visual memory accordingly preserves the type of infantile memory. (Freud 1901, 47)

By identifying visual consciousness with both dreams and infantile memory, Freud was making clear that, contra Jackson, he did not believe that the capacity to be "conscious in objects" was just as lofty a form of human introspection as the capacity to be "conscious in

[3] Freud had made a similar comment in the *Project*: "One shuts one's eyes and hallucinates; one opens them and thinks in words" (1895a, 339).

words." Although he always conceded that it was possible, particularly in talented *visuels* like Charcot, for "thought processes to become conscious through a reversion to visual residues," he was inclined to feel that, in most people, only the "concrete subject matter of the thought" could be represented to consciousness in this manner. "The relations among the various elements of this subject matter," he said, "which is what specially characterizes thoughts [i.e., propositional thoughts, in the Jacksonian sense], cannot be given visual expression. Thinking in pictures," Freud thus concluded, "is, therefore, only a very incomplete form of becoming conscious. In some way, too, it stands nearer to unconscious processes than does thinking in words, and it is unquestionably older than the latter both ontogenetically and phylogenetically" (Freud 1923, 11).

Whether or not Jackson and Freud agreed on the relative importance of the visual mode of representation in consciousness is not essential to the argument here. The more basic fact is that, within four years of hailing Jackson as his "guiding principle" in all matters relating to the functioning and pathology of speech, Freud developed a theory linking symbolic representation, specifically of the verbal mode, to the possibility of consciousness. This theory, moreover, resembles Jackson's views on object consciousness or propositional thought as much as it *fails* to resemble the theories of consciousness held by the other main figures likely to have exerted an influence on Freud at this time. In the eyes of both Meynert and Breuer, for example, consciousness was a direct function of the level of excitation in the nervous system: when latent ideas reached a certain level of stength they automatically became conscious. As Fullinwider (1983) pointed out in a slightly different context, only Jackson had made the difference between consciousness and unconsciousness turn on the word, or to be more precise, on the presence of the symbolic image. Only Jackson had seen that, in order for ideas to become conscious, they must be objectified, or as Freud would say given quality, and that this could be accompanied by linking them to "images," or as Freud would have it "mnemic residues."

8.3. Subject/Object Consciousness versus Primary/Secondary Thought

But the conceptual parallels between the two men's thought do not end there, as an examination of Freud's later psychoanalytic writings shows. It has been argued (e.g., Jones 1953, 281; Sulloway 1979, 44–45) that Freud's distinction between primary and secondary thought processes took its point of origin from Theodor Meynert's hierarchical division of mental activity into a lower (primitive, subcortical) level and a higher (voluntary, cortical) level. However, Freud's belief that the transition from the primary to the secondary process depended upon thoughts becoming associated with word-presentations makes Jackson's division of thought into subjective and objective components a plausible alternative, or complementary, source of influence. Jackson's dual conception of thought was, like Meynert's, a hierarchical one, at least in functional terms. The nervous arrangements on the right side of the brain, serving the first, subjective half of verbal thought stood, it will be recalled, at a "lower level of evolution" than the nervous arrangments on the left side serving object consciousness. Like Freud's primary process, subjective mental activity was automatic and without any capacity to moderate itself in accordance with the demands of the environment. Its associative capacity was crude, guided by the primitive principle of "association by resemblances." In contrast, the higher nervous arrangements serving object consciousness functioned voluntarily; that is, with considerable capacity to redirect or inhibit activities in response to environmental cues, much like Freud's secondary process. Association of ideas at this level of mental activity was guided by critical attention to "differences," which in turn allowed for judgment and reasoning.

In a recent paper, "Sigmund Freud, John Hughlings Jackson, and speech," Fullinwider (1983) also argued that Freud's distinction between the (conscious) secondary process and the (unconscious) primary process was inspired by Jackson's distinction between propositional thought and "speechless lower levels of ideation." "Jackson gave Freud the theory of functional levels and the notion that speech defines the functioning of the highest level. It was with this twin legacy that Freud grounded his theory of the unconscious as both orderly and active" (p. 158).

While, obviously, the argument just developed has broad points in common with Fullinwider's views, in many points of detail I am unable to follow his interpretation. He fails, for example, to realize that the key to the Jacksonian concept of propositional thought was not so much *speech* as the symbol, which could just as readily be a visual image; his view that Jackson's propositional thought was tantamount to the capacity to associate ideas is contradicted by Jackson's repeated insistence that nonpropositional thought followed the law of association by resemblances; and, most unforgivably, he seems not to have understood that Jackson's concept of propositional thought as the highest functional level in the nervous system cannot be understood in strictly vertical, hierarchical terms but must be placed within the context of his views on hemisphere functional specialization and mental duality (subject/object consciousness). Jackson's views on the double brain are not discussed at any point in his essay.

8-4. *Pathological Subject Consciousness and the Mechanism of Repression*

I now take my argument one final step and dovetail to some extent with the commentators mentioned in section 8.1. Not only is there evidence that Freud was influenced by Jackson in formulating his views on consciousness mental activity in healthy individuals; it also is possible that the Jacksonian model of mental duality in the double brain provided inspiration for certain aspects of Freud's theory of neurosis. More specifically, I suggest a Jacksonian influence for Freud's views on the psychological-physiological processes responsible for repression. In his writings, Freud distinguished between preconscious thoughts (latent ideas capable of entering into consciousness) and unconscious thoughts (latent ideas that are barred from becoming conscious). This distinction might have been modeled in broad outline after Jackson's views on how a thought can get stuck in its first subjective (unconscious) stage.

Let us see how this might be. For Freud, the distinction between the preconscious and the unconscious was not a primary one. The two systems only came to be established "after repulsion [or repression] has sprung up" (Freud 1912, 264). That is to say, the unconscious was a pathological outgrowth of the preconscious, not an innate feature of the human psychic apparatus. Certain ideas became unconscious or were primally repressed because the mind's censor, eventually identified with the unconscious part of the ego (Freud 1923), objected to their subject matter, which was invariably of a sexual nature.

Now, according to Freud, repression was brought about physically through some form of functional interference with the process whereby an objectionable idea should have entered into an assocation with word presentations. Once attached to words, it would, of course, have been in a position to make its qualities visible to consciousness. "A presentation which is not put into words remains thereafter in the Ucs [unconscious] in a state of repression" (Freud 1915, 202).

> A rough but not inadequate analogy to this supposed relation of conscious to unconscious activity might be drawn from the field of ordinary photography. The first stage of the photograph is the "negative"; every photographic picture has to pass through the "negative process," and some of the negatives which have held good in ex-

> amination [by the censor] are admitted to the "positive process" [involving linkage with word-presentations] ending in the picture. (Freud 1912, 264)

As noted by several other commentators (Edelheit 1969; Forrester 1980, 31–36), Freud's language-dependent concept of repression closely recalls his early conception of asymbolic aphasia, as described in the 1891 monograph:

> A word acquires its significance through its association with the "idea (concept) of the object," at least if we restrict our consideration to nouns. The idea, or concept, of the object is itself another complex of association composed of the most varied, auditory, tactile, kinaesthetic and other impressions. . . . In consequence, we have arrived at a division of speech disorders into two classes: (1) verbal aphasia, in which only the associations between the single elements of the word concept are disturbed; and (2) asymbolic aphasia, in which the association between word concept and object concept are disturbed. (1891, 77–78)

Repression, then, might have been envisioned by Freud as a sort of content-specific asymbolic aphasia. Where the true aphasic suffered a global loss of functioning, the neurotic was only aphasic for a specific memory or associated group of memories. And because those memories were always cut off from consciousness *at the moment* they were trying to enter, they persisted in the unconscious mind of the individual—permanently activated Jacksonian subject-propositions unable to turn themselves into object-propositions and enter consciousness, yet equally unable to disassemble themselves and disappear. The resulting neurotic symptom, whether a fetish (Katan 1969), a hysterical disorder (Forrester 1980), or some other abnormality, was thus closely analogous to the aphasic recurrent utterance, as interpreted by Jackson. It was a piece of meaning hindered in the act of trying to attach itself to its symbols and consequently retained, a chronic, unconscious, senseless echo of a traumatic idea or memory.

One last point, perhaps, needs to be made clear: if Jackson's views on recurrent utterances, the duality of thought, and the role of symbolic images, especially words, in consciousness may have influenced Freud in all the ways just outlined, one is not justified in assuming that Freud also accepted the anatomical framework within which Jackson had cast his views. After 1895, with the failure of the *Project*, Freud became reluctantly convinced that it would not be possible to develop a psychology grounded in the neurophysiology of the brain. In the end, the

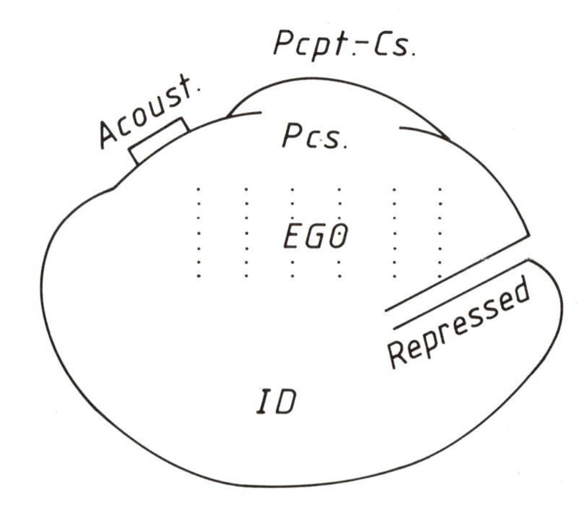

Figure 10. Freud's model of the ego, which wears its neurological cap of hearing "awry" on the left side. (Source: After Freud 1923, 24.)

great riddle of the mind/body relationship had simply proved too formidable (cf. Chapter Nine). In constructing the metapsychological framework of psychoanalysis, therefore, he stressed that "our psychical topography has *for the present* nothing to do with anatomy; it has reference not to anatomical localities, but to regions in the mental apparatus" (1915, 175). Because of this crucial development in Freud's mature thought, I would emphasize that my argument for a psychoanalytic indebtedness to Jackson's double brain simply affirms that, had Jackson *himself* not been struck by the bilateral duality and functional asymmetry of the human brain, he would not have developed those ideas on language and dual mental processing to which Freud in his turn seems intellectually indebted.

At the same time, I cannot resist pointing out that the aphasiologist in Freud naturally was aware that all the best medical knowledge of his time indicated that language, which played such a crucial role in his metapsychological system, predominantly was mediated by only one half of the brain. Is there any reason to think that Freud wondered about the implications of this fact? If not, why, in describing the conscious part of the ego as a verbal construction whose origin lay in things heard, would he bother to observe that it wore its neurological "cap of hearing" "awry" or "on one side only" (Freud, 1923, 15) (see Figure 10)? Similarly, why, in the course of explaining how the ego could be identified with the "cortical homunculus" of the anatomists, would he again think it necessary to point out that this ego had "its speech-area on the left-hand side" (1923, 16)? In making such oblique neurological allusions in the course of constructing an overtly psychological model of consciousness and the self, was he hinting that, if only the *Project* had succeeded to his satisfaction, then the doctrine of cerebral functional asymmetry would have played no insignificant role in the "psycho-physical topography" he would have gone on to develop?

CHAPTER NINE

The Fate of the Double Brain

IN PAST CHAPTERS, I reviewed, in considerable detail, how, in the second half of the nineteenth century, the duality and functional asymmetry of the human brain came to occupy an important place in the emerging sciences of mind and brain. Indeed, these two features of human neuroanatomy together managed to spawn a veritable smorgasbord of speculation. Not only were they seen as important for understanding the aphasic syndrome and other forms of neurological disorder, as might have been predicted; they also managed to insinuate themselves into nineteenth-century perspectives on religion, morality, philosophy of mind, biological and social evolution and degeneration, racial and sexual differences, criminology, education, the occult, hysteria and hypnosis, and especially psychopathology: double personality, "lucid madness," voices in the head, obsessive ideas, monomania, doppelgängers, etc.

Then, during the first years of the twentieth century, and certainly by the end of World War I, the overall tenor of speculation seems to have changed markedly. Although an exhaustive search has been impossible, I tried, and largely failed, to uncover articles and essays on man's duplex, asymmetrical brain written after 1920 and before the 1960s, which are equivalent in tone and orientation to those written in the nineteenth century. It is true that one can find isolated exceptions to this general trend, and I discuss a sampling of these later in this chapter. However, nothing is to be found that approaches the bold, wide-ranging, mutually reinforcing collection of writings on the topic offered up by the nineteenth century. From this, two questions suggest themselves. How could ways of speculating about mind and brain that were natural in one age, twenty or thirty years later, have more or less vanished from the working worldview? And, if the first half of the twentieth century was more or less a "negative" chapter in the history of the double brain, what brought about the astonishing renaissance of interest in brain duality and hemisphere asymmetry since the 1960s? Why, again, has the double brain suddenly become so interesting?

These are formidable questions, both of them, and the fact that I pose them here does not mean that I am planning in this chapter to resolve them in any sort of definitive fashion. They are guiding questions only, torchlights to help me pick my way across an intimidating social and intellectual terrain. In a single chapter, obviously, it would be impossible to survey this terrain in any comprehensive manner, and I have no intention of trying to do so. My aims are considerably more modest, or perhaps more audacious. I propose simply to try to beat out some sort of trail leading from the late nineteenth-century era of the double brain to the present day (late twentieth century). In the process, I hope to shed some preliminary light on the way modern ideas about brain laterality and "split-brain" consciousness fit into the larger ebb and flow of thinking about mind, brain, and madness, over the last 150 years.

9-1. From Hysteria to Schizophrenia: The Consequences of Clinical Cartesianism

I begin, then, by focusing on the decline of interest in Wigan-style theories linking brain duality to various forms of psychopathology or more everyday peculiarities of normal consciousness. One early signpost to a possible explanation for this phenomenon lies in a paper by the Swiss-trained American psychiatrist Adolf Meyer, published in 1904. Surveying the state of psychiatry in his time, Meyer spoke approvingly of the discipline's recent "emancipation" from its former "peculiar position of an adjunct to neurology." There were few in the nineteenth century who would have dreamed of questioning the premise that neurology ultimately offered the best hope of attaining a truly scientific understanding of mental disorder. But now Meyer argues that psychiatrists, increasingly influenced by the work of such men as Emil Kraepelin and Carl Wernicke, were learning to appreciate that mental disease was an entity whose meaning was to be found in its complex of symptoms rather than in some hypothetical brain lesion. Increasingly, it was being realized, said Meyer, that *psychology*, not neurology, must serve as the conceptual foundation for the study of psychopathology (Meyer 1904a, 221–22).

Meyer's perception at this very early stage, of a growing rift between the clinical world of brain malfunction (neurology) and the clinical world of personality disorder (psychiatry) seems, with the advantage of hindsight, remarkably acute. Over the next several decades, the view that psychiatry and neurology rightly were distinct and independent fields of inquiry was to become thoroughly entrenched in twentieth-century thought. Medical historian William Bynum drew attention to the ironical fact that, "at a time when evolutionary biology has undermined the ontological position of the human mind, we still live more or less in a Cartesian world. We recognize diseases of the brain and diseases of the mind, and if the border between them is not quite so clear as it might be in the world that Descartes constructed, the number of patients that psychiatrists and neurologists fight over is on the whole rather small" (Bynum 1981, 36).

I suggest, then, that the twentieth century's apparent resurrection of at least an operational Cartesianism within the clinic is one of the most important historical developments, if not the only relevant one, we need to notice and explain. After all, if the essence of mental pathology is no longer thought to reside in some anatomical or physiological defect of the brain, then one obvious result would be a sharp decline of interest in theories accounting for mental disorder in terms of discordant or independent action of the two hemispheres.

Consider, for example, the problem of hysteria. I have shown how the great majority of post-Broca double-brain theorists lived and worked in France and, in one way or another, had intellectual ties with, or sympathy for, the teachings of the Charcot school on hypnosis and hysteria. So long as hysteria could be conceived as a neurological disorder, it was possible for cases of hysterical double personality, like that of Louis Vivé, to be interpreted in terms of hemisphere independent action, and it was equally possible for such phenomena as transfer and hemi-hypnosis to be interpreted as experimental proof of hemisphere functional independence. To a considerable degree, then, much of the most impressive evidence available to post-Broca advocates of duality of mind depended for its coherence upon the continuing viability of Charcot's special categories of reference for hysteria.

The wave of reaction against these categories, beginning in the late 1880s, can be blamed to no small extent on the disasters to which they had led when applied uncritically to hypnosis theory and research. The devastating criticisms of the Nancy school under Bernheim (Chapter Six) not only fostered an attitude of extreme reserve among outsiders about Charcot's latest enterprise; it also left a number of Charcot's one-time loyal disciples feeling foolish and betrayed. It should come as no surprise, then, that when Joseph Babinski attacked the theoretical foundations of his old Master's approach to hysteria, in a 1901 lecture before the Société de Neurologie, his message fell on receptive ears (Babinski 1901). Through his own work in neurology, Babinski had discovered a number of signs indicative of localized neurological damage, the most widely known being the cutaneous-plantar reflex, Babinski's sign, in pyramidal disorders. Such signs, he believed, were completely absent in cases of so-called hysterical disease. Therefore, there was no justification for thinking of hysteria as a neurological disorder at all; for Babinski, the syndrome consisted of nothing more than the sum total of symptoms that could be called into existence by suggestion and eliminated though reverse-suggestion, or persuasion. He consequently recommended that the very term *hysteria* be dropped and replaced by the word *pithiatism*, from the Greek words meaning "persuasion" and "curable." The effect of this proposed change of nomenclature was to convince many of Babinski's neurological colleagues that hysteria did not "really" exist, or at least was not worth bothering about. As Guillain, Charcot's biographer put it, "the word has a powerful influence on the mind," and "the term 'pithiatism' . . . was directed to a total rejection of the clinical facts established by Charcot" (Guillain 1959, 152).

Another early, and highly influential, critic of the Charcot neurophysiological approach to hysteria was, of course, Sigmund Freud. Although he greatly admired Charcot's work, Freud's 1893 paper, "Some Points for a Comparative Study of Organic and Hysterical Motor Paralyses," identified a number of ways in which hysterical ailments failed to correspond symptomatically to their organic counterparts, as required by Charcot's theory. For example, "the [hysterical] aphasic utters not a word, whereas the organic aphasic almost always retains a few words, 'yes,' 'no,' a swear-word, etc." Similarly, "the hysteric [suffering from paralysis] drags the leg like an inert mass instead of performing a circumduction with the hip as does the ordinary hemiplegic." She does not "know" that "the proximal portion of the limb is always to some extent exempt" in organic hemiplegia. In short, Freud concluded in a now-famous passage, "hysteria behaves as though anatomy did not exist or as though it had no knowledge of it" (Freud 1893b).

Satisfactory insight into the hysterical syndrome, then, was clearly not going to come from anatomical or physiological studies alone. In his hysteria researches with Breuer, Freud was to take the crucial step of abandoning the objective, general viewpoint of psycho-physiology and transferring the problem of neurosis to the subjective, individual plane. "The problem thus posed by Freud . . . [became] a matter not of impairment in psychophysiological mechanisms that control thought, but of a personal life story" (Baruk 1978, 159). By confronting mental disorder on its own terms, Freud, and those who followed him, restored, in Michel Foucault's assessment, "the possibility of a dialogue with unreason" (Foucault 1973, 198).

This is not the place to go into the multifaceted and disputed story behind Freud's decision to abandon his neurological training, and develop a science of psychopathology, at least overtly, grounded in psychological facts alone. It will be sufficient simply to call attention again to the remarks made at the end of Chapter Eight: that, in the wake of his own unsuccessful attempt to create a "scientific" psychology rooted in the most up-to-date neurophysiology of the time (Freud 1895a), Freud had come to the conclusion that psychiatry was not yet in a position where it could profitably employ concepts from neurology to explain mental disorder:

> Research has given irrefutable proof that mental activity is bound up with the function of the brain as it is with no other organ. We are taken a step further—we do not know how much—by the discovery of the unequal importance of the different parts of the brain and their special relations to particular parts of the body and to particular mental activities. But every attempt to go on from there to discover a localization of mental processes, every endeavour to think of ideas as stored up in nerve cells and of excitations as travelling along nerve-fibres, has failed completely. The same fate would await any theory which attempted to recognize, let us say, the anatomical position of the system Cs.—conscious mental activity—as being in the cortex, and to localize the unconscious processes in the subcortical parts of the brain. Our psychical topography has *for the present* nothing to do with anatomy; it has reference not to anatomical localities, but to regions in the mental apparatus. (Freud 1915, 174–75)

For our purposes, the chief importance of the Freudian movement lies in the powerful impetus it gave to the trend toward a psychological psychiatry, first observed by Meyer in 1904. Freud held that his psychic apparatus was not just a metaphysical concept but something tangible, whose reality could be demonstrated clinically even if it could not be perceived with the senses. As a real and independent entity, the psychic apparatus was capable of becoming disordered no less than any of the organs of the body; and its diseases were quite emphatically its own, following its own laws, and having nothing to do with the diseases and disorders of the brain (Zilboorg 1959, 146).

Bromberg argued that the first World War further contributed to the growing rift between neurology and psychiatry by demonstrating that certain sorts of emotional breakdowns and neuroses could be neither understood nor treated through a narrow, brain-oriented approach. Neurologists had much to do, studying and caring for brain-injured soldiers, but they could offer little help to the shell-shocked boy who was needed on the front lines; they could not speak to the "military tragedies, social dislocations, and psychological pressures" brought on by the war (Bromberg 1982, 5–7). In the civilian sphere, changing philosophies about the proper relationship between the individual and the state, along with the institutionalization of a variety of social reform programs, also meant that psychiatrists increasingly would be expected to accept a different agenda of interests than that expected of neurologists. No longer would they focus strictly on mental illness in the hospital wards; they would concern themselves as well with personal guidance problems, criminal and delinquent behavior, industrial relations, child behavior problems, educational retardation, personality problems in schools, colleges, and the armed forces, propaganda, and the psychology of the public during wartime (Bromberg 1942, 118–19). The example of Alfred Binet might be cited as a prime case in point: he began his career in the Charcot-school tradition, studied metalloscopy and hypnosis with Charles Féré, developed a theory of hysteria as a form of double personality, but ended his life as an employee of the ministry of public education, designing tests to identify schoolchildren with learning difficulties.

Certainly, the trend toward psychologism in psychiatry was far from absolute or unproblematic. For examaple, even while the French rejected the Charcot neurophysiological view of "functional" disorders or neuroses, the psychogenetic alternative embodied in Freud's psychoanalysis initially met with much opposition in that country. France, of course, had its own answer to Sigmund Freud: the philosopher and medical psychologist Pierre Janet. Janet proposed to explain such baffling phenomena as somnambulism, trance personalities, and alternating consciousness not by reference to various hypothetical alterations in the brain, but by conceiving of the patient's mind as a hierarchical structure operating according to dynamic principles of "force" and "tension" and susceptible, under stress, to "*désagrégation.*" This flexible, psychological approach to the permutations of the neurotic personality was applauded and variously extended by such foreign workers as Morton Prince (whose famous *Dissociation of a Personality* was published in 1905), William James, William McDougall, and Boris Sidis. In his native land, however, Janet's innovations received a considerably more mixed reception. Neurologists such as Déjerine and Babinski, accused their colleague of perpetuating Charcot's errors, because he still addressed himself to the problems of hysteria and hypnosis, and there is reason to believe that these men indulged in various intrigues to reduce Janet's influence and keep him out of the main wards of the Salpêtrière (Ellenberger 1970, 343).

Despite the general revulsion in conservative French circles away from the more extravagant species of "brain mythology," of which much of the double-brain literature represented an example par excellence, it also is important to realize that organicism of a certain type continued to maintain a strong hold upon the French psychiatric imagination. Henry Ey wrote (1975, 27) that French psychiatry in the early twentieth century was marked both by a preoccupation with the constitutional and hereditary basis of mental disorder, associated with the works of men like Dupré, de Fleury, Delmas, and Logre, and by a lingering mechanistic bias inherited from the nineteenth century. One of the leaders in French psychiatry during the first half of the twentieth century, along with perhaps Jean Délay and Henri Baruk, Ey himself made a bold attempt in the 1930s to offer a corrective to sterile mechanism, which would nevertheless not represent a flight into Cartesianism. In 1938, he wrote a monograph with Julien Rouart that was largely to make his reputation. In it, he advocated a so-called organo-dynamic approach to mental disorder, one that built upon the hierarchical principles of mind and brain laid down at the end of the nineteenth century by Hughlings Jackson (the original paper is reprinted in extenso in Ey [1975]).

So much, then, for hysteria and the changing face of French psychiatry. Matters are more complicated still. One of the great psychiatric events of the prewar years was the publication in Switzerland of Eugen Bleuler's *Dementia Praecox oder Gruppe der Schizophrenie* (Bleuler 1911; see also the early papers in M. Bleuler [ed.] 1979). The disorder known as dementia praecox had been exhaustively described and catalogued by Emil Kraepelin in Germany at the end of the nineteenth century. It was now rechristened *schizophrenia* by Bleuler, and transformed into a dynamic category that focused many of the new concerns in psychiatry. Following Freud's example, Bleuler stressed that psychotics were not victims of some type of physical alteration of the cerebral structure but rather suffered from a severe disorder of the psyche, having its chief source in their life histories.

As his former student, Carl Gustav Jung explained,

> Here is a person like ourselves, beset by universal human problems, no longer merely a cerebral machine thrown out of gear. Hitherto we thought that the insane patient revealed nothing to us by his symptoms, save the senseless products of his disordered cerebral cells; but that was academic wisdom reeking of the study. When we penetrate into the human secrets of our patients, we recognise mental disease to be an unusual reaction to emotional problems which are in no way foreign to ourselves, and the delusion discloses the psychological system upon which it is based. (cited in McDougall 1926, 378)

More than any other mental disorder in the twentieth century, schizophrenia would come to exemplify "madness" in the minds of both medical men and the lay population. This conceptual triumph has a special relevance for understanding the fate of the double brain in early twentieth-century psychiatry. It may seem that nothing is more natural than that this should be so. After all, the very word *schizophrenia* means "splitting of the mind," so it is not surprising that it should end up surfacing somewhere in our story. It may be less intuitively obvious, however, that this new disease category might have played a *negative* role in the life and times of the double brain. But this is just what I wish to argue.

Bleuler had introduced the term *schizophrenia* to replace Kraepelin's *dementia praecox* because, as he explained,

> in every case we are confronted with a more or less clear-cut splitting of the psychic functions. If the disease is marked, the personality loses its unity; at different times different psychic complexes seem to represent the personality. Integration of different complexes and strivings appear insufficient or even lacking. The psychic complexes do not combine in a conglomeration of strivings with a unified resultant as they do in a healthy person; rather, one set of complexes dominates the personality for a time, while other groups of ideas or drives are "split off," and seem either partly or completely impotent.
>
> . . . Thus the process of association often works with mere fragments of ideas and concepts. This results in associations which normal individuals will regard as incorrect, bizarre, and utterly unpredictable. (Bleuler 1911, 9)

"Splitting" of the mental functions, breakdown of personal unity, competition between conflicting ideas and drives, all these phenomena had proved a rich field of inspiration for nineteenth-century double brain theorists. In delineating the wide range of symptoms (both primary and secondary) that he now wished to appropriate to schizophrenia, it is clear that Bleuler was harvesting from this same field. However, he was replanting the various clinical data in a conceptual framework where the action of two warring brain-halves was not only an unnecessary hypothesis but appeared positively implausible. Bleuler argued that, in contrast to the hysteric whose mental functions seemed to be divided into two sharply separated groups (an image at least congenial to speculation about hemispheres in conflict), the schizophrenic had experienced a fundamental mosaic-like fragmentation of his or her thinking processes, a breakdown of the alleged "associationist" pathways linking ideas to one another in a logical sequence. This primary loosening of the associational structure of the mind led in turn to various secondary forms of mental disintegration: the systematic splitting off of parts of mental life into dominating idea-complexes; the "blocking" of thought processes; the delusion of being "two people" or turning into someone else, hallucinated voices, auditory "double thinking," etc.

The diverse cognitive symptoms of schizophrenia coexisted and interacted, Bleuler believed, with a type of splitting of the will and affect. This expressed itself most strikingly in the characteristic ambivalence of the schizophrenic patient, who simultaneously loved the rose because of its beauty and hated it because of its thorns. For our purposes, one of Bleuler's remarks on the question of ambivalence is particularly noteworthy. He explained how

> [t]he ambivalence of feelings is often revealed in . . . contradictory [hallucinated] voices: there are voices which console the patient, take his part; and there are others which complain about him, annoy and torture him. They may even divide themselves as to which of the two ears they appear to speak into: the "good" voices using the right ear, and the "bad" ones using the left ear. . . . [O]ne of our patients spoke symbolically of the Holy Ghost in his right ear and the snake in his left one. (1911, 405, 100)

As discussed in Chapter Four, the curious syndrome of conflicting auditory hallucinations (good and bad voices) in the two ears had impressed no less an authority than Valentin Magnan as powerful evidence in favor of the case for hemisphere independent action (Magnan 1883). In now offering a strictly psychological-symbolic explanation for the same phenomenon, it is striking that Bleuler did not feel it necessary even to *mention* this rival physiological perspective. There is no question of his familiarity with it. He had studied under Magnan and Charcot in the 1880s, when French interest in the double brain was at its height. Moreover, in 1902, he had been quite willing to invoke hemisphere independent action to account for his case of "unilateral delirium" (Bleuler 1902–03; also Chapter Five). In now constructing the grand theoretical edifice of schizophrenia, did he fail to reflect upon the double brain because the rather static, bipolar imagery it offered was so clearly inappropriate to the conceptual framework in which he was working? Alternatively, could it be that, following Freud's example, he had decided to steer clear of all attempts to descend into the murky realms of neurophysiology and brain anatomy; maintaining, perhaps, a somewhat more open mind towards the growing and increasingly fashionable field of biochemistry? In the 1911 monograph, after all, he plainly stated his conviction that "the symptomatology of schizophrenia (including disturbances of motility) provides no definite indication of localization" in the brain (p. 381).

Be that as it may, it seems that the rise of schizophrenia as a psychogenetic, or at least not obviously neurological, disorder in the twentieth century discouraged psychiatric interest in the duality of the human brain. On this score, it reinforced the damage already wreaked by the revolt against the Charcot school's brain-oriented approach to hysteria and the rise of various alternative psychodynamic approaches to personality disorder. I believe, moreover, that both the decline of hysteria as a brain disorder *and* the rise of schizophrenia as a mind disorder need to be understood within the wider context of a growing institutional and conceptual estrangement between neurology and psychiatry, what I have been calling the resurrection of Cartesianism in the clinic.

Few intellectual and cultural trends, however, are without a stubborn countercurrent. When one begins to explore the canals off the mainstream of psychiatry, one in fact finds several authors continuing to suggest that the last word might not yet be in on the double brain. These scattered writings—and doubtless I have not unearthed them all—revolve around the same themes and speak in more or less the same accents as those of the nineteenth century. However, because they all seem to have been relatively uninfluential at the time, I do not feel they undermine my "end of an era" argument, but only place it in salutary perspective.[1]

[1] It is possible that the more flamboyant nineteenth-century–style approach to the double brain was retained somewhat longer in the popular literature of the early twentieth century than in the professional. The year 1934, for example, saw the publication of a curious book, *The Human Gyroscope. A Consideration of the Gyroscopic Rotation of Earth as Mechanism of the Evolution of Terrestrial Living Forms; Explaining the Phenomenon of Sex: Its Origin and Development and Its Significance in the Evolutionary Process.* Written by the once well-known antifeminist and antivivisectionist Arabella Kenealy, the greater part of this work is an opaque mixture of scientism and mystical indulgence not germane to the present discussion. What *is* relevant, from our point of view, is the way in which Kenealy's theory relating the earth's gyroscopic rotation to the evolution of sex led her to consider the differential functioning of the two halves of the brain. Medical science had clearly shown, she felt, that the left brain-half was the "positive male-objective and leading half, in respect of cerebral process and faculty," while the right brain-half was the "negative-female and subjective" side of the brain. We saw how a number of nineteenth-century writers had argued variations on this theme. Kenealy, though, differed from most of these in her emphatic insistence that such functional differences did not mean that the left hemisphere was superior and the right hemisphere inferior. On the contrary, the differing activities of the male and

In 1926, psychoanalyst Sandor Ferenczi raised a specter of the old nineteenth-century idea of a link between the right hemisphere and unconscious (subconscious, sleeping, automatic, subliminal) mental activity. Ferenczi was one of Freud's more creative disciples, though later effectively disowned by the master for his "wild" therapeutic techniques. Discussing the nineteenth century belief that in hysteria "the hemianesthetic stigmata occurs more frequently on the left than on the right," he proposed, "It is possible that—in right-handed people—the sensational sphere for the left side shows from the first a dispensation for unconscious impulses, so that it is more easily robbed of its normal functions and placed at the service of unconscious libidinal fantasies" (cited in Galin 1974, 577). Ferenczi did not say explicitly that these libidinal fantasies therefore might have a home in the right side of the brain, but at least he was willing to grant credence to some of the reported observations of the Charcot school and he did not exclude the possibility of an ultimate rapprochement on some level between that school's psycho-physiological orientation and his own interpersonal, psychoanalytic one.

In the United States, the now-forgotten "experimental psychoanalyst" Werner Wolff was much more explicit in his argument for a link among unconscious thought-processes, psychopathology, and the double brain. His interest in these matters seems to have been first aroused by a series of experiments he carried out on some students of his in the early 1940s. These consisted, quite simply, of asking subjects to identify photographs either of their own faces or of those of various friends and acquaintances. Unknown to the subjects, however, all the photographs were reconstructed chimeras, made up either of the two left-halves or the two right-halves of a single face. Wolff found that subjects had significantly more difficulty recognizing the left-left faces

female sides of the brain perfectly balanced and complemented each other.

As she explained, "Male-Objective [on the left side]—exact, critical, analytic, mathematical—defines, dissects, disintegrates, and resolves things into their ultimate factors—atoms, protons, neutrons, electrons, and these even to dispersion in radiation. . . . But the subjective faculties [on the right side], aesthetic and intuitive, are integrative and constructive. They synthetize and unify; and supplementing the realist with the visionary viewpoint, reveal the vital in the formulae, the spiritual thread of Purpose gleaming through all phenomena" (Kenealy 1934, 216).

Neurology, then, offered an analogy to, and an implicit biological sanctioning of, Kenealy's ideal society. She had long argued that men and women were made to complement each other, each in his or her necessary but different realm. It was sheer folly for the sensitive, intuitive fair sex, taking advantage of the social upheavals wrought by the Great War, to seek employment for which it was physically and mentally unsuited (see Kenealy 1920).

of their own photographs than the right-right faces. Asked to make a judgment about an unrecognized portrait of themselves, they tended to project "wish images upon the form of expression to be judged. . . . The most remarkable finding was that wishful tendencies appeared more frequently in the left-left than in the right-right faces (the proportion was more than 3 to 2)" (Wolff 1943, 154). Other experiments suggested that "the left side seems to represent the traits which are opposite to those of the sex and fundamental nature; perhaps in general it represents certain underlying wishes of the personality" (p. 167).

By 1952, Wolff was committed to the view, only hinted at in his 1943 work, that the right hemisphere served as the locus for Freudian unconscious thought processes. This discovery opened the door, he believed, to "a new hypothesis on the structure of the abnormal."

> One observation leads the way, namely that injuries to the left hemisphere tend to produce patterns of primitive and of symbolic thinking (see Kurt Goldstein) whose characteristics are known to us from primitive peoples and from dream activity. If an injury weakens the left brain dominance, the right hemisphere can become active, releasing primitive and symbolic processes characteristic of the unconscious, which thus seems to be centered in the non-dominant hemisphere. Hence, our theory suggests the hypothesis of a dual brain function. . . . Neurosis may result from brain rivalry with respect to conscious and unconscious realities and schizophrenia may be the consequence of splitting both approaches to reality so far apart that the integrative and balancing functions of the organism were unable to work. (Wolff 1952)

In 1955, clinical neurologist N. Dorin Ischlondsky reported on a case in *The Journal of Nervous and Mental Diseases* that, both in terms of clinical manifestations and author interpretation, is strikingly reminiscent of the Bourru and Burot Louis V. case of the 1880s and the "Welsh" case reported by Bruce in the 1890s (Chapter Five). The patient described was suffering from a form of double personality in which she alternated between a state of behavior characterized by submissiveness, shyness, and obedience (her Mickey personality) and a state of behavior characterized by violence, abusiveness, and obscenity (her Flossie personality).

> The physicians soon noticed that each of the two opposed mental states was associated with a very specific spatial distribution of certain neurological manifestations, sensory and motor. During the "Flossie" phase of the patient's behavior, a routine neurological

> examination revealed that the left and right side of her body responded differently to sensory stimuli: while the right side was *hypo*-sensitive the left side displayed a *hyper*-sensitivity. Thus vision and hearing were unclear and far away on the right side but very clear and close on the left side. Her response to touch and pain showed a high threshold on the right and a low threshold on the left side. . . . When the patient changed from the "Flossie" pattern of behavior to the "Mickey" type, the picture changed also with regard to the neurological manifestations. Just as fast as the psyche of the patient switched to the antipode of the preceding personality pattern . . . so also did the sensory and other neurological manifestations switch rapidly from one side of the body to the opposite side. (Ischlondsky 1955, 8–9)

It seems likely that Ischlondsky believed his case to be unprecedented; at least he did not indicate any awareness of similar reports in the older literature. His interpretation of the case was twentieth century in the sense that it drew upon concepts of reflex inhibition derived from S-R psychophysiology, although one senses also a distant debt to Hughlings Jackson. Nevertheless, the lesson this physician ultimately drew from his study would not have been out of place in Arthur Wigan's *Duality of Mind*. It was his view that the "amoral" personality, Flossie, had been able to come into existence only because there had been a weakening of the inhibitory capacity of the true, morally upright personality, Mickey, who had her focus in "the highest and most integrated" parts of the brain, those associated with the speech centers. "A conclusion of great import may be reached: mathematical precision of cortical dynamics corresponds to, and coincides with, justness and morality of attitude. This conclusion . . . implies that the systematic training of well-differentiated neuropsychic responses in the growing individual contributes to the development of his moral conscience and to the moral rehabilitation of society as a whole. It is needless," he concluded gravely, "to stress the significance of this conclusion for the activity of the educator" (p. 14).

9-2. *Neurology's Rediscovery of the "Whole"*

Up to this point, I have focused on the fate of the double brain in psychiatry and related its decline to a general growing disenchantment with nineteenth-century–style organicism and the rise of alternative perspectives on mental disorder. I now must explain why the neurologists, who continued after all to be interested in the functioning and malfunctioning of the human brain, also would lose sight of much of the nineteenth-century work both on brain duality and hemisphere differences. How is it that, when attention was again drawn to these problems in the 1940s and 1950s, culminating in the laterality explosion of the late 1960s and 1970s, most researchers would be imbued with a sense that they were pushing back frontiers earlier generations had never suspected (section 9-5)?

One plausible answer to this question may be found in neurology's "rediscovery of the whole" early in the twentieth century (Riese 1959, 118). By this, we refer to a substantial though *not* unequivocable reaction, early in the new century, against the classical conception of cerebral localization developed by such workers as Broca, Fritsch and Hitzig, Ferrier, and especially Wernicke and Lichtheim (the Germans having largely seized ascendency by this time). A new generation of researchers, led by such men as Pierre Marie in France, Carl von Monakow in Switzerland, Henry Head in England, Kurt Goldstein in Germany, and Karl Lashley in the United States, instead increasingly were inclined to interpret brain/mind, or brain/behavior, relations in holistic, hierarchical, developmental, and dynamic terms. Although it should be stressed that the concept of hemisphere functional asymmetry was not fundamentally rejected by the holists, as a research problem it tended to be placed on a back burner. Certainly, the bold nineteenth-century style of speculation about opposing faculties in two hemispheres and "two minds" poised within two brains does not seem to have interested this particular group of neurologists.

The holistic trend in neurology left its mark on both the aphasiology clinic and the laboratory, but the extent and nature of the mark was rather different in each case and affects our story in different ways. In this section, I restrict myself to the clinical developments; the next section is devoted to related developments in the laboratory.

On the clinical front, then, although aphasiologists like Jackson and, to a certain extent, Trousseau were criticizing certain localizationist assumptions as early as the 1860s, the first truly effective challenge to the classic view of language disorders was not launched until 1906. In that year, Pierre Marie in France, "struck," as his student A. Souques would later put it, "by the discordance between the facts and the reigning theories . . . made up his mind to drive a pickaxe into the edifice of aphasia" (Souques 1928, 362). In three iconoclastic papers, Marie subjected the entire theoretical foundations of aphasia to a scathing critique (Marie 1906a, 1906b, 1906c). Ten years of clinical and postmortem study had convinced him that the standard view of the day, which recognized the existence of a wide variety of distinct aphasias, because it held that there were separable centers in the cortex dealing with different aspects of language, was simply untenable. All the evidence pointed instead towards a view of aphasia as a single, unitary disorder. This disorder was characterized, first and foremost, by a disturbance of intelligence; a sort of dementia particularly concerned with that part of thought dependent upon language. It was not a question of loss of some hypothetical verbal images, as Wernicke had taught.

Marie went on to affirm that brain-lesions associated with aphasia always lay in the zone of Wernicke, which Marie held to be a considerably larger area than Wernicke himself recognized. In certain cases, aphasia (that is, Wernicke's aphasia) was *complicated* by a disturbance of articulation, *anarthria*, resulting from a lesion in the "lenticular zone," in the corpus striatum. Such subcortical damage led to the clinical syndrome known misleadingly as Broca's "aphasia." All cases of so-called Broca's aphasia were really cases of Wernicke's aphasia complicated by anarthria. Postmortem anatomical examination of such cases, including Broca's own famous case of "Tan," invariably revealed damage to the zone of Wernicke in the temporal region of the brain. There was *no* evidence, and this was Marie's most stunning declaration, that Broca's area (the third frontal convolution) played any special role in the production of speech!

Marie's work caused a sensation, splintering the aphasiologists into angry factions that debated each other without ever really managing to understand each other's point of view. Neurology's faith in the paradigm that had guided its research for four decades was badly shaken, and ultimately this threw the field into a state of chaos, as Henry Head would later describe it. "The whole problem of disorders of speech was thrown into the melting pot and each worker was free to take up an individual position" (Head 1926, 77).

In his 1959 *History of Neurology*, Riese argues that Marie's "revision of aphasia" was not the only factor responsible for growing doubts in these years about the adequacy of the classical localizationist paradigm. He identifies the phenomenon of recovery from brain damage as another anomaly facing a new skeptical generation of aphasiologists. Increasingly, it was said that recovery of functions was simply incompatible with the idea that the nervous system was a purely mechanical apparatus operating according to the laws of reflex and association. A machine could not repair itself after it had suffered damage and become disordered, but a living human being was very capable of regaining speech and movement even after suffering severe damage to the brain (Riese 1959, 127).

Thus, in the second decade of the new century, the Swiss neurologist

Carl von Monakow offered an explanation for recovery in terms of something he called *diaschisis*. This semi-philosophical, semi-physiological dynamic principle was intended to open the door to a wholly new way of thinking about cerebral localization. In von Monakow's words,

> diaschisis is the basic dynamic principle, it forms a bridge between those phenomena which can be localized distinctly and those which cannot. Thus, in reality, it is nothing but an irruption in the nerve cells which are not anatomically changed, which may be far away from the focus both spatially and temporally. This irruption leads to a war for the maintenance of function, a war which can end by victory or defeat of the involved elements or connection. (von Monakow 1911, 250)

The central idea here is that functional disorders could arise from areas not directly affected by lesions because all the parts of the brain are in a state of dynamic interdependence. Consequently, the relationship between the size and location of a particular lesion and the symptoms that result was extremely complex and fluid and could not be understood without taking into account the history of the injury (its onset and the various stages of functional loss and recovery). That history in turn needed to be understood in terms of the total phylogenetic and ontogenetic development of the brain. "We should never think of localization of function as represented by geometrical lines in certain groups of gyri in the adult man," von Monakow warned. He added, in an attempt to make his views more clear,

> The connections between a local anatomical lesion and the residual functional disturbance are similar to that of a music box out of whose cylinder a locally circumscribed series of pegs has been taken (local defect) and the disturbance of the melody. . . . The error of tunes will be, even by the experienced person, deduced only with difficulty from the number and place of the lacking pegs. . . . Certainly, nobody will put the melody (or some bars of it) . . . into locally circumscribed parts of the cylinder. (von Monakow 1911)

For the historian, it is intriguing to note that von Monakow's influential interpretation of brain malfunction and recovery in terms of diaschisis bears considerable family resemblance to Brown-Séquard's argument, largely ridiculed back in the 1870s, that sensory and motor disturbances resulting from localized brain lesions were due not to the lesions themselves but to "inhibitory action" caused by other, distant parts of the brain. Certainly, Brown-Séquard's argument was presented in a rather less cogent and considerably more strident fashion than von Monakow's. It seems most unlikely, though, that this alone can explain why the former only managed to make himself unpopular with his colleagues, while the latter became one of the guiding lights of a new approach to brain functioning in aphasiology.

In a similar way, we can acknowledge Riese's documentation of the rise of attention to the problem of recovery, but we should be wary of accepting that matters were quite so simple as he paints them. It is important to remember that the twentieth century in no sense "discovered" the problem of recovery. The nineteenth century also was perfectly aware of this phenomenon and had developed a number of models to account for it, models that functioned *within* the general framework of localization theory; hindsight judgements concerning the adequacy or inadequacy of these models, of course, are irrelevant here. The point is, if people half a century later suddenly felt that recovery challenged classical localization theory as a guiding principle in clinical neurology, it is hardly likely that the behavior of the brains they were examining was the single, or most important, reason for their change of heart. To do justice to the rise of holism in clinical neurology, a more generous, socio-cultural approach would seem to be required.

Matters took a new turn with the declaration of World War I in 1914. Cynical as it sounds, there can be no denying that the war was an exciting time for clinical neurologists. As Henry Head frankly admitted, the combat

> was producing a series of cases unique in the history of the subject. Young men with local wounds of the brain, but otherwise in perfect health, anxious to be examined and interested in their own condition, presented material for investigation that had never before passed through the hands of the neurologist. (Head 1926, 140)

Among this army of investigators was the German neurologist Kurt Goldstein, whose contribution to the corpus callosum debate was described in Chapter Five. Working now with a psychologist, Adhémar Gelb, Goldstein carried out a study of the effects of gunshot wounds on the brain. To his surprise, he found that slight injuries inflicted on the visual center did not lead to localized or function-specific disorders, as the localizationist paradigm suggested should be the case. Instead, he and Gelb found that there seemed to be a compensatory recasting of the whole visual field. Casting about for a way of interpreting this finding, the two men turned to *Gestalttheorie*, pioneered in 1890 by the German psychologist Christian von Ehrenfels. The word *gestalt* means "form" or "configuration," and von Ehrenfels had argued that the essence of psychological experience inhered not in the specific elements of which it was composed—as insisted by the associationist psychology favored by the German localizers—but in the relationship among those elements.

Goldstein and Gelb's World War I patient, *Schn.*, suffering from visual agnosia, served as the first example of Gestalt ideas applied to clinical neurology, though years later their diagnosis and interpretation would be challenged by the Freiburg neurologist Richard Jung, who re-examined the patient. On the basis of this case, it began to be argued that, instead of reacting mosaic-fashion to specific brain injuries, the individual compensated for his disability by adjusting his total set of responses to a reduced level of functioning (Murphy 1949, 379).

The principles learned from a study of perceptual disorders were ultimately to be applied by Goldstein equally to language and cognitive disorders. In his magnum opus, *Der Aufbau des Organismus*, written in the 1930s, Goldstein taught that language must be conceived not as a specialized skill but as a mode of functioning that permeated all aspects of the individual's mental orientation. Language allowed for the possibility of operating on the level of abstract thought; what Goldstein called categorical behavior. Categorical behavior was contrasted with a more primitive mental attitude characterized by a preoccupation with momentary sense impressions and concrete things in their immediate uniqueness. On this lower level, words were used less as symbolic representatives of categories than as assumed inherent properties of the objects in question. Because of his brain damage, the (amnesic) aphasic was only able to operate on the more concrete level, seen ultimately by Goldstein as a predominance of action over thought. It seemed to him that this same basic concreteness might explain a wide range of clinical syndromes generally considered as different entities: aphasia, apraxia, agnosia, and defects of attention and orientation (Riese 1968, 26–27).

After the war, Goldstein joined with Gelb in founding a hospital and research center for the study of brain injuries, the *Institut zur Erforschung der Folgeerscheinungen von Hirnverletzungen*. The new institute attracted such men as neurologist and historian Walther Riese, German psychologist Egon Weigl, psychoanalyst Frieda Fromm-Reichmann, and philosopher-linguist Ernst Cassirer, Goldstein's cousin. The idea that Gestalt theory could serve as a model for understanding brain disorder began to spread. In 1920, for example, neurologist W. Fuchs addressed himself to the curious fact that, when certain brain-injured patients with visual field defects were shown a geometric figure through a tachistoscope, so that part of the image fell in their blind field, they would nevertheless report "seeing" a complete figure. Such a phenomenon was inexplicable in terms of classical localization but made perfect sense within the framework of Gestalt theory. Gestalt psychologists had long taught that regular or symmetrical geometric figures were endowed with a high degree of internal cohesion, which had the effect of producing closure if the figure were presented in an incomplete form, the Gestalt law of Prägnanz. Thus, the completion effect observed in the brain-injured patient was simply a special case of a general tendency in perception (Zangwill 1963, 274–75).

In England, the publication of Henry Head's *Aphasia and Kindred Disorders of Speech*, in 1926, represents a landmark in the history of clinical neurology. Indebted to the pioneering work of both Pierre Marie and Kurt Goldstein, Head is particularly significant from the point of view of this study for his role in resurrecting and promoting the theories of Hughlings Jackson.

> He was one of the most remarkable pioneers in this field of research. . . . But even amongst the younger men his aphoristic dicta fell upon deaf ears. Until Arnold Pick dedicated *Die agrammatischen Sprachstörungen* to "Hughlings Jackson, the deepest thinker in neuropathology of the past century," no one attempted to understand his contribution to the subject. (Head 1926, 30)

The lucid overview Head then proceeded to present of Jackson's essential thought relieved many neurologists of the task of going back and struggling with the original texts themselves. One wonders about the consequences of this for future perceptions of Jackson's "contribution to the subject," for there seems little doubt that Henry Head's Jackson is an ideologically distorted version of the Jackson one finds in the original articles. "Some of his most valuable conceptions appear amongst a good deal that is no longer of importance to us," Head told his readers offhandedly (p. 52). This remark apparently was intended to secure him license to weed out those features of Jackson's thought that undermined the image he was trying to create of the man: a brave, lonely visionary who anticipated just the sorts of ideas that Head was now championing.

Particularly rankling, at least so far as this study is concerned, was Head's conspicuous failure to discuss Jackson's physiological/philosophical views on hemisphere functional differences and brain duality. Now, it is quite possible that Head did not agree with Jackson's perspective on the double brain. The fact remains, however, that this perspective represents an essential cornerstone of Jackson's model of higher nervous functioning in man. To omit all reference to it because one has decided that it is "no longer of importance to us" borders on intellectual dishonesty.

But Head did not simply fail to discuss the double brain. He devoted a section each to Jackson's theories of mental duality and imperception and not only failed to refer to the anatomical ideas upon which those theories were grounded but also hopelessly misrepresented, or perhaps misunderstood, the theories themselves. Subject-consciousness was confounded with (conscious) imagistic thought rather than understood as the unconscious preparatory stage of all thought. Imperception was identified with agnosia, recognized since Lissauer (1889) as a bilateral or dominant-lobe defect. No mention was made of the fact that Jackson personally regarded imperception as a right-brain disorder (a disorder of the highest sensory centers) anatomically and clinically complementary to aphasia (a disorder of the highest motor centers).

His treatment of Jackson aside, Head is an important figure in our inquiry because of his role in spreading the new holistic ideas within the English-speaking world.

> the whole conception of "amnesic" or "sensory" forms of aphasia has undergone a profound change, especially under the influence of recent developments in psychology. The older views, based on theories of combination and synthesis of elementary functions, have given place to the idea of a single unitary reaction in response to multiple factors. This principle underlies what I have called the "psychological attitude." It was foreshadowed by Hughlings-Jackson more than fifty years ago and is gaining adherents day by day as aphasia and kindred disorders of speech come to be considered as manifestations of the mental processes, thinking and speaking. (Head 1926, 133)

Head distinguished among four clinical varieties of aphasia (verbal, syntactical, nominal, and semantic), each with its characteristic complex of symptoms. He was the first clinician to introduce a system of serialized testing into the study of aphasia. He also is notable for his recognition that the concept of a symbolic disorder should be broad enough to encompass, in addition to disturbances of speech and language per se, disturbances in perceptual judgments that require internal verbal mediation, such as left-right confusion. Another of his early innovations was the concept of the "body scheme"—that is, the idea that, within each of our brains, there was a representation of our *perception* of our own body and how its different parts related to one another (and this would include artificial bodily extensions such as canes and even feather hats). It was argued that destruction of portions of the body scheme could cause that part of the body to drop out of the individual's awareness, rendering impossible "recognition of posture or of the locality of a stimulated spot in the affected part of the body" (Head & Holmes 1911).

9-3. Complementary Trends in the Laboratory

While the holists were revising the localizationist paradigm in the clinic, psychology and physiology researchers working in laboratories had begun to challenge it on quite different grounds. In England, Sir Charles Scott Sherrington (1857–1952) devoted the first decades of the twentieth century to laying out the principles for a new understanding of reflex action based on a neuronal (that is, cellular) conception of nervous organization. The idea that the nervous system was composed of physically discrete cells, separated from one another by gaps that Sherrington called *synapses*, had been extensively developed in the 1880s by the Spanish histologist Santiago Ramón y Cajal, and Freud adopted it for his *Project* in the 1890s (Chapter Eight). It was only under Sherrington, though, that the new theory effectively superseded the previously popular "reticular hypothesis," which had looked upon the nervous system as a physically connected network. Neuron theory offered a way to understand how nerve impulses involved in various simple reflexes might combine selectively to give rise to more complex behavior. In his experimental work and theoretical writings, Sherrington focused on interaction between reflex levels in the spinal cord. He showed how input at one level could modify input at another, giving rise to a hierarchical pattern of integration and control. His basic message was that behavior was more than the sum of individual reflexes and that the organism must ultimately be viewed as an integrated functional unity (Allen 1975, 88–92).

In the years following World War I, the world's intellectual center of gravity shifted from Germany and France to the English-speaking countries, and a movement arose within these countries to turn psychology into a laboratory science. The aim behind this movement was to rid psychology of the cobwebs that still clung to it and to establish it as a "pure" experimental science, comparable in its alleged objectivity to experimental chemistry or physics. As historian Daniel Robinson pointed out, a metaphysical stance had been adopted within Anglo-American psychology, not on the question of *truth* but on the nature of *psychology* as a discipline. If psychology was to be an experimental science in the manner of the physical sciences, then its subject matter could only contain those pieces of truth amenable to the experimental method (Robinson 1976, 357–58).

As early as 1913, the American John B. Watson had declared that the new experimental psychology aimed to eliminate "states of consciousness as proper objects of investigation in themselves" as a first step in the crusade to "remove the barrier from psychology which exists between it and the other sciences" (Watson 1913, 177). The goal of Watson's psychology was simply and strictly "the prediction and control of behavior." Once the decision was made to recognize "no dividing line between man and brute," a step philosophically justified by Darwinism, research became almost wholly animal oriented. Animal psychology effectively eliminated the subjective factor (no one knew or cared anything for the contents of a rat's mind) and provided an appropriate arena for the introduction of methods by which behavior could be measured and described under controlled conditions.

In the United States, one of the most important intellects to emerge from Watson's animal-oriented, behaviorist psychology was a young neurophysiologist, Karl Spencer Lashley. With his rates and mazes, Lashley would demonstrate that physiological psychology, no less than psychology proper, could get by without resorting to "unscientific" concepts, like mind and consciousness. But Lashley would do more than this. In 1926, he wrote a paper, "The Relation Between Cerebral Mass, Learning and Retention," which argued that loss of memory (failure to perform learned tasks) in the rat was not dependent upon removal of one or another cortical area but simply was proportional to the total amount of brain matter removed (Lashley 1926). Such a finding represented a major challenge to the rigid "connectionist" theory of conditioned reflex taught by Lashley's mentor, Watson; and it also seemed to be inconsistent with the localization principles now being challenged in the medical wards.

As he continued to explore the relationship between learned behavior and the brain (using mostly laboratory rats, some monkeys, and later a few chimpanzees), Lashley became more and more convinced that there was little, if any, relationship between the locus of a brain lesion and the type of postoperative difficulties experienced by the animal. In *Brain Mechanisms and Intelligence* (1929), he wrote,

> Specialization of function of different parts of the cortex occurs in all forms, but at best this is only a gross affair, involving general categories of activity rather than specific reactions. The more complicated and difficult the activity, the less the evidence for its limi-

> tation to any single part of the nervous system, and the less the likelihood of its disintegration into subordinate physiological elements.

Then, in a bold statement meant to silence anyone presumptuous enough to suggest that what was true for rats might not be *generally* applicable in cerebral neurophysiology, Lashley added,

> there is little evidence of finer cortical differentiation in man than in the rat. (Lashley 1929, 155–56)

Much like Flourens a century before him, Lashley may be seen as the intellectual leader of a movement in experimental neurophysiology away from a localizationist approach to cerebral functioning and toward a field theory of brain activity, a theory neatly encapsulated in two key terms: *equipotentiality* and *mass action.* In the laboratory, psychologists and neurologists began to make the Lashley emphasis on unitary functioning "a cardinal touchstone for modernness of thinking" (Murphy 1968, 34). The fundamental alliance between the clinic and the laboratory in their joint championship of holism is plainly demonstrated by the fact that Lashley wrote the foreword to the English translation of Kurt Goldstein's major work, published in the United States as *The Organism: A Holistic Approach to Biology Derived from Pathological Data in Man* (Goldstein 1939).

The relationship between lab and clinic, however, also was not without its tensions. Confronted every day with the complexities of the disordered human brain, the clinic was never so radical nor so united in its antilocalization stance as the laboratory. Lashley's pronouncements notwithstanding, most clinical neurologists knew, as C. Judson Herrick had put it bluntly, that "rats are not men. . . . Men are bigger and better than rats" (Herrick 1926, 347, 365; also cf. Geschwind 1964). This is at least part of the reason why, in spite of the theoretical upheavals and dissensions within its ranks, a modest but respected section of clinical neurology quietly could start developing a new conception of differential left-brain/right-brain functioning; one that would set the stage for the 1960s revival of the double brain (see section 9-4).

It is useful to realize that the holistic reorientation in neurology, both in its experimental and its more equivocable clinical manifestations, was paralleled, as Allen (1975) points out, by a similar metamorphosis within society at large. The Bismarckian ideal of politics as a balance of fixed power relationships began to give way to a philosophy of dynamic equilibrium, with constantly changing spheres of political influence. The spread of Marxism, with its dynamic view of material development, and of Keynesian economics, with its concept of deficit-spending based on the notion of internal regulation, were challenging the static pay-as-you-go, laissez-faire ideal of economic organization. The general philosophical mood seems to have been changing throughout the Western world; mechanism was on the decline. Vitalistic and organicist philosophers, such as Henri Bergson and Alfred North Whitehead, were finding receptive audiences. Allen argues that these wide cultural changes may have influenced the questions that began to be asked even in a seemingly insular discipline like physiology (p. 111). Along these same lines, it is probably relevant that the changes in neurophysiology also coincided historically with the swing in psychiatry from a static neurologism to a dynamic psychologism, as discussed in section 9-1. This fact at least raises the possibility that *both* developments can be related to a wider shift in what Watson has called the prescriptive orientation of the human sciences: disciplines which seem to move pendulumlike between the poles of objectivism and subjectivism, staticism and dynamicism, mechanism and vitalism, molecularism and molarism, etc. (Watson 1975).

9-4. The Problem of Hemisphere Differences: New Trends in the Clinic

In the past two sections, I attempted to explain why the problems of hemisphere functional independence and lateralized brain functioning ceased to be high-priority issues in the brain sciences during the early decades of the twentieth century. I laid special stress on the intellectual upheavals wrought by the rise of a holistic trend within the brain sciences, both clinical and experimental, and drew attention to a number of developments that interacted with the main theme: neurology's role in the first World War, the rise of psychology as a laboratory science of behavior, etc.

However, in the clinic at least, not every neurologist accepted holism, and these workers from the old localizationist school continued to be published in mainstream scholarly journals, to receive university appointments, and to attract disciples. Why, then, did these people, no less than the holists, also apparently lose touch with the nineteenth-century tradition of speculation about the double brain?

My reply is a preliminary and still largely impressionistic one. I would like to suggest that, in breaking away from psychiatry and various associated areas of inquiry (philosophy, criminology, the occult), neurology simply lost touch altogether with big, bold traditions of speculation. For this reason, much of the nineteenth-century literature on hemisphere differences and brain duality came to be quite remote from its immediate concerns. I do not know the extent to which early twentieth-century localizers were familiar with nineteenth-century perspectives on the double brain, but I suspect they would have regarded a fair portion of it (not all certainly) somewhat askance. Generally speaking, their own research had become more piecemeal and more focused. So far as they continued to treat problems on the double brain with roots going back to the nineteenth century, they tended to stick to those solidly in their field of expertise: the nature and localization of the various left-brain dysfunctions (aphasia, agnosia, apraxia), the relationship between speech dominance and handedness, the problems of the left-hander and so-called crossed aphasia, etc.

In dealing with such issues, moreover, these men, by and large, were working within a conceptual framework derived from the Wernicke-Lichtheim tradition not from the Broca faculty school (cf. Head's 1926 historical review of this era). It will be recalled that the German approach to localization, which conceived of higher brain functions as an outcome of diverse associative processes between various functional centers, simply did not lend itself readily to romantic French-style ideas about various self-sufficient mental "faculties" in a state of polar opposition in the two hemispheres of the brain (cf. Chapter Three, section 3-4). Also Germans, from Wernicke to Liepmann, had taken the lead in promoting what was to become the mid twentieth-century concept of left cerebral "dominance," which essentially held that intellect was tied up solely with the left hemisphere and that the right side of the brain made no appreciable contribution to mental functioning (cf. Dandy 1930; Chapter Three, section 3-1). Growing acceptance of this way of thinking naturally would have the effect of further discouraging interest in the possible functional duality of the brain. We may conclude, then, that to the extent that German values dominated in early twentieth-century neurology, this would have had the effect of generally muting awareness of the nineteenth-century literature on brain duality, where the French had played a leading role.

On this point, it is significant that most leading neurology textbooks in the early twentieth century were written by Germans or modeled after German examples. As Murray (1985) has pointed out, it is largely the textbook writers, after all, who decide how much of a discipline's past will be remembered and which parts. In a related argument, Leary (1985) has noted that, because the norms of science encourage people to study nature instead of books, scientists constantly tend to lose sight of those parts of their discipline's history upon which they are no longer actively building. In other words, given an ethic that separates science from scholarship, it is almost inevitable that the old literature on the double brain largely would fail to retain a place in neurology's collective memory once people were preoccupied with other research problems and theoretical approaches.

I now want to complicate matters by suggesting that, ironically, all these negative developments in the history of the double brain may have had an unexpected positive consequence. It is my impression that the decline and general obscuring of much of the wide-ranging, cross-disciplinary nineteenth-century work on the double brain cleared the intellectual field for, and thus helped make *possible*, the rise, in the 1930s and 1940s of a new, more focused body of clinical thought roughly concerned with at least some of the same questions with which the forgotten, older literature had been grappling. Although the general thrust of clinical neurology was firmly toward a quite severe interpretation of left-brain dominance, a quiet but respected band of clinical researchers had begun to accumulate evidence at least hinting that a more democratic view of the brain might be in order. One of the first challenges had been issued with the growing documentation that certain clinical syndromes caused by brain damage were dependent upon bilateral, rather than strictly left-sided, lesions. The English neurologists Holmes and Horrax, for example, implicated bilateral lesions in a variety of disturbances of visual disorientation: inability to localize an object accurately in space; inability to determine the relative lengths of two lines, or the relative sizes of two objects; disturbances of topographical memory that caused patients to get lost in their own homes; impaired spatial orientation causing patients to run into objects they could see clearly (Holmes & Horrax 1919).

The syndrome of simultaneous agnosia, first described in 1924 by the German neurologist I. Wolpert, also seemed to depend upon damage to the right hemisphere as well as the left. Wolpert defined it as a condition in which a patient could identify specific details of an action or picture but was unable to integrate the separate elements into a meaningful whole (Wolpert 1924). As the century progressed, still more clinical data was reported that suggested at least some right hemisphere involvement in such interesting disorders as prosopagnosia (inability to recognize and remember familiar faces), chronognosia (disturbance of one's subjective sense of time), and sensory amusia (loss of ability to distinguish musical sounds and patterns of rhythm).

In 1914, Joseph Babinski in France introduced the term *anosognosia* to denote a peculiar condition in which a patient suffering from left hemiplegia either denied or appeared to be completely unaware of his disability (Babinski 1914). Babinski's description of this disorder before the Paris Société de Neurologie struck a responsive chord in a number of his colleagues. In the discussion following the paper, Souques referred to a doctor suffering from left hemiplegia and hemianaesthesia who "forgot" his left side and who, when reference to his disability was made to him, seemed not to hear. Ballet told of patients with occipital-lobe brain tumors who, as their illness progressed, stopped complaining of headaches, and finally declared they had no more problems with their vision, after they in fact had become blind.

Babinski, however, had been struck by the fact that the disorder seemed to occur only in conjunction with left hemiplegia; i.e., without comparable symptoms accompanying right hemiplegia. He could not help wondering, "Could anosognosia particularly concern the right hemisphere?" For us, familiar with the nineteenth century literature linking hysteria to the right hemisphere, this is an intriguing remark. After all, unawareness or denial of disability (or, alternatively, *la belle indifférence*) was a recognized dimension of the hysterical syndrome throughout the late nineteenth century. One cannot help but wonder whether Babinski, who had spent so many years studying hysterical patients under Charcot, also noticed this similarity between classical hysteria and his newly christened organic disorder. Perhaps he did, but kept his thoughts to himself. After all, in 1901, he had set out to discredit Charcot's neurological approach to hysteria (section 9-1). There would have been no sense muddying the waters at this point.

In any event, during the decades following, the question of a link between anosognosia and right-brain damage would continue to be discussed, and the disorder itself would come to be regarded by some as a peculiar disturbance of Head's body scheme, one distinct from any body scheme disorder associated with left-brain damage. As J. M. Nielsen summed up matters in 1938,

> There are certain peculiar differences . . . in the types of disturbance of the body scheme resulting from lesions of the two sides. Destruction of the major areas, usually the left, by lesions of certain types or locations produces defects on both sides of the body; destruction of the minor area produces defects on the opposite side only.
>
> . . . lesions on the major side, in the areas necessary, cause disturbance of the knowledge of the two hands, especially the fingers. . . . There is in the same area, a laterality-coordinating area, a lesion of which causes loss of the sense of right and left, and even entire loss of sense of direction. . . . The more closely the lesion approaches

> Wernicke's zone, the more of a language or symbolic element one finds. . . . On the other hand, a lesion of the minor parietal region causes, when disturbance of body scheme appears, a dropping out of the opposite side of the body, first from the attention and then from consciousness.
>
> . . . It is indeed strange that on the minor side there should exist a group of cells the duty of which is to keep the opposite side of the body in attention or consciousness without a similar mechanism being present on the other side. . . . It seems possible that through phylogenetic training of the left side of the brain for finer differentiation, as for the use of fingers and symbolization, the cruder function has been abandoned. (Nielsen 1938, 536, 556–57, 559)

In 1941, W. Russell Brain, in England, described three patients suffering from unilateral lesions in the right hemisphere who experienced disorientation in their left visual half-field, even though there was no apparent loss of visual acuity. Brain attributed the disorder to "inattention" of the left half of external space and felt it bore a relationship to the anosognosia syndrome.

> The visual localization of an object in the external world implies in the first instance the power to relate its position to that of the body also perceived as a visual object. . . . Visual localization is thus a process involving the body scheme and the scheme of the external world. This interrelatedness of spatial orientation and awareness of the body implies an anatomical relationship. (Brain 1941)

By linking left-sided "hemi-neglect" with disorders of the body scheme, Brain found himself able to offer an interpretation of a rather curious defect he had observed in two of his patients, which he called *apraxia for dressing*. He argued that the key element underlying this disability was a type of left-right confusion, different from that described by Head several decades earlier,

> When the patients said that they were unable to recognize the right and left sides of a garment, they meant that seeing the garment no longer evoked a knowledge of its spatial relationship to the body, and so did not lead to the appropriate actions. This disorder . . . is independent of the inability to name the right and left sides of the body. . . . [In fact] No. 4 . . . was able to use these verbal symbols to escape from his difficulties. (p. 266)

In 1944, Paterson and Zangwill elaborated on Brain's 1941 description of left-sided hemi-neglect, observing that patients suffering from this disorder were not only disoriented on the left side but tended to leave out material on the left side of drawings, or else crowd parts into the right side and to "forget" food on the left side of their plates. They also observed in these patients a disturbance in such activities as assembling, building, and drawing similar to a dominant-lobe disorder called *constructional apraxia* (Kleist 1923) but lacking a semantic component (Paterson & Zangwill 1944).

In France, Henry Hécaen and J. de Ajuriaguerra contributed a great deal to the elaboration of disorders associated with unilateral right-brain damage. In the 1950s, they called attention to the fact that certain forms of dyslexia, dysgraphia, and dyscalculia could form a component of hemi-neglect. A patient, for example, might neglect the left side of a written text. In writing, he might tend to add extra strokes to certain graphemes; lines of script might slant at various levels of inclination; spaces might be inserted between graphemes that destroyed the unity of the word; and the writing might be crowded onto the right side of the paper (Hécaen & Marcie 1979, 120). Hécaen and his colleagues were also among the first, in the twentieth century at least, to contrast the "catastrophic" reaction accompanying left-brain damage with the oddly dismissive, even euphoric, attitude of right-hemisphere patients toward their disabilities (Hécaen & Ajuriaguerra 1945). This seemed to suggest that the affective peculiarities accompanying unilateral brain disorder might be more complex than Goldstein, with his emphasis on the patient's inevitable defensive reaction to upheavals in his world, had thought. Here, moreover, one sees another, albeit unrecognized, line of continuity with nineteenth-century thinking, specifically Luys' work comparing the emotional states of left and right hemiplegics (Chapter Three).

In their 1955 work *Denial of Illness*, Edwin Weinstein and Robert Kahn noted further (pp. 4–5) that the so-called denial syndrome associated with left-sided hemi-neglect was not just a question of inadequate or inappropriate emotional response to one's plight. In certain patients, it was accompanied by disorientation for place and time; "reduplication" for places, times, and people (the patient confabulated the existence of two or more people, places, or events where only one existed); alterations in sexual behavior (generally a shift from pre-illness modesty to a preoccupation with sexual topics, often of a lewd nature); and an over-literalness in interpreting even quite everyday questions (a reply to the doctor's routine "How do you feel?" might be "With my hands").

Meanwhile, during this same general period, the Canadian neurosurgeon Wilder Penfield was involved in a systematic exploration of the human cortex, using focused electrical stimulation on the exposed brains of (conscious) epileptics awaiting surgical removal of diseased tissue. He and his colleagues extensively "mapped" the speech areas of the left hemisphere in this fashion, noting where stimulation produced speech arrest, word distortion, slurring and hesitancy, loss of naming capacity, and vocalization changes (Penfield & Roberts 1959). He found that the right hemisphere had a capacity for vocalization; and he placed spatial orientation and awareness of body scheme within that hemisphere as well. He also found that, when he stimulated the temporal lobes, he was able to elicit certain subjective signals whose function seemed to be "to interpret the relationship of the individual to his immediate environment": "Examples of such signals are: these things are 'familiar' or 'frightening,' they are 'coming nearer' or 'going away'; and so on" (Penfield 1975, 442). Although Penfield's "interpretive" signals appeared for the most part to be bilaterally represented, there were two notable exceptions. In the great majority of cases, visual illusions and illusions bearing on the passing of time, notably the "feeling of familiarity" or "dreamy mental state" described by such nineteenth-century workers as Jackson and Crichton-Browne, could be evoked by stimulating the right hemisphere only (Mullan & Penfield 1959, 283).

By the 1950s, there was an increasing openness to the idea that right hemisphere might not be quite so insignificant after all, and a few scattered attempts were made to sum up the new, still nascent perception of the relationship between the two sides of the brain in a functional formula. All these attempts were based on an assessment of the clinical work of the previous several decades and included such left/right dichotomies as "symbolic or propositional" versus "visual or imaginative"; "education of relations" versus "education of correlates"; "verbal" versus "perceptual"; and "symbolic" versus "visuospatial" (Bogen 1969).

9-5. *"Split-Brain" Man, and the Launching of a New Era*

In spite of this emerging new consensus within certain clinical circles about the different functional capacities of the two hemispheres, the late twentieth-century explosion of interest in the bilateral functioning of the human brain would only be made possible by a decline in the fortunes of the Lashley plasticity/equipotentiality paradigm and by a related dramatic reversal in the Akelaitis-inspired perception of the corpus callosum as a big but essentially "useless" structure. The manifest side of this still-unfolding story can be rather quickly sketched out in rough; I do not intend to say much about the story's more intractable subterranean layers. That such layers exist, I have no doubt, but many have not yet hardened sufficiently in time to permit a balanced appraisal.

Let me begin instead by returning to a key actor in our drama: the corpus callosum. During the first half of the twentieth century, the controversy over the corpus callosum had continued unabated. Around 1940, Warren McCulloch summed up the state of scientific knowledge with his wry remark that the only demonstrated function for the structure seemed to be that of aiding in the transmission of epileptic seizures from one side of the body to the other. Ten years later, Karl Lashley still apparently found ample justification to adhere to his own facetious proposal that the callosum's principle function was probably mechanical in nature—to keep the two hemispheres from sagging (Sperry 1962, 43).

By the third decade of the century, Hugo Liepmann's concept of a callosal syndrome (Chapter Five, section 5-8) (left-handed apraxia resulting from a lesion in the corpus callosum) was widely felt to have been discredited. It was pointed out that callosal lesions were rarely, if ever, isolated, so it was impossible to decide whether deficits attributed to such a lesion, at least in part, might not be due to associated damage; that the signs attributed to callosal lesions often subsided or disappeared altogether; that, in numerous cases of callosal disease, no perceptible "disconnection" signs could be found; and that patients with agenesis of the corpus callosum failed to manifest most, if any, of the so-called callosal signs (Bogen 1979, 315–18).

An authoritative report published in 1943 seemed to settle the matter. A group of epileptics who, for therapeutic reasons, had undergone extensive sectioning of the corpus callosum were systematically tested by Andrew Akelaitis for evidence of "disturbance of the voluntary higher integrative activities in the subordinate side of the body." Results were unequivocably negative:

> Language functions studied unilaterally as visual lexia in each homonymous visual field, tactile lexia (including recognition of skin writing and wooden letters) and graphia (handwriting) were not disturbed in the subordinate or dominant side of the body.
>
> These findings suggest that commisural systems other than the corpus callosum are utilized for the activities tested in the interhemispheral connections between the dominant and subordinate hemispheres. (Akelaitis 1943, 261)

In the laboratories, the Akelaitis study became a major stronghold supporting the radical anticonnectionist view associated with Lashley and his followers. If two million fibers in the brain could be severed without noticeable consequences, then even the flexible-seeming Sherringtonian concepts of nerve specificity of function and central nervous integration were clearly naive. Rather, the brain seemed, in the words of American neuropsychologist Roger Sperry, "to possess an almost mystical plasticity in its ability to achieve proper orderly function in spite of radical disruptions in its normal wiring plan" (Sperry 1975, 427).

Roger Sperry, a graduate student under Lashley in the late 1930s, was at that time, as he tells it, as much under the spell of the plasticity-equipotentiality paradigm as everyone else. Nevertheless, to establish a more solid basis for further research, he took on the task of reproducing a variety of experiments generally believed to support that paradigm. He found that his own findings sharply contradicted previous accounts. In fact, he felt they tended to support a distinctly localizationist/connectionist approach to the brain (Sperry 1975).

One of the last brain structures Sperry submitted to experimental re-examination was the corpus callosum. Early in the 1950s, he and a colleague, Ronald E. Myers, surgically divided both the corpus callosum and the optic chiasma in a cat so that visual information from the animal's left eye made its way to the left hemisphere alone and information from the right eye to the right hemisphere alone. In its everyday behavior, the cat appeared to be relatively unaffected by this surgery. In specially designed tests for unilateral learning capacity, however, matters got more interesting. Working on a problem with one eye, the animal responded normally and learned to perform correctly. When, however, that eye was covered and the same problem was then presented to the other eye, the creature showed no signs of understanding what it was meant to do and had to learn the task all over again with the other half of its brain (Gazzaniga 1967; Myers & Sperry 1953). Sperry and Myers were startled by these results. Had the surgeon's knife produced two independent minds within a single cat skull? Later, still more dramatic experiments with chiasm-callosum sectioned monkeys did much to fan conviction that something of this nature had indeed occurred (see, e.g., Downer's 1962 description of witnessing a unilateral Klüver-Bucy syndrome in a "split-brain" rhesus monkey: a veritable Dr. Jekyll/Mr. Hyde transformation).[2]

[2] The Klüver-Bucy syndrome, first described in rhesus monkeys in the late 1930s, is associated with damage to the temporal lobes of the brain, especially the amygdala and hippocampus below the cerebral cortex. Formerly aggressive and intimidating laboratory animals subjected to such damage are said to become much more tame and manageable. The switch from preoperative aggression to postoperative passivity is usually

Not long after, in the early 1960s, two California neurosurgeons, Joseph E. Bogen and Philip J. Vogel, performed a radical sectioning of the corpus callosum in a series of patients suffering from intractable epilepsy. From a therapeutic point of view, the operations seemed remarkably successful, limiting the epileptic convulsions to one half of the body and, unexpectedly, significantly reducing the overall number and severity of attacks. Moreover, on casual examination, these so-called split-brain patients seemed to confirm earlier assertions that corpus callosum transection has a negligible effect on human mental functioning. Following surgery, Bogen and Vogel's patients appeared unchanged in temperament, personality, and general intelligence. One patient, for example, a twelve-year-old boy, "on awakening from surgery . . . quipped that he had a 'splitting headache,' and in his still drowsy state . . . was able to repeat the tongue twister 'Peter Piper picked a peck of pickled peppers' " (Gazzaniga 1967, 24).

The publicity of the split-brain experiments with laboratory animals, however, had alerted Bogen and Vogel to the possibility that, in spite of their patients' superficial normalcy, symptoms of a callosal syndrome might be detectable, given appropriate testing. There was some feeling that the largely negative test impression obtained by mid twentieth-century workers, such as Akelaitis, might have been caused by the application of inadequate or sloppy postoperative testing procedures.

At Bogen's request, Sperry, with several of his students, agreed to design a series of tests for use on the split-brain patients. Results could hardly have been more dramatic. In the testing environment, as Sperry put it, "instead of the normally unified single stream of consciousness, these patients behave in many ways as if they have two independent streams of conscious awareness, one in each hemisphere, each of which is cut off from and out of contact with the mental experiences of the other" (Sperry 1968, 724). It was found, for example, that

> if two objects are placed simultaneously one in each hand [out of sight behind a screen], and then are removed and hidden for retrieval in a scrambled pile of test items, each hand will hunt through the pile and search out selectively its own object. In the process each hand may explore, identify, and reject an item for which the other hand is searching. It is like two separate individuals working over

emphasized in discussions of the syndrome, even though the animals in question also tend to become hypersexual, show a compulsive tendency to place objects in the mouth, and have trouble recognizing objects presented visually (Valenstein 1973, 131–35).

the collection of test items with no cooperation between them. (1968, 727)

The cumulative effect of a wide range of tests like this was to convince Sperry that "in the minor [nonspeaking, right] hemisphere we deal with a second conscious entity that is characteristically human and runs along in parallel with the more dominant stream of consciousness in the major [left] hemisphere" (1968, 732; see also the similar, earlier conclusions of Geschwind 1965). He pointed out that the right hemisphere seemed able to generate emotional responses independent of those of the left, expressing displeasure (through facial grimaces) when it heard the left hemisphere making verbal mistakes to questions for which *it* "knew" the answers; blushing and giggling when a slide of a nude pin-up was unexpectedly flashed at it, even though the patient insisted verbally that she had seen nothing, "just a flash of light." There was even some anecdotal evidence that the right hemisphere was the more impulsive and violent of the two brain halves, much as the nineteenth century had feared. Michael Gazzaniga tells the story of a patient who once "grabbed his wife with his left hand and shook her violently, while with the right hand trying to come to his wife's aid in bringing the left belligerent hand under control" (Gazzaniga 1970, 573).

Almost as exciting as the discovery that these split-brain patients had, apparently, "two minds" was the growing realization that each of these minds seemed to have its own distinctive cognitive strengths. Very roughly, it appeared that the left, talking, side of the brain was superior in writing, calculation, and tasks requiring analytic skills; the right side of the brain seemed to be better at recognizing spatial relationships, constructing novel figures, and making sense of patterned material, including human faces. None of these findings would not have come as much of a surprise to the more alert observers of recent trends in the clinic (section 9-4). However, to a rising generation of, mostly American, academic psychologists during the 1960s, in the process of freeing themselves from the iron grip of an increasingly uninspiring behaviorism and developing a new psychology focusing on human cognition, the concept of differential hemisphere functioning had a fascination and aura of novelty that seem together to have proved irresistible.

In 1969, Joseph Bogen set the flavor for much further speculation in this area by proposing a classification formula for the two brain-halves that ran counter to the mid-twentieth-century habit of treating the left side of the brain as the dominant half and the right side as minor, with no specific functions of its own. In Bogen's view, man's propositional left hemisphere, which Bogen christened in honor of Hughlings Jackson, was complemented by an appositional mind on the right side of the brain.[3] This "other" brain processed information in a nonlinear, synthetic mode characteristic of musical and artistic expression, and possibly of dreaming as well. In Bogen's opinion, it was high time that neurology outgrew its bias against the right side of the brain and began to subject the "rules of appositional thought" to the same sort of intensive scrutiny that the left hemisphere had so long enjoyed. Bogen went on to speculate that the age-old tendency for societies to dichotomize human experience and consciousness (pitting reason against intuition, science against art, yang against yin) might reflect the fact that, in the human animal, "there are two types of thinking generated in the same cerebrum." He felt it likely that, not just split-brains, but *all* people pass through life with two functionally independent and cognitively complementary brains. In another gesture to history, he proposed to call his perspective *neowiganism*, a tribute to the author of *Duality of Mind*, whose work he had discovered and read with delight in the early 1960s (Bogen 1969; see also the elegant new edition of Wigan's work issued by Bogen & Simon 1985).

By the end of the 1960s, the psychological and neurological community had begun to cast its sights beyond the world of the split-brain and set itself the task of discovering how the two hemispheres interact in subjects with an intact corpus callosum. As the data kept spiraling to ever more heady heights, virtually no area of inquiry was untouched that impinged on mind, brain, and their relations. Did the new research cast light on some of the neurological conditions underlying certain dynamic unconscious processes described by Freud (Galin 1974)? Had it any implications for attempts by philosophers to reconcile a personal, mentalist view of human consciousness with an understanding of human beings as biological systems (Nagel 1979; Puccetti 1981)? What did it mean for science's understanding of personality differences (Bakan 1969) or differences in the cognitive capacities of the sexes (Witelson 1976)? What light could it throw on the neurobiology of mental disorder, from schizophrenia to manic-depression (Marin & Tucker 1981)? What were its educational implications (Bogen 1977)? Did it have anything to teach us about the neurophysiology of such

[3] In fact, Jackson believed that both the left *and* the right side of the brain engaged in propositional thought, by means of words and images, respectively (see Chapter Seven).

altered states of consciousness as dreaming and hypnosis (Huppe 1977; Frumkin, Ripley, & Cox 1978)? Could we find in it an explanation for that most intractable of philosophical and scientific problems, the origins of human consciousness (Jaynes 1976)? Did it offer any insights into the evolution of human culture (Sagan 1977)? Did it have anything to say about the mysterious process by which the human "self" interacted with "its" brain (Popper & Eccles 1977)?

All these new and, more often, not-so-new themes in the history of the double brain interacted in various ways with a number of wider socio-cultural currents. The rediscovery of the double brain in the 1960s coincided with a period in the English-speaking world marked by extreme ambivalence toward science and technology and, by extension, toward rational, analytic thinking in general. An escalating nuclear arms race, overpopulation, contamination of the natural environment, and industrial overgrowth had bred distrust and disillusionment, which in turn had led to the rise of alternative value systems that stressed the virtues of other ways of knowing. The emerging notion of the brain's "bimodal consciousness" was to prove itself admirably suited to the ideological needs of this counterculture movement. A 1970 book by Robert Ornstein, *The Psychology of Consciousness*, argued that Western society, with its emphasis on logical thinking and language, had overdeveloped its left-brain at the expense of the allegedly intuitive, mystical right-brain, whose functions Ornstein believed to be far more developed in the cultures of the Orient (Ornstein 1970). In the wake of Ornstein's book one finds a growing tendency to equate the left hemisphere with the evils of modern, technological society, and to look upon the right hemisphere as a victim of heartless discrimination and oppression. In 1976, for example, Jungian analyst June Singer declared that

> the cutting of this connecting tissue between the two hemispheres of the brain can be viewed as the ultimate symbolic act of Pisces, the Age of Polarities. It gives credence to the idea that the Day World can be separated from the Night World, and that rational functioning can be separated from intuitive functioning, and that masculine consciousness can be separated from feminine consciousness. . . . Of course this comes as no surprise to the people who have survived the unimaginable horrors of Nazi concentration camps, the leaders of which justified some of their most nefarious activities in the names of "scientific research" and "objectivity." Nor should it surprise readers of *The Gulag Archipelago*, who discovered how human feeling could be eliminated in the service of

> "efficient government." Hiroshima and Nagasaki belong to the same genre, but we do not like to dwell on these monstrosities or to consider what rationales were being consulted in the left hemisphere while the emotions of moral indignation were being segregated and isolated in the right. (Singer 1976, 222–23)

In Southern California, a system of mind development began making the rounds that involved binding the right arm in a cast and sling so as to prod the right hemisphere into action and thereby stimulate the intuitive faculties (Weil 1977, 50). A 1978 article proposing "ten ways to develop your right brain" included such exercises as "one day a week, mak[ing] it a rule that no one in the office or plant can use the word no" because "the right hemisphere has no equivalent of no" (cited in Springer & Deutsch 1985, 247). Paperback books explained how to use the right hemisphere to enhance one's athletic abilities ("inner skiing") or become more artistic (e.g., Betty Edward's 1980 *Drawing on the Right Side of the Brain*). Some writers went so far as to suggest that psychic (psi) abilities as well as intuition and art were bound up with the right hemisphere, as Myers and a few others in the last century had proposed, and they hinted that unbelievers in the paranormal were blinded to its reality by their left-brain dominance (Gardner 1979, 76).

9-6. *On the Relations between Old Views and New: Some Closing Questions*

Responding to my 1985 article "Nineteenth century ideas on hemisphere differences and 'duality of mind,' " Bradshaw (1985), Cernāček (1985), and Gruzelier (1985) were struck by the range of recurring themes and preoccupations in the nineteenth- and twentieth-century literature on brain duality and hemisphere differences. That the two traditions are essentially discontinuous with each other naturally makes such parallels all the more intriguing. Bradshaw has argued (1985, 635) that we really are in no position to decide how far this phenomenon is due to similarities in the late nineteenth- and late twentieth-century socio-cultural milieu, the constraints of external reality (i.e., the fact that brains a century ago and brains today probably behave more or less the same under similar conditions), or the modeling propensities of the human mind (e.g., a possible tendency for human beings to impose dichotomous categories of organization onto data and experience).

It is true enough that we cannot construct a balance sheet of factors that account in a definitive fashion for the existence of two bodies of thought that, like Old and New World monkeys, have much in common but seem largely to have evolved independently of each other. It seems to me, however, that we might at least ask why two similar traditions should have emerged *when* and *where* they did, and whether certain historical conditions might be conducive to certain ways of thinking about mind and brain.

Is it significant, for example, that we again are living in a time when the explanatory possibilities of the brain sciences are widely perceived as almost limitless; and if so, what does it mean? That neurologists and psychologists again are feeling increasingly free to draw social, moral, and philosophical conclusions from their work? That the walls separating various diverse fields of social and biological inquiry (psychology, psychiatry, neurology, anthropology, philosophy, etc.) again have become increasingly fluid? That traditional barriers between elite-scientific and so-called popular culture again are being widely eschewed?

I ask these questions. I do not feel able at this point properly to answer them. Nor am I suggesting that any amount of musing on the relationship between the two research traditions would ever allow one to predict where present trends in the history of the double brain are leading: whether the double brain will live up to at least some of the high hopes invested in it; whether intellectual fatigue and frustration will ultimately set in and encourage a swing back to perspectives on mind and brain currently out of favor; or whether some novel finding or cultural development will change fundamentally the parameters of discourse.

At the same time, even if it cannot help one predict the future, the history of the double brain is not therefore a subject of antiquarian interest only. Naturally, it is of antiquarian, or historical, interest as *well*; that is to say, I believe that the theme of this study represents a fresh and potentially fruitful avenue for exploring a number of wider issues currently of interest to historians of psychology and brain functioning. My narrative, for example, has made points of contact with evolutionary theory (cf. Young 1970; Block 1984), the problem of mind-body relations (cf. Smith 1973), the dilemma of free will and moral responsibility (cf. Daston 1978); the relationship between ideas of brain functioning and the social and political order (cf. Jacyna 1981, 1982; Pauly 1983); and the generally polemical function of science and medicine in an increasingly secular society (cf. Turner 1975; Nye 1984). It has examined the way in which ideas about asymmetry and hemisphere differences were employed in the service of an ideology of white male supremacy (cf. Gould 1981). It even has asked how far science's capacity to come to terms with the double brain might be constrained not only by changing socio-cultural influences but also the enduring mytho-symbolic propensities of the scientists' own minds and brains (cf. Corballis 1980).

At the same time, I would reiterate a remark I made at the very beginning of this study: that the history of the double brain may not only be in a position to flesh out, modify, or generally offer a new perspective on the scientific and medical culture of an earlier age, it also may be in a position to comment usefully upon the present effort in neurology to make sense of the human animal's duplex, asymmetrical brain. Taken seriously, reflection upon such a history could encourage the modern-day scientist, above all interested in how the human mind and brain "really" works, to give some thought as well to the problem of how *science* also "really" works. Virtually everything I have learned in the course of writing this book has reinforced my conviction that the fashioning of "truths" in science is never so straightforward an affair as some people perhaps are still inclined to think it should be. This perhaps applies most of all to those truths in science that purport to tell us something about ourselves. To a greater or lesser extent, I believe, all scientific truths necessarily bear the mark of human fingerprints.

APPENDIX

Guide to the Major Structures of the Human Brain Discussed in This Study

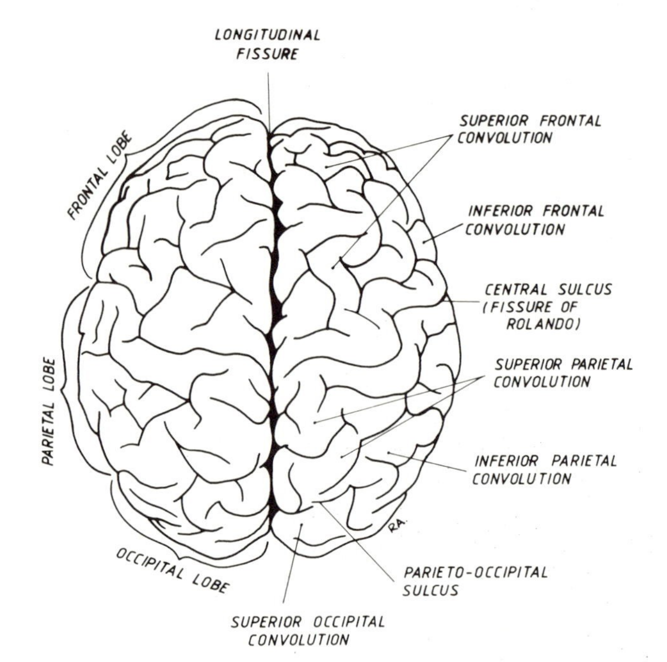

Figure I. The brain seen from above.

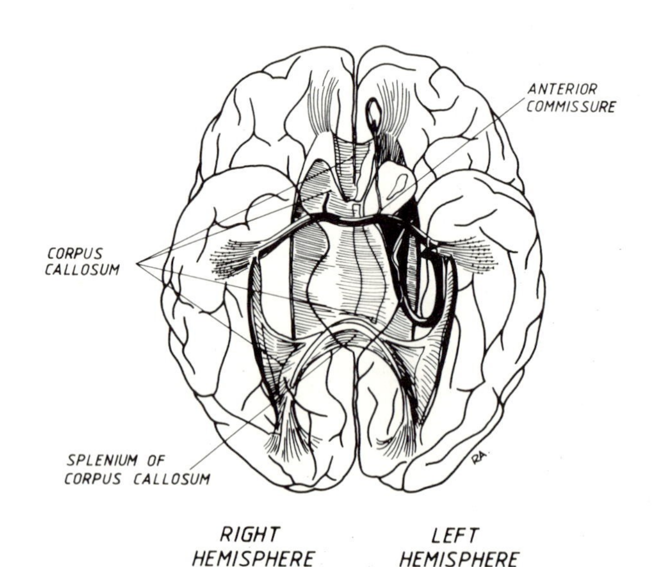

Figure II. Basal view of the brain, showing commissural connections.

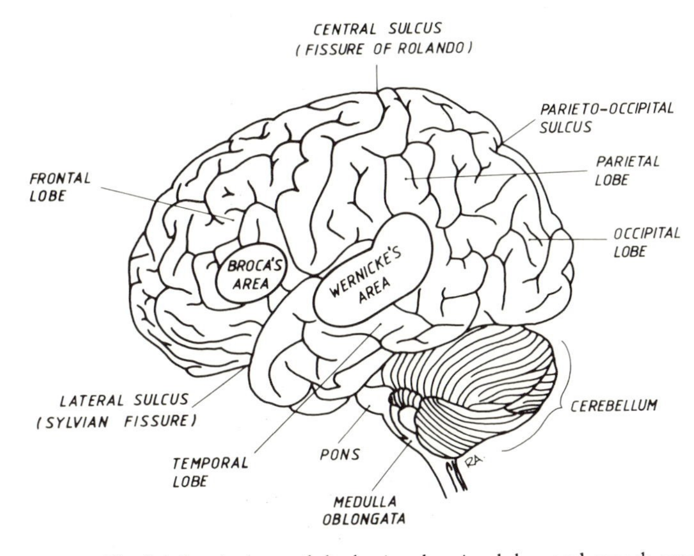

Figure III. The left hemisphere of the brain, showing lobes and speech areas.

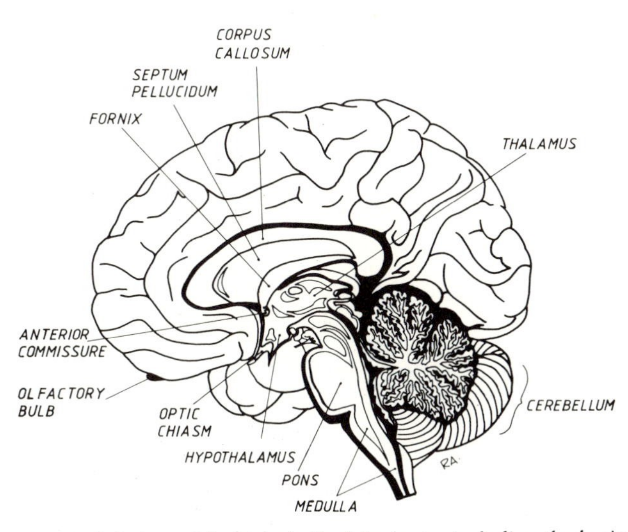

Figure IV. Medial view of the right half of the brain, including the brain stem.

References

Abercombie, J. 1831. *Inquiries Concerning the Intellectual Powers and the Investigation of Truth.* 2nd ed. Edinburgh: Waugh and Innes.

Adamkiewicz, A. 1880. Ueber bilaterale Functionen. *Archiv für Physiologie* (Leipzig): 159–62.

Adler, A. S. 1879. *Ein Beitrag zur Lehre von den "bilateralen Functionen" im Anschlusse an Erfahrungen der Metalloscopie.* (Inaugural dissertation, University Frederic-Guilhume of Berlin) Berlin: Gutmann.

Adler, A. S. 1880–81. A contribution to the doctrine of bilateral functions after experiences of metaloscopy [sic]. *San Francisco Western Lancet* 9:536–52.

Akelaitis, A. 1943. Studies on the corpus callosum VII. Study of language functions (tactile and visual lexia and graphia) unilaterally following section of the corpus callosum. *Journal of Neuropathology and Experimental Neurology* 2:226–62.

Albutt [no initial]. 1868. *Review IX*: Medical ophthalmoscopy. *The British and Foreign Medico-Chirurgical Review* 41:126–50.

Allen, G. 1975. *Life Science in the Twentieth Century.* Cambridge: Cambridge Univ. Press, 1979.

Alliot, E. 1886. *Contribution à l'étude de la suggestion mentale et de l'action des médicaments à distance.* Paris: J.-B. Baillière et fils.

Alston, W. P. 1967. Emotion and feeling. In *The Encyclopedia of Philosophy*, Vol. 2:479–86. New York & London: Macmillan Publishing Co. & The Free Press.

Alvarado, C. S. 1986. Early speculation about psychic phenomena and the brain's hemispheres. Unpublished paper, Dept. Behavioral Medicine and Psychiatry (division of parapsychology), University of Virginia.

Anatomy of the brain. 1849. *Buchanan's Journal of Man* 1 (Nos. 6 & 7):240–51, 289–30.

Atrophie complète du lobule de l'insula et de la troisième circonvolution du lobe frontale avec conservation de l'intelligence et de la faculté du langage articulé—Observation par M. le Dr. Parrot. 1863. *Bulletins de la Société Anatomique* 38:372–401.

Audenino, E. 1908. Mancinismo e destrismo. *Archivio di Psichiatria,*

Neuropatologia, Antropologia Criminale et Medicina Legale (Torino) 29:292–301.

Azam, E. G. 1887. *Hypnotisme, double conscience et altération de la personalité*, preface by J. M. Charcot. Paris: J. B. Baillière.

Babinski, J. 1886. *Recherches servant à établir que certaines manifestations hystériques peuvent être transférées d'un sujet à un autre sous l'influence de l'aimant.* Paris: Delahaye et Lecrosnier.

Babinski, J. 1892. Hypnotism and hysteria, translated by W. J. Herdman. *Journal of Nervous and Mental Diseases* 19:383–95, 495–511, 584–98.

Babinski, J. 1901. Définition de l'hystérie. *Revue Neurologique* 9:1074–80.

Babinski, J. 1914. Contribution à l'étude des troubles mentaux dans l'hémiplégie organique cérébrale (anosognosie). *Revue Neurologique* 27:845–48.

Bakan, P. 1969. Hypnotizability, laterality of eye-movements, and functional brain asymmetry. *Perceptual and Motor Skills* 28:927–32.

Baker, S. 1893. Etiological significance of heterogeneous personality. *Journal of Nervous and Mental Diseases* 20:664–74.

Baldwin, J. M. 1889. Dr. Maudsley on the double brain. *Mind* 14:545–50.

Ball, B. 1884. Le dualisme cérébral. *Revue Scientifique*, 3rd ser., 7:33–37.

Ballet, G. 1880. Nouveau fait à l'appui de la localisation de Broca (Demonstration expérimentale de la localisation de la faculté du langage dans l'hémisphère gauche du cerveau. *Le Progrès Médical* 8:739–41.

Banquet (le) en l'honneur du Docteur Bérillon. 1906–07. *Revue de l'hypnotisme et de la psychologie physiologique* 21:33–50, 98–116.

Bárety, A. 1887. *Le magnétisme animal étudié sous le nom de force neurique rayonnante et circulante dans ses propriétés physiques, physiologiques et thérapeutiques.* Paris: Octave Doin.

Barkow, H. C. L. 1864. *Bemerkungen zur Pathologischen Osteologie.* Breslau: Ferdinand Hirt's Königliche Universitäts-Buchhandlung.

Barlow, J. 1843. *Man's power over himself to prevent or control insanity.* London: William Pickering.

Baruk, H. 1978. Modern neuropsychiatry and Freudian psychoanalysis. In *Historical Explorations in Medicine and Psychiatry*, edited by H. Riese. New York: Springer Publishing Co.

Bastian, H. C. 1869. On the various forms of loss of speech in cerebral disease. *The British and Foreign Medico-Chirurgical Review* 43:209–36, 470–92.

Bastian, H. C. 1869–70. Consciousness. *The Journal of Mental Science* 15:501–23.

Bastian, H. C. 1880. *The Brain as an Organ of Mind.* London: Kegan Paul.

Bateman, F. 1869–70. On aphasia, or loss of speech in cerebral disease. *The Journal of Mental Science* 15:367–93.

Bateman, F. 1890a. *On Aphasia or Loss of Speech and the Localisation of the Faculty of Articulate Language.* London: Churchill.

Bateman, F. 1890b. Hypnotism: With a criticism on some recent experiments at la Salpêtrière. *Journal of Nervous and Mental Diseases* 15:287–94.

Bell, C. 1811. Idea of a new anatomy of the brain. In *Readings in the History of Psychology*, edited by W. Dennis. New York: Appleton-Century-Crofts, Inc., 1948.

Bender, D. 1965. The development of French anthropology. *Journal of the History of the Behavioral Sciences* 1:139–51.

[Obituary] Benjamin Ball, M.D. 1893. *British Medical Journal* 1:613.

Bentley, M. 1916. The psychological antecedents of phrenology. *Psychological Monographs* 21:102–15.

Benton, A. 1972. The "minor" hemisphere. *Journal of the History of Medicine and Allied Sciences* 27:5–14.

Berger, O. 1880. Experimentelle Katalepsie (Hypnotismus). Neue Beiträge. *Deutsche Medicinische Wochenschrift* (Berlin) 6:116–18.

Bérillon, E. 1884. *De l'indépendance fonctionnelle des deux hémisphères cérébraux* (Doctoral dissertation.) Paris: A Parent.

Bérillon, E. 1899. *L'oeuvre scientifique de Dumontpallier.* Paris: A. Quelquejeu.

Bérillon, E. 1914. *L'hypnotisme et la psychothérapie.* [Address given at the inauguration of a bust of Amédée Dumontpallier, June 24, 1913.] Paris: Aux Bureaux de la Revue de Psychothérapie.

Bérillon, E., and P. Magnin. 1901–02. Hypnotisme de degré différend [*sic*] pour chaque côté du corps. *Revue de l'hypnotisme expérimental et thérapeutique* 16:282–83.

Berjon, A. 1886. *La grande hystérie chez l'homme. Phénomènes d'inhibition et de dynamogénie. Changements de la personnalité. Action des médicaments à distance.* Paris: Baillière et fils.

Bernheim, H. 1881a. De la magnétothérapie, historique et faits nouveaux. *Revue médicale de l'Est* (Nancy) 13:305–308.

Bernheim, H. 1881b. Magnétothérapie. Historique et faits nouveaux.

Revue médicale de l'Est (Nancy) 13:547–53, 579–88, 620–28, 654–59, 688–95, 728–31.

Bernheim, H. 1882. Nouvelles observations de magnétothérapie. *Revue médicale de l'Est* 14:626–32, 662–66.

Bernheim, H. 1883. De la suggestion dans l'état hypnotique et dans l'état de veille. *Revue médicale de l'Est* 15:513–20, 545–59, 577–92, 610–19, 641–58, 674–85, 712–21.

Bernheim, H. 1884. De la suggestion dans l'état hypnotique et dans l'état de veille. *Revue médicale de l'Est* 16:7–20.

Bernheim, H. 1885. Notes et discussions. L'hypnotisme chez les hystériques. *Revue Philosophique de la France et de l'Étranger* 19:311–16.

Bernheim, H. 1886. Reponse à l'article de M. Binet sur le livre de M. Bernheim: De la suggestion et de ses applications thérapeutiques. *Revue de l'hypnotisme et de la psychologie physiologique* 1:213–18.

Bianchi, A. 1883. Changes in handwriting in relation to pathology. In review of the works of the following authors, viz: Marcé, Poincaré, Charcot, Buchwald, Erlenmeyer, Vogt, Swortzoff, Grasset, Ireland, and Durand, 1863–1882. *The Alienist and Neurologist* 4:566–90 (Translated by J. Workmann from *Il Pansani Gazetta Sicula*, Palermo, 1882.)

Bichat, X. 1805. *Recherches physiologiques sur la vie et la mort.* 3rd ed. Paris: Brosson/Gabon.

van Biervliet, J. J. 1899. L'homme droit et l'homme gauche. *Revue philosophique* 47:113–43, 277–96, 371–89.

Binet, A. 1889. *On Double Consciousness.* Chicago: Open Court Publishing Co., 1905.

Binet, A. 1891. *Alterations of Personality*, translated by H. G. Baldwin. London: Chapman & Hill, Ltd., 1896.

Binet, A., and C. Féré. 1885. L'hypnotisme chez les hystériques. I. Le transfert psychique. *Revue Philosophique de la France et de l'Étranger* 19:1–25.

Binet, A., and C. Féré. 1887. *Animal Magnetism* (International Science Series). London: Kegan Paul, Trench & Co.

Bleuler, E. 1902–1903. Halbseitiges Delirium. *Psychiatrisch-Neurologische Wochenschrift* 4(34):361–67.

Bleuler, E. 1911. *Dementia Praecox oder die Gruppe der Schizophrenie.* English translation as *Dementia praecox or the group of schizophrenias*, by J. Zinkin (Monograph series on schizophrenia, no. 1). New York: International Universities Press, 1950.

Bleuler, M., ed. 1979. *Beiträge zur Schizophrenielehre der Zürcher*

Psychiatrischen Universitätsklinik Burghölzl (1902–1971). Darmstadt: Wissenschaftliche Buchgesellschaft.

Block, E., Jr. 1984. James Sully, Robert Louis Stevenson, and evolutionary psychology in the Cornhill Magazine 1875–1880. *Journal of the History of Ideas* 45(3):465–75.

[Review] *The Blot upon the Brain, etc.* by W. W. Ireland. 1885–86. *The Journal of Mental Science* 31:540–49.

Bogen, J. E. 1969. The other side of the brain 2: An appositional mind. *Bulletin of Los Angeles Neurological Societies* 34:135–62.

Bogen, J. E. 1977. Some educational implications of hemispheric specialization. In *The Human Brain*, edited by M. C. Wittrock. New York: Prentice-Hall.

Bogen, J. E. 1979. The callosal syndrome. In *Clinical Neuropsychology*, edited by K. M. Heilman and E. Valenstein. New York/ Oxford: Oxford Univ. Press.

de Boismont, A. B. 1854. Revue de *Considérations pratiques et théoriques sur l'oblitération et l'altération de l'esprit, déduites de trois cents autopsies faites à l'asile public Saint-Athanase de 1833 à 1854*, par le docteur Follet. *Annales Médico-Psychologiques* 6:655–56.

le Bon, G. 1878a. Sur l'inégalité des régions correspondantes du crâne. *Bulletins de la Société d'Anthropologie*, 3rd ser., 1:104–106.

le Bon, G. 1878b. Recherches expérimentales sur les variations de volume du cerveau et du crâne. *Bulletins de la Société d'Anthropologie*, 3rd ser., 1:310–15.

le Bon, G. 1879. Variations du volume du cerveau et sur leurs relations avec l'intelligence. *Revue d'Anthropologie* 8:27–104.

Bonhoeffer, K. 1914. Klinischer und anatomischer Befund zur Lehre von der Apraxie und der "motorischen Sprachbahn." *Monatsschrift für Psychiatrie und Neurologie* 35:113–28.

Bottey, F. 1884. *Le "magnétisme animal": Étude critique et expérimentale sur l'hypnotisme ou sommeil nerveux provoqué chez les sujets sains (léthargie, catalepsie, somnambulisme, suggestions, etc.)* Paris: E. Plon, Nourrit et Cie.

Bouillaud, M. J. 1825. *Traité clinique et physiologique de l'encéphalité ou inflammation du cerveau, et de ses suites.* Paris: J.-B. Baillière.

Bourneville, D. M., and P. Regnard. 1877–80. *Iconographie photographique de la Salpêtrière.* 3 vols. Paris: Progrès médical.

Bourru, H., and F. Burot. 1886–87. Les variations de la personnalité. *Revue de l'Hypnotisme Expérimental et Thérapeutique* 1:193–99, 261–65.

Bourru, H., and F. Burot. 1887. *La suggestion mentale et l'action à distance des substances toxiques et médicamenteuses.* Paris: J.-B. Baillière et fils.

Bourru, H., and F. Burot. 1888. *Variations de la personnalité.* Paris: J.-B. Baillière.

Boyd, R. 1861. Tables of the weights of the human body and internal organs in the sane and insane of both sexes and of various ages, arranged from 2614 post-mortem examinations. *Royal Society of London Philosophical Transactions* 151:241–62.

Bradshaw, J. L. 1985. Reinventing hemisphere differences. *The Behavioral and Brain Sciences* 8(4):635.

Braid, J. 1843. *Neurypnology; or the Rationale of Nervous Sleep, Considered in Relation with Animal Magnetism.* London: John Churchill.

Brain, W. R. 1941. Visual disorientation with special reference to lesions of the right cerebral hemisphere. *Brain* 64:244–72.

Breuer, J., and S. Freud. 1893–95. Studies on hysteria. In *The Standard Edition of the Complete Psychological Works of Sigmund Freud.* Vol. 2. London: Hogarth Press, 1955.

Broadbent, W. H. 1866, An attempt to remove the difficulties attending the application of Dr. Carpenter's theory of the function of the sensori-motor ganglia to the common form of hemiplegia. *The British and Foreign Medico-Chirurgical Review* 37:468–81.

Broadbent, W. H. 1872. On the cerebral mechanisms of speech and thought. *Medico-Chirurgical Transactions*, 2nd ser., 37:145–91.

Broca, P. 1860. Discussion sur la perfectibilité des races. *Bulletins de la Société d'Anthropologie* 1:337–42.

Broca, P. 1861a. Sur le volume et la forme du cerveau suivant les individus et suivant les races. *Bulletins de la Société d'Anthropologie* 2:139–204, 301–21.

Broca, P. 1861b. Perte de la parole. Ramollisement chronique et destruction partielle du lobe antérieur gauche du cerveau. *Bulletins de la Société d'Anthropologie* 2:235–38.

Broca, P. 1861c. Remarques sur le siège de la faculté du langage articulé, suivies d'une observation d'aphémie (perte de la parole). *Bulletins de la Société Anatomique* 36:330–57.

Broca, P. 1861d. Nouvelle observation d'aphémie produite par une lésion de la moitié postérieure des deuxième et troisième circonvolutions frontales. *Bulletins de la Société Anatomique*, 2nd ser., 6:398–407.

Broca, P. 1863. Localisations des fonctions cérébrales.—Siège du langage articulé. *Bulletins de la Société d'Anthropologie* 4:200–204.

Broca, P. 1865. Sur la faculté du langage articulé. *Bulletins de la Société d'Anthropologie* 6:493–94.

Broca, P. 1866a. Discours sur l'homme et les animaux. *Bulletins de la Société d'Anthropologie*, 2nd ser., 1:53–79.

Broca, P. 1866b. [index title] Poids comparé des lobes frontaux et occipitaux, et des deux hémisphères. In "Correspondance, *Bulletins de la Société d'Anthropologie*, 2nd ser., 1:195–96 (note page error in index).

Broca, P. 1869. L'ordre des primates. Parallèle anatomique de l'homme et des singes. XI. *Le cerveau. Bulletin de la Société d'Anthropologie*, 2nd ser., 4:374–95.

Broca, P. 1872. De l'influence de l'éducation sur le volume et la forme de la tête. *Bulletins de la Société d'Anthropologie*, 2nd ser., 7:879–96.

Broca, P. 1875. Sur les poids relatifs des deux hémisphères cérébraux et de leur lobes frontaux. *Bulletins de la Société d'Anthropologie* 10:534.

Broca, P. 1877. Rapport sur un mémoire de M. Armand de Fleury intitulé: *De l'inégalité dynamique des deux hémisphères cérébraux. Bulletins de l'Académie de Médecine* 6:508–39.

Bromberg, W. 1942. Some social aspects of the history of psychiatry. *Bulletin of the History of Medicine* 11:117–32.

Bromberg, W. 1982. *Psychiatry between the wars, 1918–1945. A recollection.* Westport, Conn./London: Greenwood Press.

Brown, E. M. 1983. Neurology and spiritualism in the 1870s. *Bulletin of the History of Medicine* 57(4):563–77.

Brown-Séquard, C.-E. 1870. Symptômes variables suivant le côté de l'encéphale qui est le siège des lésions. *Comptes Rendus de la Société de Biologie*, 5th ser., 2:27–28, 96–97.

Brown-Séquard, C.-E. 1871. Parallèle des phénomènes différents observés dans les lésions des hémisphères droits et gauches du cerveau. *Comptes Rendus de la Société de Biologie*, 5th ser., 3:96.

Brown-Séquard, C.-E. 1874a. Dual character of the brain (Toner lecture). *Smithsonian Miscellaneous Collections* (Washington, D.C., 1878) 15:1–21.

Brown-Séquard, C.-E. 1874b. The brain power of man: Has he two brains or has he only one? *Cincinnati Lancet and Observer* 17:330–33.

Brown-Séquard, C.-E. 1882. *Recherches expérimentales et cliniques sur l'inhibition et la dynamogénie. Application des connaissances fournies par ces recherches aux phénomènes principaux de l'hypnotisme et du transfert*. Paris: G. Masson. [Extract from the *Gazette hebdomadaire de médecine et de chirurgie*.]

Brown-Séquard, C.-E. 1884. Persistance de la parole dans la chant, dans les rêves et dans le délire, chez des aphasiques. *Comptes Rendus de la Société de Biologie*, 8th ser., 1:256–57.

Brown-Séquard, C.-E. 1887. Dualité du cerveau et de la moelle épinière, d'après des faits montrant que l'anesthésie, l'hyperethésie, la paralysie, et des états variés d'hypothermie et d'hyperthermie dus à des lésions organiques du centre cérébro-spinal, peuvent être transférés d'un côté à l'autre du corps. *Comptes Rendus de l'Académie des Sciences* (Paris) 105:646–52.

Brown-Séquard, C.-E. 1890. Have we two brains or one? *The Forum* 9:627–43.

Bruce, A. 1889–90. On the absence of the corpus callosum in the human brain, with the description of a new case. *Brain* 12:171–90.

Bruce, L. 1895. Notes of a case of dual brain action. *Brain* 18:54–65.

Bruce, L. 1897. On dual brain action and its relation to certain epileptic states. *Transactions of the Medical-Chirurgical Society of Edinburgh* 16:114–19.

Brunton, T. L. 1874. On the physiology of vomiting and the action of antiemetics and emetics. *The Practitioner* 13:409–29.

Burq, V. 1853. *Métallothérapie, traitement des maladies nerveuses, paralysies, rhumatisme chronique, spasmes . . . par les applications métalliques. Abrégé historique, théorique et pratique, extrait de vingt-deux mémoires ou notes aux deux Académies*. Paris: G. Baillière.

Burq, V. 1882. *Des origines de la métallothérapie. Part qui doit être faite au magnétisme animal dans sa découverte. Le Burquisme et le Perkinisme*. Paris: A Delahaye et Lecrosnier.

Butterfield, H. 1931. *The Whig Interpretation of History*. London: G. Bell & Sons.

Bynum, W. F. 1981. Rationales for therapy in British psychiatry, 1780–1835. In *Madhouses, Mad-Doctors, and Madmen*, edited by A. Scull. London: The Athlone Press.

Caldwell, C. 1824. *Elements of Phrenology*. Cited in American elements of phrenology, *The Phrenological Journal and Miscellany* (1825) 2:113–20.

Carlson, E. T. 1984. The history of multiple personality in the United States: Mary Reynolds and her subsequent reputation. *Bulletin for the History of Medicine* 58:72–82.

Carrier, A. [de Lyons]. 1867. *Étude sur la localisation dans le cerveau de la faculté du langage articulé.* Doctoral dissertation. Paris: A. Parent.

Cernăček, J. 1985. Hemisphere asymmetry: Old views in new light. *The Behavioral and Brain Sciences* 8(4):636.

Chambard, E. 1881. Revue générale. Le hypnose hémi-cérébrale. *L'Encéphale* 1:242–50.

Charcot, J.-M. 1875a. Hémiopie latérale et amblyopie croisée. *Le Progrès Médical* 3:481–82.

Charcot, J.-M. 1875b. De l'hémianesthésie hystérique (dixième leçon) *Leçons sur les maladies du système nerveux faites à la Salpêtrière,* rec. et pub. par Bourneville. Paris: V. Adrien Delahaye. Vol. 1:300–19.

Charcot, J.-M. 1878a. Des troubles de la vision chez les hystériques. *Le Progrès Médical* 6:37–39.

Charcot, J.-M. 1878b. Phénomènes divers de l'hystéro-epilepsie.—Catalepsie provoquée artificiellement. *La Lancette française: Gazette des hôpitaux* 51:1075–77.

Charcot, J.-M. 1878c. A lecture on certain phenomena of hysteria major (delivered at the Salpêtrière Nov. 17), translated by G. Sigerson. *The British Medical Journal* 2:789–91.

Charcot, J.-M. 1878d. De la métalloscopie et la métallothérapie. *La Lancette française: Gazette des Hôpitaux Civils et Militaires* 51:217–19, 235–36, 241–43.

Charcot, J.-M. 1882. Sur les divers états nerveux déterminés par l'hypnotisation chez les hystériques. *Comptes-Rendus hebdomadaires de l'Académie des Sciences* 94:403–405.

Charcot, J.-M. 1883. *Exposé des titres scientifiques du M. J.-M. Charcot.* Paris: Victor Goupy et Jourdan.

Charcot, J., and P. Richer. 1878. Catalepsie et somnambulisme hystériques provoqués. *Le Progrès Médical* 6:973–75.

[Obituary] Charles Féré. 1907. *The British Medical Journal* 1:1281.

Chazarain, L.-T., and C. Dècle. 1887–88. Les courants de la polarité dans l'aimant et dans le corps humain. *Revue de l'Hypnotisme,* 141–51.

Clapham, C. 1881–82. The noble forehead. *Journal of Mental Science* 27:623–24.

Clapham, C. 1892. Head, size and shape of, in the insane. In *A Dictionary of Psychological Medicine,* edited by D. H. Tuke. London: J. & A. Churchill. Vol. 1:574–80.

Clark, M. J. 1981. The rejection of psychological approaches to mental disorder in late nineteenth-century British psychiatry. In *Mad-*

houses, Mad-Doctors, and Madmen, edited by A. Scull. London: The Athlone Press.

Clarke, B. In press. Arthur Wigan and the "Duality of the Mind." *Psychological Medicine*.

Cobban, A. 1965. *A History of Modern France*. Rev. ed. Vol. 2. Middlesex, England: Penguin Books.

Combe, A. 1824. On the effects of injuries of the brain upon the manifestations of the mind. *Transactions of the Phrenological Society* (Edinburgh), 183–208.

Congress of psychiatry and neuro-pathology at Antwerp [Notes and News]. 1885–86. *Journal of Mental Science* 31:621–22.

Cooter, R. 1984. *The Cultural Meaning of Popular Science: Phrenology and the Organization of Consent in Nineteenth-Century Britain*. Cambridge/New York: Cambridge Univ. Press.

Corballis, M. 1980. Laterality and myth. *American Psychologist* 35:284–95.

Corraze, J. 1983. La question de l'hystérie. *Nouvelle histoire de la psychiatrie*, edited by J. Postel & C. Quétel, 401–11. Toulouse: Editions Private (Société Française d'Histoire de la Psychiatrie).

de Courmelles, F. 1891. *Hypnotism*, translated by L. Ensor. London: George Routledge & Sons.

Crichton-Browne, J. 1878. On the weight of the brain and its component parts in the insane. *Brain* 1:504–18.

Crichton-Browne, Sir J. 1895. *The Cavendish Lecture on Dreamy Mental States*. London: Baillière, Tindall & Cox.

Crichton-Browne, Sir J. 1907. Dexterity and the bend sinister. *Proceedings of the Royal Institution of Great Britain* 18:623–52.

Critchley, M. 1928. *Mirror-Writing*. London: Kegan Paul, Trench, Trubner.

Cunningham, D. 1892. *Contributions to the Surface Anatomy of the Cerebral Hemispheres*. Dublin: Royal Irish Academy.

Cunningham, D. 1902. Right-handedness and left-brainedness [the Huxley lexture for 1902]. *Journal of the Royal Anthropological Institute of Great Britain and Ireland* 32:273–96.

Dandy, W. E. 1930. Changes in our conception of the localization of certain functions in the brain. *American Journal of Physiology* 93:643.

Daston, L. J. 1978. British responses to psycho-physiology, 1860–1900. *Isis* 69:192–208.

Davey, J. G. 1844. The duality of the mind known to the early writers on medicine. *The Lancet* 1 (March–Sept.):377–78.

Dax, M. 1836. Lésions de la moitié gauche de l'encéphale coincident avec l'oubli des signes de la pensée. In English translation in *The Roots of Psychology*, edited by S. Diamond. New York: Basic Books, 1974.

Delaunay, G. 1874. *Biologie comparée du côté droit et du côté gauche chez l'homme et chez les êtres vivants*. Doctoral dissertation. Paris: A Parent.

Delaunay, G. 1878–79. *Études de biologie comparée*. 2 vols. Paris: V. Adrien Delahaye.

Delaunay, G. 1879. De la tendance des individus à se diriger à gauche ou à droite. *Lancette française: Gazette des hôpitaux* 52:253–54.

Delaunay, J. [*sic*] 1882. Sur deux nouveaux procédés d'investigation psychologique. *Lancette française: Gazette des hôpitaux* 55:22–23.

Delaunay, G. 1883. De la rotation. *Lancette française: Gazette des hôpitaux* 56:970–71.

Delaunay, G. 1884. Du croisement des membres et de la façon de s'asseoir. *Lancette française: Gazette des hôpitaux* 57:332–33.

Descourtis, G. 1881. Contribution à l'étude du dédoublement des opérations cérébrales. *L'Éncéphale* 1:126–28.

Descourtis, G. 1882. *Du fractionnement des opérations cérébrales et en particulier de leur dédoublement dans les psychopathies*. Doctoral dissertation. Paris: A Parent.

Descourtis, G. 1890. Les deux cerveaux de l'homme. *Revue d'hypnologie théorique et pratique* 1:97–106.

Dessoir, M. 1887. Hypnotism in France. *Science* 9:541–45.

Dessoir, M. 1888. *Bibliographie des modernen Hypnotismus*. Berlin: Carl Duncker's Verlag.

Dewey, R. 1896. Report of a case of "dreamy mental state." *Journal of Nervous and Mental Diseases* 23:763–69.

Dewhurst, K. 1982. *Hughlings Jackson on Psychiatry*. Oxford: Sandford Publications.

Didi-Huberman, G. 1982. *Invention de l'hystérie: Charcot et l'iconographie photographique de la Salpêtrière*. Paris: Macula.

Discussion sur l'aphémie. 1865. *Bulletin de la Société d'Anthropologie* 6:412–17.

Discussion sur la faculté du langage articulé 1864-65. *Bulletin de L'Académie Imperiale de Médecine*, 1st ser., 30:575–600, 604–38, 647–56, 659–75, 679–703, 713–18, 724–81, 787–803, 816–32, 840–68, 888–90.

Downer, J. 1962. Interhemispheric integration in the visual system. In *Interhemispheric Relations and Cerebral Dominance*, edited by V. B. Mountcastle. Baltimore: Johns Hopkins Univ. Press.

Dreams. 1892. *The Spectator* 69:189–90.

Dr. Wigan on the Duality of Mind. 1845. *The Phrenological Journal* 17:168–81.

Duality and decussation (Explanation of the mutual relations of the double brain and double body). 1850. *Buchanan's Journal of Man* 1(2):513–28.

Dumontpallier, A. 1882a. Léthargie incomplète avec conservation de l'ouie et de la mémoire—De l'indépendance fonctionnelle de chaque hémisphère cérébral. *Comptes rendus de la société de biologie*, 7th ser., 4:393–97.

Dumontpallier, A. 1882b. Indépendance fonctionnelle de chaque hémisphère cérébral.—illusions, hallucinations unilatérales ou bilatérales provoquées," *Comptes rendus de la société de biologie*, 7th ser., 4:786–97. Also *Lancette française: Gazette des hôpitaux* (1882) 55:1177–79.

Dumontpallier, A. 1889. Discours de M. Dumontpallier au congrès de l'hypnotisme. *Revue de l'hypnotisme et de la psychologie physiologique* 4:79–82.

Dumontpallier, A., and E. Bérillon. 1884. Indépendance fonctionnelle des hémisphères cérébraux.—Hallucinations bilatérales simultanées dans l'hypnotisme.—persistance à l'état de veille. *Comptes rendus de la société de biologie*, 8th ser., 1:408–409.

Dumontpallier, A., J.-M. Charcot, and J. B. Luys. 1877. Rapport fait à la Société de Biologie sur la métalloscopie du docteur Burq. *Comptes rendus de la société de biologie (sec. mémoires)* 6th ser., 4:1–24.

Dumontpallier, A., and V. Magnan. 1883. Des hallucinations bilatérales à caractère différent suivant le côté affecté, dans le délire chronique; leçon clinique de M. Magnan, et démonstration expérimentale du siège hémilatéral ou bilatéral cérébral des hallucinations. *Union médicale*, 3rd ser., 35:845–48, 869–75.

Dumontpallier, A., and P. Magnin. 1882. Étude experimentale sur la métalloscopie, l'hypnotisme et l'action de divers agents physiques dans l'hystérie. *Comptes rendus hebdomadaires des séances de l'académie des sciences* 94:60–63.

Dunn, T. D. 1895. Double hemiplegia with double hemianopsia and loss of geographical center. *Transactions of the College of Physicians of Philadelphia*, 3rd ser., 17:45–55.

Durant, J. R. 1981. The beast in man: An historical perspective on the biology of human aggression. In *Multidisciplinary approaches to aggression research*, edited by P. F. Brain and D. Benton, 17–

46. Amsterdam/New York: Elsevier/North Holland Biomedical Press.

Duval, A. 1882. [Review of] *Du fractionnement des opérations cérébrales et en particulier de leur dédoublement dans les psychopathies*, par M. le docteur Descourtis. *L'Encéphale* 2:691–93.

Ecker, A. 1868. Zur Entwicklungsgeschichte der Furchen und Windungen des Grosshirn-Hemisphären im Foetus des Menschen. *Archiv für Anthropologie* 3:203–23.

Edelheit, H. 1969. Speech and psychic structure: The vocal-auditory organization of the ego. *Journal of American Psychoanalytic Association* 17:381–412.

Edwards, B. 1980. *Drawing on the Right Side of the Brain*. Los Angeles: J. P. Tarcher.

Ellenberger, H. F. 1970. *The Discovery of the Unconscious*. New York: Basic Books.

Eling, P. 1984. Broca on the relation between handedness and cerebral speech dominance. *Brain and Language* 22:158–59.

Elliotson, J. 1845–46a. Cure of hysterical epilepsy, somnambulism &c., with mesmerism. *The Zoist* 3:39–79.

Elliotson, J. 1845–46b. Cure of a contracted foot with severe pain, cured with mesmerism. *The Zoist* 3:446–85.

Elliotson, J. 1847. On the joint operation of the two halves of the brain: with a notice of Dr. Wigan's work, entitled *The Duality of the Mind*, etc. *The Zoist* 15:209–34.

Engelhardt, H. T. 1975. John Hughlings Jackson and the mind-body relation. *Bulletin of the History of Medicine* 49:137–51.

Engelhardt, H. T. 1976. Reflections on our condition: The geography of embodiment. In *Philosophical Dimensions of the Neuro-Medical Sciences*, edited by S. F. Spicker and H. T. Engelhardt, 59–68. Dordrecht: D. Reidel.

Erb, [no initial]. 1885. A case of hemorrhage in the corpus callosum. *The Journal of Nervous and Mental Diseases* 12:121.

Esquirol, J.E.D. 1838. *Mental Maladies*, English translation (1845) by E. K. Hunt. Facsimile edition, New York and London: Hafner, 1965.

Exner, S. 1881. *Untersuchungen über die Localisation der Functionen in der Grosshirnrinde des Menschen*. Vienna: Wilhelm Braumuller.

Ey, H. 1975. *Des idées de Jackson à un modèle organo-dynamique en psychiatrie*. (Collection "Rhadamanthe.") Toulouse: Edouard Privat.

Fechner, G. T. 1860. *Elemente der Psychophysik*. Vol. 2. Leipzig: Breitkopf & Hartel, 1907.

Féré, C. 1883. Note sur un cas d'anomalie asymétrique du cerveau. *Archives de neurologie* 5:59–66.

Féré, C. 1901a. Note sur le travail alternatif des deux mains. *L'année psychologique* 7:130–42.

Féré, C. 1901b. L'excitabilité comparée des deux hémisphères cérébraux chez l'homme. *L'année psychologique* 7:143–60.

Féré, C. 1902. L'alternance de l'activité des deux hémisphères cérébraux. *L'année psychologique* 8:107–49.

Ferrier, D. 1876. *The Functions of the Brain*. London: Dawsons of Pall Mall, 1966.

de Fleury, A. 1865. Mémoire sur la pathogénie du langage articulé. *Gazette hebdomadaire de médecine et de chirurgie*, 2nd ser., 2:228–32, 244–50.

de Fleury, A. 1872. Du dynamisme comparé des hémisphères cérébraux dans l'homme. *Association française pour l'avancement des sciences* 1:834–45.

Fliess, R. 1953. On the "spoken word" in the dream. In *The Revival of Interest in the Dream: A Critical Study of Post-Freudian Psychoanalytic Contributions*, edited by R. Fliess. New York: International Univ. Press.

Fliess, W. 1906. *Der Ablauf des Lebens*. Leipzig und Vienna: Franz Deuticke, 1923.

Fliess, W. 1914. *Vom Leben und vom Tod: Biologische Vorträge*. Jena: Eugen Diederichs.

Flourens, J.P.M. 1846. *Phrenology Examined*, translated by C. L. Meigs. Philadelphia: Hogan & Thompson.

Forrester, J. 1980. *Language and the Origins of Psychoanalysis*. London and Basingstoke: The Macmillan Press.

Foucault, M. 1973. *Madness and Civilization: A History of Insanity in the Age of Reason*, translated by R. Howard. New York: Vintage Books.

Freud, S. 1886. Report on my studies in Paris and Berlin. In *Standard edition of the complete psychological works of Sigmund Freud*, translated by J. Strachey. Vol. 1:3–15. London: Hogarth Press, 1966.

Freud, S. 1887–1902. *The Origins of Psychoanalysis, Letters to Wilhelm Fliess, Drafts, and Notes: 1887–1902*, introduction by E. Kris; edited by M. Bonaparte, A. Freud, and E. Kris; translated by E. Mosbacher and J. Strachey. New York: Basic Books. London: Imago Publishing Co., 1954.

Freud, S. 1891. *On Aphasia: A Critical Study*. London: Imago Publishing Co., 1953.

Freud, S. 1893. Some points for a comparative study of organic and hysterical motor paralyses. In *Standard Edition*. Vol. 1:157–72.

Freud, S. 1895a. Project for a scientific psychology. In *Standard Edition*. Vol. 1:283–387.

Freud, S. 1895b. Obsessions and phobias. Their psychical mechanism and aetiology. In *Standard Edition*. Vol. 3:71–82. London: Hogarth Press, 1962.

Freud, S. 1896. The aetiology of hysteria. In *Standard Edition*. Vol. 3:189–221.

Freud, S. 1900. The interpretation of dreams. In *Standard Edition*. Vols. 4 and 5. London: Hogarth Press, 1953.

Freud, S. 1901. The psychopathology of everyday life. In *Standard Edition*. Vol. 6. London: Hogarth Press, 1958.

Freud, S. 1905. Three essays on the theory of sexuality. In *Standard Edition*. Vol. 7:125–243. London: Hogarth Press, 1953.

Freud, S. 1911. Formulations on the two principles of mental functioning. In *Standard Edition*. Vol. 12:21–226. London: Hogarth Press, 1958.

Freud, S. 1912. A note on the unconscious in psycho-analysis. In *Standard Edition*. Vol. 12:255–66.

Freud, S. 1915. The unconscious. VII. Assessment of the unconscious. In *Standard Edition*. Vol. 14. London: Hogarth Press, 1957.

Freud, S. 1919. "A Child Is Being Beaten": A contribution to the study of the origin of sexual perversions. In *Standard Edition*. Vol. 17:177–204. London: Hogarth Press, 1955.

Freud, S. 1923. *The Ego and the Id*, translated by J. Riviere, rev. ed. by J. Strachey. London: Hogarth Press, 1962.

Freud, S. 1937. Constructions in analysis. In *Standard Edition*. Vol. 23:256–69. London: Hogarth Press, 1964.

Fritsch, G., and E. Hitzig. 1870. Ueber die elektrische Erregbarkeit des Grosshirns. English translation in *Some Papers on the Cerebral Cortex* by Gerhardt von Bonin. Springfield, Ill.: Charles C. Thomas, 1960.

Frumkin, K., H. Ripely, and G. Cox. 1978. Changes in cerebral hemisphere lateralization with hypnosis. *Biological Psychiatry* 13:741–50.

Fullinwider, S. P. 1983. Sigmund Freud, John Hughlings Jackson, and speech. *Journal of the History of Ideas* 44(1):151–58.

Galin, D. 1974. Implications for psychiatry of left and right cerebral specialization. *Archives of General Psychiatry* 31:572–83.

Galin, D., P. Diamond, and D. Braff. 1977. Lateralization of conversion symptoms: More frequent on the left. *American Journal of Psychiatry* 134:578–79.

Gamgee, A. 1878. An account of a demonstration on the phenomena of hystero-epilepsy: and on the modification which they undergo under the influence of magnets and solenoids, given by Professor Charcot at the Salpêtrière. *The British Medical Journal* 2:545–48.

Gardner, M. 1979. *The Ambidextrous Universe*. rev. ed. New York: Charles Scribner's & Sons.

Gauld, A. 1968. *The Founders of Psychical Research*. London: Routledge & Kegan Paul.

Gaussin [no initial]. 1865. Sur la faculté d'expression (with discussion). *Bulletins de la Société d'Anthropologie* 6:398–417.

Gazzaniga, M. 1967 (August). The split brain in man. *Scientific American* 27:24–29.

Gazzaniga, M. 1970. *The Bisected Brain*. New York: Appleton-Century-Crofts.

Geison, G. L. 1978. *Michael Foster and the Cambridge School of Physiology*. Princeton, N.J.: Princeton Univ. Press.

Geschwind, N. 1964. The paradoxical position of Kurt Goldstein in the history of aphasia. In *Selected Papers on Language and the Brain: Boston Studies in the Philosophy of Science*. Vol. 16:62–72. Edited by R. S. Cohen and M. W. Wartofsky. Dordrecht, Holland: D. Reidel Publishing Co., 1974.

Geschwind, N. 1965. Disconnexion syndrome in animals and man. In *Selected Papers on Language and the Brain: Boston Studies in the Philosophy of Science*. Vol. 16:106–236. Edited by R. S. Cohen and M. W. Wartofsky. Dordrecht, Holland: D. Reidel Publishing Co., 1974.

Geschwind, N., and W. Levitsky. 1968. Human brain; left-right asymmetries in temporal speech region. *Science* 161:186–87.

Gilman, S. L. 1983. Photography of the insane. *History of Science* 21:432–34.

Gley, E. 1899. La Société de Biologie de 1849 à 1900. Rapport presenté à la séance du cinquantenaire de la Société. *Comptes rendus et mémoires de la Société de Biologie* 51 (11th ser., 1):1011–80.

Goldstein, J. 1982. The hysteria diagnosis and the politics of anticlericalism in late nineteenth-century France. *Journal of Modern History* 54:209–39.

Goldstein, K. 1908. Zur Lehre von der motorischen Apraxie. *Journal für Psychologie und Neurologie* 11:169, 270.

Goldstein, K. 1939. *The Organism: A Holistic Approach to Biology derived from Pathological Data in Man*, with a foreword by K. S. Lashley (American Psychology Series). New York: American Book Company.

Goltz, F. 1888. Ueber die Verrichtungen des Grosshirns. In *Some Papers on the Cerebral Cortex*, English translation by G. von Bonin, 118–58. Springfield, Ill.: Charles C. Thomas, 1960.

Goltz, F. 1892. Der Hund ohne Grosshirn. *Pflügers Archives* 51:570–614.

Gould, G. M. 1907. The origin of right-handedness. *The Boston Medical and Surgical Journal* (July–Dec.) 157:597–601.

Gould, G. M. 1908. *Righthandedness and Lefthandedness*. Philadelphia & London: J. B. Lippincott Co.

Gould, S. J. 1977. *Ontogeny and Phylogeny*. Cambridge, Mass.: The Belknap Press of Harvard Univ. Press.

Gould, S. J. 1981. *The Mismeasure of Man*. New York: W. W. Norton.

Gowers, W. R. 1855. *Lectures on the Diagnosis of Diseases of the Brain*. London: J. & A. Churchill.

Grasset, J. 1903. *L'hypnotisme et la suggestion*. Paris: Octave Doin.

Gratiolet, P., and F. Leuret. 1839–57. *Anatomie comparée du système nerveux, considérée dans ses rapports avec l'intelligence*. 2 vols. (2nd vol. by Gratiolet alone). Paris: J. B. Baillière et Fils.

Greenblatt, S. H. 1965. The major influences on the early life and work of John Hughlings Jackson. *Bulletin of the History of Medicine* 39:346–76.

Greenblatt, S. H. 1970. Hughlings Jackson's first encounter with the work of Paul Broca: The physiological and philosophical background. *Bulletin of the History of Medicine* 44:555–70.

Greenblatt, S. H. 1977. The development of Hughlings Jackson's approach to diseases of the nervous system, 1863–1866: Unilateral seizures, hemiplegia and aphasia. *Bulletin of the History of Medicine* 51:412–30.

Greenwood, F. 1892. Imagination in dreams. *The Contemporary Review* 62:165–82.

Griesinger, W. 1861. *Mental Pathology and Therapeutics*, translated by G. L. Robertson and J. Rutherford, a facsimile of the English edition of 1867. New York & London: Hafner Publishing Co., 1965.

Gruzelier, J. 1985. Nineteenth-century views on madness and hypnosis: A 1985 perspective. *The Behavioral and Brain Sciences* 8(4):638–39.

Guillain, G. 1959. *J.-M. Charcot: His Life—His Work*, translated and edited by P. Bailey. London: Putman Medical Pub. Co.

Hall, G. S. 1881. Recent researches on hypnotism. *Mind* 6:98–104.

Hall, G. S., and E. M. Hartwell. 1884. Research and discussion, bilateral asymmetry of function. *Mind* 9:93–109.

Hall, T. S. 1969. *Ideas of life and matter*, Vol. 2: *From the Enlightenment to the End of the Nineteenth Century*. Chicago and London: Univ. of Chicago Press.

Hamilton, D. J. 1884–5. On the corpus callosum in the human brain. *Journal of Anatomy and Physiology* 19:385–414.

Hammond, M. 1980. Anthropology as a weapon of social combat in late-nineteenth-century France. *Journal of the History of the Behavioral Sciences* 16:118–32.

Hammond, W. A. 1870. The physics and physiology of spiritualism. *North American Review* 110:233–60.

Hammond, W. A. 1885. Unilateral hallucinations. *New York Medical Journal* 42:649–52.

Harrington, A. 1985. Nineteenth century ideas on hemisphere differences and "duality of mind." *The Behavioral and Brain Sciences* 8(4):617–59.

Harrington, A. 1986. Models of the mind and the double brain: Some historical and contemporary reflections. *Cognitive Neuropsychology* 3(4):411–27.

Harrington, A. In press. Metals and magnets in medicine: Hysteria, hypnosis, and medical culture in fin-de-siècle Paris. In *Psychological Medicine* (London).

Harris, L. J. 1980. Left-handedness: Early theories, facts, and fancies. In *Neuropsychology of Left-Handedness*, edited by J. Herran. New York/London: Academic Press.

Harris, R. 1985. Murder under hypnosis in the case of Gabrielle Bompard: Psychiatry in the courtroom in belle époque Paris. In *The Anatomy of Madness: Essays in the History of Psychiatry*, edited by W. F. Bynum, R. Porter, and M. Sheperd. Vol. 2:197–241. London: Tavistock Pub.

Hart, E. 1893. *Hypnotism, Mesmerism, and the New Witchcraft*. New York: D. Appleton & Co.

von Hartmann, E. 1869. *Philosophy of the Unconscious: Speculative Results According to the Inductive Method of Physical Science*, translated by W. C. Coupland. London: Kegan, Paul, Trench, Trubner & Co. Ltd., 1893.

Hartmann, F. 1907. Beiträge zur Apraxielehre. *Monatsschrift für Psychiatrie und Neurologie* 21 (Jan.–June):97–118, 248–70.

Harvey, J. 1983. Races specified, evolution transformed: The social context of scientific debates originating in the société d'anthropologie de Paris, 1859–1902. Ph.D. dissertation, History of Science Department, Harvard University.

Haymaker, W. 1970. Friedrich Goltz (1834–1902). In *The Founders of Neurology*, edited by W. Haymaker and F. Schiller, 217–21. 2nd ed. Springfield, Ill.: Charles C. Thomas.

Head, H. 1926. *Aphasia and Kindred Disorders of Speech*. Vol. 1. Cambridge: Cambridge Univ. Press.

Head, H., and G. Holmes. 1911. Sensory disturbances from cerebral lesions. In *Studies in neurology*, by H. Head. London: Oxford Univ. Press, 1920.

Hécaen, H., ed. 1978. *La dominance cérébrale: une anthologie*. Paris: E. E. S. S. & Mouton & Co.

Hécaen, H., and J. de Ajuriaguerra. 1945. L'apraxie de l'habillage; ses rapports avec la planotopokinésie et les troubles de la somatognosie. *Encéphale* 34:113–44.

Hécaen, H., and G. Lanteri-Laura. 1977. *Evolution des connaissances et des doctrines sur les localisations cérébrales.* (Bibliothèque Neuro-Psychiatrique de Langue Française) Desclée de Brouer.

Hécaen, H., and P. Marcie. 1979. Agraphia: Writing disorders associated with unilateral cortical lesions. In *Clinical Neuropsychology*, edited by K. M. Heilman and E. Valenstein. New York/ Oxford: Oxford Univ. Press.

Heidenhain, R. 1880. *Animal Magnetism*, 4th German edition translated by I. C. Wooldridge. Preface by G. J. Romanes. London: Kegan Paul & Co.

Herrick, C. J. 1926. *Brains of Rats and Men*. Chicago: Univ. of Chicago Press.

Hertz, R. 1909. Death and the right hand. In *Right and Left: Essays on Dual Symbolic Classification*, edited by R. Needham. Chicago and London: Univ. of Chicago Press, 1973.

Hillman, R. G. 1965. A scientific study of mystery: The role of the medical and popular press in the Nancy-Salpêtrière controversy on hypnotism. *Bulletin of the History of Medicine* 39:163–83.

Hirtl, L. 1884. Ueber das Auftreten von Transfert-Erscheiningen während der Behandlung der partiellen Epilepsie. *Neurologisches Centralblatt* (Leipzig) 3:9–12.

Holland, H. 1840. On the brain as a double organ. In *Chapters on Mental Physiology*. London: Longman, Brown, Green & Longmans, 1852.

Holmes, G., and G. Horrax. 1919. Disturbances of spatial orientation and visual attention, with loss of stereoscopic vision. *Archives of Neurology and Psychiatry* 1:385–407.

Holmes, O. W. 1842. Homeopathy and its kindred delusions (two lectures delivered before the Boston Society for the Diffusion of Useful Knowledge). In *Medical Essays: 1842–82*. 2nd ed. Boston: Houghton, Mifflin, & Co., 1883.

Hood, A. 1825–26. Continuation of the singular and important case of R. W. (organ of language diseased). *Phrenological Journal and Miscellany* 3:26–36.

Huppe, K. 1977. Split brains and psychoanalysis. *The Psychoanalytic Quarterly* 40:220–24.

Huppert, M. 1869. Doppelwahrnehmung und Doppeldenken. Eine psychologische Studie. *Allgemeine Zeitschrift für Psychiatrie* 26:529–50.

Huppert, M. 1872. Ueber das Vorkommen von Doppelvorstellungen, eine formale Elementarstörung. *Archiv für Psychiatrie und Nervenkrankheiten* 3 (3rd part):66–110.

Ireland, W. 1881. On mirror-writing and its relation to left-handedness and cerebral disease. *Brain* 4:361–67.

Ireland, W. 1886. *The Blot upon the Brain: Studies in History and Psychology*. New York & London: G. P. Putnam's Sons.

Ireland, W. 1891. On the discordant action of the double brain. *British Medical Journal* 1:1167–69.

Ireland, W. 1892. Double brain. In *A Dictionary of Psychological Medicine*, edited by D. H. Tuke, 397–401. London: J. & A. Churchill.

Ischlondsky, N. D. 1955. The inhibitory process in the cerebrophysiological laboratory and in the clinic. *Journal of Nervous and Mental Diseases* 121:5–18.

Jackson, J. 1895. *Upright versus Sloping Writing*. New York: Pocket Pedagogical Library.

Jackson, J. 1903–04. The physiology of simultaneous ambidextral work. *Hospital* (London) 35:461–62.

Jackson, J. 1905. *Ambidexterity or Two-Handedness and Two-brainedness: An Argument for Natural Development and Rational Education*. London: Kegan, Paul, Trench, Trubner.

Jackson, J. 1909. Ambidexterity and recent criticism: being a reply to Sir James Crichton Browne's lecture on "dexterity and the bend sinister." *The General Practitioner* (Feb. 6 and 13). Reprint by the Ambidextral Cultural Society.

Jackson, J. 1910. *Hand-Writing and Brain-Building*. London: [publisher not stated].

Jackson, J. H. 1864a. Hemiplegia on the right side, with loss of speech [Letter to the Editor]. *The British Medical Journal* 1:572–73.
Jackson, J. H. 1864b. Clinical remarks on cases of defects of expression (by words, writing, signs, etc.) in diseases of the nervous system. (Under the care of Dr. Hughlings Jackson.) *The Lancet* 2:604–605.
Jackson, J. H. 1864c. Clinical remarks on cases of defects of sight in diseases of the nervous system. *The Medical Times and Gazette* 1:480–82.
Jackson, J. H. 1866a. Clinical remarks on emotional and intellectual language in some cases of disease of the nervous system. *The Lancet* 1:174–76.
Jackson, J. H. 1866b. Clinical remarks on cases of temporary loss of speech and of power of expression (epileptic aphemia? aphrasia? aphasia?) and on epilepsies. *The Medical Times and Gazette* 1:442–43.
Jackson, J. H. 1866c. Remarks on those cases of diseases of the nervous system, in which defect of expression is the most striking symptom. *The Medical Times and Gazette* 1:659–62.
Jackson, J. H. 1866d. Hemiplegia of the left side, with defect of speech [Letter to the Editor]. *The Medical Times and Gazette* 2:210.
Jackson, J. H. 1867. Remarks on the disorderly movements of chorea and convulsion, and on localisation. *The Medical Times and Gazette* 2:642–43, 669–70.
Jackson, J. H. 1868a. On the physiology of language. [Abstract of presentation to the British Association for Advancement of Science. Norwich, 1868.] *The Medical Times and Gazette* 2:275–76.
Jackson, J. H. 1868b. Hemispheral coordination. *The Medical Times and Gazette* 2:208–209.
Jackson, J. H. 1868c. Hemispheral coordination. *The Medical Times and Gazette* 2:358–59.
Jackson, J. H. 1868d. Notes on the physiology and pathology of the nervous system—Remarks on Broadbent's hypothesis. In *Selected Writings of John Hughlings Jackson*, edited by J. Taylor. Vol. 2:220. London: Hodder & Stoughton, 1932.
Jackson, J. H. 1869. Notes on the physiology and pathology of the nervous system—the unit of constitution of the nervous system. In *Selected Writings of John Hughlings Jackson*. Vol. 2:234–37.
Jackson, J. H. 1870. Notes on cases of disease of the nervous system: Speech-defects in healthy persons. *The British Medical Journal* 2:460.

Jackson, J. H. 1871. On voluntary and automatic movements. *The British Medical Journal* 2:641–42.

Jackson, J. H. 1872. Case of disease of the brain—left hemiplegia—mental affection. *The Medical Times and Gazette* 1:513–14.

Jackson, J. H. 1873. On the anatomical and physiological localisation of movements in the brain. *The Lancet* 1:84–85, 162–64.

Jackson, J. H. 1874a. Clinical lecture on a case of hemiplegia. *The British Medical Journal* 2:69–71, 99–101.

Jackson, J. H. 1874b. On the scientific and empirical investigation of epilepsies. *Medical Press and Circular* 2:325–27, 347–52, 389–92, 409–12, 475–78, 497–99, 519–21.

Jackson, J. H. 1874c. Remarks on systematic sensations in epilepsies. *The British Medical Journal* 1:174.

Jackson, J. H. 1874d. On the nature of the duality of the brain. In *Selected Writings of John Hughlings Jackson.* Vol. 2:129–45.

Jackson, J. H. 1876a. On the scientific and empirical investigation of epilepsies. *Medical Press and Circular* 1:63–65, 129–31, 173–76, 313–16.

Jackson, J. H. 1876b. Case of large cerebral tumour without optic neuritis and with left hemiplegia and imperception. In *Selected Writings of John Hughlings Jackson.* Vol. 2:146–52.

Jackson, J. H. 1876c. Notes on cases of diseases of the nervous system (under the care of Dr. Hughlings Jackson). *The Medical Times and Gazette* 2:700–702.

Jackson, J. H. 1878–79. On affections of speech from disease of the brain. In *Selected Writings of John Hughlings Jackson.* Vol. 2:155–70.

Jackson, J. H. 1879. Psychology and the nervous system. *Medical Press and Circular* 2:199–201, 239–41, 283–85, 409–11, 429–30.

Jackson, J. H. 1879–80. On affections of speech from disease of the brain. In *Selected Writings of John Hughlings Jackson.* Vol. 2:171–204.

Jackson, J. H. 1880. On affections of speech from disease of the brain. *Brain* 2:324–56.

Jackson, J. H. 1880–81. On right or left-sided spasm at the onset of epileptic paroxysms, and on crude sensation warnings, and elaborate mental states. *Brain* 3:192–205.

Jackson, J. H. 1882. On some implications of dissolution of the nervous system. *Medical Press and Circular* 2:411–14, 433–34.

Jackson, J. H. 1887. Remarks on evolution and dissolution of the nervous system. In *Selected Writings of John Hughlings Jackson.* Vol. 2:92–118.

Jackson, J. H. 1887–88. Remarks on evolution and dissolution of the nervous system. *The Journal of Mental Science* 33:25–48.

Jackson, J. H. 1893a. Words and other symbols in mentation. *Medical Press and Circular* 2:205–208.

Jackson, J. H. 1893b. Cerebral paroxysms (epileptic attacks) with an auditory warning; In slight seizures the special imperceptions called "Word-Deafness" (Wernicke) and "Word-Blindness" (Kussmaul); Inability to speak and spectral words (auditory and visual). *The Lancet* 2:252–53.

Jackson, J. H. 1897. On the relations of different divisions of the central nervous system to one another and to parts of the body [first Hughlings Jackson lecture, delivered before the Neurological Society, Dec. 8, 1897]. In *Selected Writings of John Hughlings Jackson*. Vol. 2:422–443.

Jackson, S. W. 1969. The history of Freud's concepts of regression. *The Journal of the American Psychoanalytic Association* 17(3):743–84.

Jacyna, L. S. 1981. The physiology of mind, the unity of nature, and the moral order in Victorian thought. *British Journal History of Science* 14:109–32.

Jacyna, L. S. 1982. Somatic theories of mind and the interests of medicine in Britain, 1850–79. *Medical History* 26:233–58.

James, W. 1889. Notes on automatic writing. *Proceedings of the American Society for Psychical Research* 1:548–64.

James, W. 1890. *Principles of Psychology*, 2 vols. New York: Dover Publications, 1950.

James, W. 1973. *William James on Psychical Research*, edited by G. Murphy and R. O. Ballou. Clifton, N.J.: A. M. Kelley.

Janet, P. 1889. *L'automatisme psychologique*. 7th ed. Paris: Librairie Felix Alcan, 1913.

Janet, P. 1925. *Psychological Healing: A Historical and Clinical Study*, translated by Eden and Cedar Paul. 2 Vols. London: George Allen & Unwin, Ltd.

Janet, P., and F. Raymond. 1899. Note sur l'hystérie droite et sur l'hystérie gauche. *Revue neurologique* 7:851–55.

Jaynes, J. 1970. The problem of animate motion in the seventeenth century. *Journal of the History of Ideas* 3:219–34.

Jaynes, J. 1976. *The Origin of Consciousness in the Breakdown of the Bicameral Mind*. Boston: Houghton Mifflin Co.

Jeannerod, M. 1985. *The Brain Machine: The development of neurophysiological thought*, translated by D. Urion. Cambridge, Mass.: Harvard Univ. Press.

Jensen [no first initial]. 1868. Ueber Doppelwahrnehmungen in der gesunden, wie in der kranken Psyche. *Allgemeine Zeitschrift für Psychiatrie* 25 (Supplement):48–64.

Jones, E. 1908. Le côté affecté par l'hémiplégie hystérique. *Revue neurologique* 16:193–96.

Jones, E. 1953. *The Life and Work of Sigmund Freud.* Vol. 1: *The Formative Years and the Great Discoveries, 1856–1900.* New York: Basic Books. London: Hogarth Press.

Joynt, R. J., and A. L. Benton. 1964. The memoir of Marc Dax on aphasia. *Neurology* 14:851–54.

Katan, M. 1969. The link between Freud's work on aphasia, fetishism, and constructions in analysis. *International Journal of Psychoanalysis* 50:547–53.

Kenealy, A. 1920. *Feminism and Sex-Extinction.* London: T. Fisher Unwin, Ltd.

Kenealy, A. 1934. *The Human Gyroscope: A Consideration of the Gyroscopic Rotation of Earth as Mechanism of the Evolution of Terrestrial Living Forms; Explaining the Phenomenon of Sex: Its Origin and Development and Its Significance in the Evolutionary Process.* London: John Bale, Sons & Danielsson, Ltd.

Kiernan, J. G. 1883. Different unilateral auditory hallucinations on opposite sides [abstract of report by Valentin Magnan]. *Journal of Nervous and Mental Diseases* 10:697.

Kiernan, J. G. 1896. Dual action of the cerebral hemispheres. *Medicine* (Detroit) 2:31–33.

Kleist, K. 1923. Kriegverletzungen des Gehirns in ihrer Bedeutung für die Hirnlokalisation und Hirnpathologie, cited in A. Benton, Visuoperceptive, visuospatial, and visuoconstructive disorders. In *Clinical neuropsychology*, edited by K. M. Heilman and E. Valenstein. New York/Oxford: Oxford Univ. Press.

Klippel, M. 1898. La non-équivalence des deux hémisphères cérébraux. *Revue de Psychiatrie*, 52–57.

Knorr-Cetina, K. D., and M. Mulkay, eds. 1983. *Science Observed: Perspectives on the Social Study of Science.* London: Sage Publications.

Knox, D. N. 1875. Description of a case of defective corpus callosum. *The Glasgow Medical Journal*, new ser., 7:227–37.

Kurella, H. 1911. *Césare Lombroso: A Modern Man of Science*, translated from German by M. Eden Paul. London: Rebman Co.

Lacapra, D. 1984. Is everyone a *mentalité* case? Transference and the "culture" concept. *History and Theory: Studies in the Philosophy of History* 23(3):296–311.

Lange, F. A. 1881. *History of Materialism and Criticism of Its Present Importance*, translated by E. C. Thomas. Vol. 3. London: Trubner & Co.

Lashley, K. S. 1926. The relation between cerebral mass, learning and retention. *Journal of Comparative Neurology* 41:1–58.

Lashley, K. S. 1929. *Brain Mechanisms and Intelligence*. Chicago: Univ. of Chicago Press.

Lattes, L. 1907. Asimmetrie cerebrali nei normale e nei delinquenti. *Archivio di Psichiatria, Neuropatologia, Antropologia Criminale e Medicina Legale* (Torino) 28:1–22.

Laycock, T. 1845. On the reflex function of the brain. *British and Foreign Medical Review* 19:298–311.

Laycock, T. 1876. Reflex, automatic, and unconscious cerebration: A history and a criticism. *The Journal of Mental Science* 21:477–98; 22:1–17.

Leary, D. E. 1985. Scientific amnesia. *The Behavioral and Brain Sciences* 8(4):641.

Lélut, L. F. 1864–65. Rapport sur le mémoire de M. Dax, relatif aux fonctions de l'hémisphère gauche du cerveau. *Bulletin de l'Académie Impériale de Médecine*, 1st ser., 30:173–75.

Lesky, E. 1970. Structure and function in Gall. *Bulletin of the History of Medicine* 44:297–314.

Lewes, G. H. 1877. *The Physical Basis of Mind*. London: Trubner & Co.

Lichtheim, L. 1886. On aphasia. *Brain* 7:433–84.

Liepmann, H. 1900. Das Krankheitsbild der Apraxia (motorischen Asymbolie) auf Grund eines Falles von einseitiger Apraxie. *Monatsschrift für Psychiatrie und Neurologie* 8:15–44, 102–32, 182–97.

Liepmann, H. 1905. Die linke Hemisphäre und das Handeln. [Originally published in *Münchener med. Wochenschrift*, Nos. 48, 49.] Reprinted in *Drei Aufsätze aus dem Apraxiegebiet* (neu durchgesehen und mit Zusatzen versehen), 17–50. Berlin: Von Karger, 1908.

Liepmann, H. 1906. Der weitere Kranksheitsverlauf bei dem einseitig Apraktischen und der Gehirnbefund auf Grund von Serienschnitten. *Monatsschrift für Psychiatrie und Neurologie* 19 (Jan.–June):217–43.

Liepmann, H. 1907. Ueber die Funktion des Balkens beim Handeln und die Beziehungen von Aphasie und Apraxie zur Intelligenz. [Orig. pub. in *Med. Klinik*, Nos. 25, 26.] Reprinted in *Drei Auf-*

sätze aus dem Apraxiegebiet (neu durchgesehen und mit Zusatzen versehen), 51–80. Berlin: Von Karger, 1908.

Liepmann, H., and O. Maas. 1907. Fall von linksseitiger Agraphie und Apraxie bei rechtsseitiger Lähmung. *Journal für Psychologie und Neurologie* 10:214–27.

Lissauer, H. 1889. Ein Fall von Seelenblindheit nebst einem Beiträge zur Theorie derselben. *Archiv für Psychiatrie und Nervenkrankheiten* 21:222–70.

Lombard, J. S. 1878. On the effect of intellectual and emotional activity on the temperature of the head. *Proceedings of the Royal Society* 27:462–65.

Lombroso, C. 1884. Sul mancinismo e destrismo tattile nei sani, nei pazzi, nei ciechi e nei sordomuti. *Archivio di Psichiatria, Scienze Penali, ed Antropologia Criminale* 5:187–97.

Lombroso, C. 1903. Left-handedness and left-sidedness. *North American Review* 177:440–44.

Lombroso, C. 1908. Psychology and spiritism [review of Enrico Morselli's *Psicologia e "Spiritismo"*] *The Annals of Psychical Science* 7:376–80.

Lombroso, C. 1909. *After Death—What?*, translated by W. S. Kennedy. Boston: Small, Maynard & Co.

Lusanna, F. 1854. Della duplicità indipendente degli emisferi cerebrali. *Gazetta Medica Italiana* (Lombardia) [Appendice Psychiatrica], 3rd ser., 5:421–22.

Luys, J. B. 1879. Études sur le dédoublement des opérations cérébrales et sur le rôle isolé de chaque hémisphère dans les phénomènes de la pathologie mentale. *Bulletin de l'académie de médecine*, 2nd ser., 8:516–34, 547–65. (See 1888, below.)

Luys, J. B. 1881a. Recherches nouvelles sur les hémiplégies émotives. *Encéphale* 1:378–98.

Luys, J. B. 1881b. Contribution à l'étude d'une statistique sur les poids des hémisphères cérébraux à l'état normal et à l'état pathologique. *Encéphale* 1:644–46.

Luys, J. B. 1881c. *The Brain and its functions.* London: Kegan Paul.

Luys, J. B. 1888 [reprint of 1879 article]. *Encéphale* 8:404–24, 516–37.

Luys, J. B. 1889. Le dédoublement cérébral du pianiste. *Revue de l'hypnotisme et de la psychologie physiologique* 3:282–84.

Luys, J. B. 1890a. Faits tendant à démontrer que le lobe droit joue un rôle dans l'expression du langage articulé. *Revue d'hypnologie théorique et pratique* 1:134–46.

Luys, J. B. 1890b. *Hôpital de la Charité. Leçons cliniques sur les principaux phénomènes de l'hypnotisme dans leurs rapports avec la pathologie mentale*. Paris: Georges Carré.

Luys, J. B. 1891. De la sollicitation isolée du lobe gauche et du lobe droit dans l'état hypnotique, au point de vue des manifestations de la parole. *Comptes rendus de la Société de Biologie* 3:201–205.

[Obituary] Luys (Dr. J. B.). 1897. *The British Medical Journal* 2:619.

Lyon, S. B. 1895. Dual action of the brain. *New York Medical Journal* 62:107–10.

Maas, O. 1907. Ein Fall von linksseitiger Apraxie und Agraphie. *Neurologisches Centralblatt* 26:789–92.

Magnan, V. 1883. Des hallucinations bilatérales de caractère différent suivant le côté affecté. *Archives de neurologie* 6:336–55.

Mahoney, P. 1979. Friendship and its discontents. *Contemporary Psychoanalysis* 15(1):55–109.

de Manacéïne, M. 1894. Suppléance d'un hémisphère cérébral par l'autre. *Archives italiennes de biologie* 21:326–32.

de Manacéïne, M. 1897. *Sleep: Its Physiology, Pathology, Hygiene and Psychology*. London: Walter Scott, Ltd.

Mandelbaum, M. 1955. Societal facts. *The British Journal of Sociology* 6:305–17.

Mandelbaum, M. 1971. *History, Man, and Reason: A Study in Nineteenth-Century Thought*. Baltimore & London: Johns Hopkins Univ. Press.

Manouvrier, L. 1883. Étude craniométrique sur la plagiocéphalie. *Bulletins de la société d'anthropologie de Paris*, 3rd ser., 6:526–42; (discussion) 542–53.

Marie, P. 1906a. Revision de la question de l'aphasie: la troisième circonvolution frontale gauche ne joue aucun rôle special dans la fonction du langage. *Semaine médicale*, 241–47.

Marie, P. 1906b. Revision de la question de l'aphasie: que faut-il penser des aphasies sous-corticales (aphasies pures)? *Semaine médicale*, 493–500.

Marie, P. 1906c. Revision de la question de l'aphasie: l'aphasie de 1861 à 1866; essai de critique historique sur la genèse de la doctrine de Broca. *Semaine médicale*, 565–71.

Marin, P. S., and G. J. Tucker. 1981. Psychopathology and hemisphere dysfunction: A review. *Journal of Nervous and Mental Diseases* 169:546–57.

Marro, A., and C. Lombroso. 1883. Ambidestrismo nei pazzi e nei criminali. *Archivio di psichiatria, antropologia criminale e scienze penali* 4:229–30.

Marshall, J. C. 1984. Multiple perspectives on modularity. *Cognition* 17:209–42.

Marx, O. M. 1966. Aphasia studies and language theory in the 19th century. *Bulletin History of Medicine* 40:328–49.

Maudsley, H. 1868. Concerning aphasia. *The Lancet* 2:690–92.

Maudsley, H. 1870. *Body and Mind: An Inquiry into Their Connection and Mutual Influence, Specially in Reference to Mental Disorders* (Gulstonian lectures for 1870). London: Macmillan & Co.

Maudsley, H. 1889. The double brain. *Mind* 14:161–87.

Maudsley, H. 1895. *The Pathology of Mind: A Study of its Distempers, Deformities, and Disorders.* London & New York: Macmillan & Co.

McDougall, W. 1911. *Body and Mind: A History and Defense of Animism.* New York: The Macmillan Company.

McDougall, W. 1926. *An outline of abnormal psychology.* 5th ed. London: Metheun & Co., 1946.

Mercier, C. A. 1901. *Psychology, Normal and Morbid.* London: George Allen & Unwin.

la Mettrie, J. O. 1747. *Man a Machine (L'homme machine).* French-English ed. La Salle, Ill.: Open Court, 1912.

Meyer, A. 1904a. A few trends in modern psychiatry. *The Psychological Bulletin* 1:217–40.

Meyer, A. 1904b. [Review of] "Halbseitiges Delirium" by Prof. Bleuler, in "Psychological Literature." *The Psychological Bulletin* 1:285–86.

Miloche, P. 1982. *Un méconnu de l'hystérie. Victor Dumont Pallier (1826–1899).* Doctoral dissertation, Faculté de Médecine, Université de Caen.

Mitchell, S. W. 1888. Mary Reynolds: A case of double consciousness [with discussion of the double brain hypothesis]. *Transactions of the College of Physicians of Philadelphia*, 3rd ser., 10:366–89.

Miyoshi, M. 1969. *The Divided Self: A Perspective on the Literature of the Victorians.* New York Univ./Univ. of London Press.

von Monakow, C. 1911. Lokalisation der Hirnfunktionen. English translation in *Some Papers on the Cerebral Cortex*, by G. von Bonin, 231–50. Springfield, Ill.: Charles C. Thomas, 1960.

Monard, J. 1880. La métallothérapie en 1820. *Lyon médical* 34:411–20, 449–54, 485–89.

de Montyel, M. 1884. Contribution à l'étude de l'inégalité de poids des hémisphères cérébraux dans la folie nevrosique et la démence paralytique. *Encéphale* 4:574–89.

de Montyel, M. 1887. Contribution à l'étude du poids des hémisphères cérébraux chez les aliénés. *Annales Médico-Psychologiques*, 1st ser., 6:364–82.

Morel, P. 1983. Dictionnaire biographique. In *Nouvelle histoire de la psychiatrie*, edited by J. Postel & C. Quètel, 564–735. Toulouse: Editions Private (Société Française d'Histoire de la Psychiatrie).

Moutier, F. 1908. *L'aphasie de Broca*. Paris: G. Steinheil.

Moxon, W. 1866. On the connexion between loss of speech and paralysis of the right side. *The British and Foreign Medico-Chirurgical Review* 37:481–89.

Mullan, S., and W. Penfield. 1959. Illusions of comparative interpretation and emotion. *Archives of Neurology and Psychiatry* 81:269–84.

Murphy, G. 1949. *Historical Introduction to Modern Psychology*. New York: Harcourt, Brace & Co.

Murphy, G. 1968. Personal impressions of Kurt Goldstein. In *The Reach of Mind: Essays in Memory of Kurt Goldstein*, edited by M. L. Simmel, 31–34. New York: Springer Publishing Co.

Murray, D. J. 1985. What textbooks between 1887 and 1911 said about hemisphere differences. *The Behavioral and Brain Sciences* 8(4):544–645.

Myers, A. T. 1885–86. Psychological retrospect: The life-history of a case of double or multiple personality. *The Journal of Mental Science* 31:596–605.

Myers, A. T. 1889–90. Note on hypnotism and hysteria. *Proceedings of the Society for Psychical Research* 15–17 (supplement):199.

Myers, F.W.H. 1885. Automatic writing—II. *Proceedings of the Society for Psychical Research* 3:23–63.

Myers, F.W.H. 1886. Multiplex personality. *Nineteenth Century* 20:648–66.

Myers, F.W.H. 1889–90. Professor Pierre Janet's "Automatisme Psychologique." *Proceedings of the Society for Psychical Research* 15–17 (supplement):186–98.

Myers, F.W.H. 1892. The Subliminal Consciousness. Chapter 3. The mechanism of genius. *Proceedings of the Society for Psychical Research* 21–23.

Myers, F.W.H. 1898–99. Review of "Some Peculiarities of the Secondary Personality," by G.T.W. Patrick. *Proceedings of the Society for Psychical Research* 34–35 (supplement):382–86.

Myers, F.W.H. 1915. *Human Personality and Its Survival of Physical Death*. 2 Vols. London: Longman, Green.

Myers, R., and R. Sperry. 1953. Interocular transfer of a visual form discrimination habit in cats after section of the optic chiasma and corpus callosum. *Anatomical Record* 115:351–52.

Nagel, T. 1979. Brain-bisection and the unity of consciousness. In *Mortal Questions*. Cambridge: Cambridge Univ. Press.

Needham, R., ed. 1973. *Right and Left: Essays on Dual Symbolic Classification*. Chicago and London: Univ. of Chicago Press.

Needham, R. 1979. *Symbolic Classification*. Santa Monica, Calif.: Goodyear Perspectives in Anthropology Series.

[Review of] *New View of Insanity. The Duality of Mind . . .* (1848) *Journal of Psychological Medicine and Mental Pathology* I:218–29.

Nielsen, J. M. 1938. Gerstmann syndrome: Finger agnosia, agraphia, confusion of right and left, and acalculia; comparison of this syndrome with disturbance of body scheme resulting from lesions of the right side of the brain. *Archives of Neurology and Psychiatry*, 535–60.

Notes of a case of dual brain action [abstract of Bruce report (1895)] 1896. *Journal of Nervous and Mental Diseases* 23:217.

Nye, R. 1984. *Crime, Madness, and Politics in Modern France: The Medical Concept of National Decline*. Princeton, N.J.: Princeton Univ. Press.

Ogle, W. 1871. On dextral pre-eminence. *Medico-Chirurgical Transactions* 54 (2nd ser., 36):279–301.

Olmsted, J.M.D. 1946. *Charles-Edouard Brown-Séquard: A Nineteenth-Century Neurologist and Endocrinologist*. Baltimore: Johns Hopkins Univ. Press.

Oppenheimer, J. M. 1976. Studies of brain asymmetry: Historical perspective. *Annals of the New York Academy Sciences* 299:4–17.

Ornstein, R. 1970. *The Psychology of Consciousness*. San Francisco: W. H. Freeman & Co., 1972.

Ott, I. 1883. Functional independence of each hemisphere [abstract of report by Dumontpallier]. *Journal of Nervous and Mental Diseases* 10:323–24.

Owen, A.R.G. 1971. *Hysteria, Hypnosis, and Healing: The Work of J.-M. Charcot*. London: Dennis Dobson.

Parant, V. 1875. *De la possibilité des suppléances cérébrales*. Doctoral dissertation. Paris: A Parent.

Parrot, J. 1879. Sur le développement du cerveau chez les enfants du premier âge. *Archives de Physiologie Normale et Pathologique*, 2nd ser., 6:505–21.

Paterson, A., and O. L. Zangwill. 1944. Disorders of visual space perception associated with lesions of the right cerebral hemisphere. *Brain* 67:331–58.

Pauly, P. J. 1983. The political structure of the brain: cerebral localization in Bismarckian Germany. *International Journal of Neuroscience* 21:145–50.

Penfield, W. 1975. The mind and the brain. In *The Neurosciences: Paths of Discovery*, edited by F. G. Worden, J. P. Swazey, and G. Adelman. Cambridge, Mass.: MIT Press.

Penfield, W., and L. Roberts. 1959. *Speech and Brain Mechanisms.* Princeton, N.J.: Princeton Univ. Press.

Perkins, B. D. 1798. *The influence of metallic tractors on the human body, in removing various painful inflammatory diseases, such as Rheumatism, Pleurisy, some Gouty, Affections &c, &c. lately discovered by Dr. Perkins, of North America; and demonstrated in a series of experiments and observations, by Professors Meigs, Woodward, Rogers, &c, &c. by which the importance of the discovery is fully ascertained, and a new field of enquiry opened in the modern science of galvinism or animal electricity.* J. Johnson/Ogilvy & Son.

Petit, L.-H. 1880 (title page date 1879). *Sur la métallothérapie, ses origines et les procédés thérapeutiques qui en derivent.* Paris: Octave Doin.

Pick, A. 1903. On "dreamy mental states" as a permanent condition in epileptics. *Brain* 26:242–51.

Pierret, [no initial]. 1895. Le dualisme cérébrale. *La province médicale* (Lyons) 9:268–73.

Plumer, W. S. 1860. *Mary Reynolds: A Case of Double Consciousness.* [Reprinted from *Harper's Magazine* for May 1860.] Chicago: Religio-Philosophical Publishing House.

Poincelot, A. 1886. L'hypnotisme et le magnétisme (Nouvelle conference faite à la salle des Capucines). *Revue de l'hypnotisme et de la psychologie physiologique* 1:184–85.

Popper, K., and J. Eccles. 1977. *The Self and Its Brain.* London/New York: Springer Verlag.

The posterior lobes of the brain and the seat of intellectuality. 1883. *The Alienist and Neurologist* 4:322.

Prince, M. 1891. Hughlings-Jackson on the connection between the mind and the brain. *Brain* 14:250–69.

Prince, M. 1906. *The Dissociation of a Personality.* New York/London: Longman's, Green & Co.

Pruner-Bey, [no initial]. 1865. L'homme et l'animal. *Bulletins de la Société d'Anthropologie* 6:522–62.

Puccetti, R. 1981. The case for mental duality: Evidence from split-brain data and other considerations. *The Behavioral and Brain Sciences* 4:93–123.

Ransom, W. B. 1895. On tumours of the corpus callosum, with an account of a case. *Brain* 18:531–50.

Régis, E. 1881. Des hallucinations unilatérales. *Encéphale* 1:43–74.

Report of the general meeting. 1886. *Journal of the Society for Psychical Research* 2:218–29.

Ribot, T. 1877. Philosophy in France. *Mind* 2:366–86.

Ribot, T. 1891. *The Diseases of Personality*, authorized translation. 4th ed. Chicago: The Open Court Publishing Co., 1910.

Richer, P. 1881. *Études cliniques sur l'hystéro-épilepsie ou grande hystérie*. Paris: Delahaye et Lecrosnier.

Riese, W. 1947. The early history of aphasia. *Bulletin of the History Medicine* 21:322–34.

Riese, W. 1949. An outline of a history of ideas in neurology. *Bulletin of the History of Medicine* 23:111–36.

Riese, W. 1958. Freudian concepts of brain function and brain disease. *The Journal of Nervous and Mental Diseases* 127:287–307.

Riese, W. 1959. *A History of Neurology*. MD Monographs on Medical History. New York: MD Publications, Inc.

Riese, W. 1968. Kurt Goldstein—The man and his work. In *The Reach of Mind: Essays in Memory of Kurt Goldstein*, edited by M. L. Simmel, 17–29. New York: Springer Publishing Co.

Riese, W., and W. Gooddy. 1955. An original clinical record of Hughlings Jackson with an interpretation. *Bulletin of the History of Medicine* 29:230–38.

Riese, W., and E. C. Hoff. 1950. A History of the doctrine of cerebral localization: Sources, anticipations, and basic reasoning. *Journal of the History of Medicine* 5:50–71.

Robertson, A. 1867. The pathology of aphasia. *The Journal of Mental Science* 13:520–21.

Robertson, A. 1875. On the unilateral phenomena of mental and nervous disorders. *The Glasgow Medical Journal*, new ser., 7:496–516.

Robertson, G. M. 1892. Hypnotism at Paris and Nancy. Notes of a visit. *Journal of Mental Science* 38:494–531.

Robinson, D. 1976. *An Intellectual History of Psychology*. London/New York: Macmillan Publishing Co.

de Rochas, A. 1904–05. *Les Frontières de la Science* (2nd ser.). 2 Vols. Paris: Librairie des Sciences Psychologiques.

Rolleston, J. D. 1930–31. Jean-Baptiste Bouillaud (1796–1881). A pioneer in cardiology and neurology. *Proceedings of the Royal Society of Medicine* 24:1253–62.

Roques, F. 1869. Sur un cas d'asymétrie de l'encéphale, de la moelle, du sternum et des ovaires. *Bulletins de la société d'anthropologie de Paris*, 2nd ser., 4:727–32.

Rosenthal, M. 1882. Untersuchungen und Beobachtungen ueber Hysterie und Transfert. *Archives für Psychiatrie und Nervenkrankheit* 12 (part 1):201–31.

Rosse, I. C. 1892. Triple personality. *Journal of Nervous and Mental Diseases* 19:186–91.

Rush, B. 1981. *Lectures on the Mind*, edited by E. T. Carlson, J. L. Wollock, and P. S. Noel. Philadelphia: American Philosophical Society.

Ryalls, J. 1984. Where does the term "aphasia" come from. *Brain and Language* 21:358–63.

Ryan, M. 1844. The non-duality of the brain. *The Lancet* 1:154.

Sabatier, C. 1908. L'homme est-il simple, double, ou multiple? *Revue générale des sciences pures et appliquées* 19 (May):358–67.

Sabatier, C. 1918. *Le duplicisme humain.* 2nd ed. preface by M. le Dr. Abelous. Paris: Librairie Felix Alcan.

Sagan, C. 1977. *The Dragons of Eden.* New York: Random House.

Sarbin, T. R., and W. C. Coe. 1979. Hypnosis and psychopathology: Replacing old myths with fresh metaphors. *Journal of Abnormal Psychology* 88:506–26.

Sarda, J. H. 1908. Sommeil alternatif des deux hémisphères cérébraux. *Congrès français de médecine* (10th session, Geneva) 2:186–92.

Savage, G. H. 1892. Handwriting of the insane. In *A Dictionary of Psychological Medicine*, edited by D. H. Tuke. Vol. 1:568–74. London: J. & A. Churchill.

Schiff, E. 1879. Prof. Schiff ueber Metalloskopie und Metallotherapie. *Wiener Medizinische Press* 20:1379–82.

Schiller, F. 1979. *Paul Broca: Founder of French Anthropology, Explorer of the Brain.* Berkeley & Los Angeles: Univ. of California Press.

Schur, M. 1972. *Freud; Living and Dying.* International Psycho-Analysis Library. London: Hogarth Press.

Scull, A., ed. 1981. *Madhouses, Mad-Doctors, and Madmen: The Social History of Psychiatry in the Victorian Era.* London: The Athlone Press.

Shapin, S. 1979a. Homo phrenologicus: Anthropological perspectives on an historical problem. In *Natural Order: Historical Studies of Scientific Culture.* Beverly Hills/London: Sage Publications.

Shapin, S. 1979b. The politics of observation: Cerebral anatomy and social interests in the Edinburgh phrenology disputes. In *On the Margins of Science*, edited by R. Wallis. Univ. of Keele.

Singer, J. 1976. *Androgyny*. Garden City, N.Y.: Anchor Press.

Smith, C.V.M. 1982. Evolution and the problem of mind. Part II. John Hughlings Jackson. *Journal of the History of Biology* 15 (2):241–62.

Smith, E. N. 1900. Ambidexterity: A plea for its general adoption. *British Medical Journal* 2:579–80.

Smith, R. 1973. The background of physiological psychology in natural philosophy. *History of Science* 11:75–124.

Sollier, P. 1897. *Genèse et nature de l'hystérie*. Paris: Ancienne Librairie Germer Baillière.

Sollier, P. 1900. De la localisation cérébrale des troubles hystériques. *Revue neurologique* 8:102–107, 364–71.

Souques, A. 1928. Quelques cases d'anarthie de Pierre Marie: Aperçu historique sur la localisation du langage. *Revue neurologique* 2:319–368.

Soury, J. 1896. Le lobe occipital et la vision mentale. I. Les hémianopsies. *Revue philosophique de la France et de l'étranger* 41:145–68.

Soury, J. 1899. *Le système nerveux central. Structure et functions. Histoire critique des théories et des doctrines*. Paris: George Carré et C. Naud.

Spencer, H. 1862. *First Principles*. 6th ed. London: Williams & Norgate, 1928.

Sperry, R. 1962. Some general aspects of interhemispheric integration. In *Interhemispheric Relations and Cerebral Dominance*, edited by V. B. Mountcastle. Baltimore: Johns Hopkins Univ. Press.

Sperry, R. 1968. Hemisphere disconnection and unity in conscious awareness. *American Psychologist* 23:723–33.

Sperry, R. 1975. In search of psyche. In *The Neurosciences: Paths of Discovery*, edited by F. G. Worden, J. P. Swazey, and G. Adelman. Cambridge, Mass.: MIT Press.

Springer, S. P., and G. Deutsch. 1985. *Left brain, right brain*. rev. ed. New York: W. H. Freeman & Co.

Spurzheim, J. G. 1833. *Phrenology or the Doctrine of the Mental Phenomena*. rev. 2nd American ed. Reprinted, Philadelphia/London: J. B. Lippincott Co., 1908.

Stengel, E. 1953. Introduction to *On Aphasia: A Critical Study*, by Sigmund Freud, translated by E. Stengel. London: Imago Publishing Co.

Stengel, E. 1954. A re-evaluation of Freud's book "On Aphasia." Its significance for psycho-analysis. *International Journal of Psycho-analysis* 35:85–89.

Stengel, E. 1963. Hughlings Jackson's influence on psychiatry. *British Journal of Psychiatry* 109:348–55.

Stern, D. B. 1977. Handedness and the lateral distribution of conversion reactions. *Journal of Nervous and Mental Diseases* 164–65:122–28.

Stewart, W. A. 1969. *Psychoanalysis: The First Ten Years, 1888–1898*. London: Allen & Unwin.

Sulloway, F. J. 1979. *Freud: Biologist of the Mind*. Bungay, Suffolk: Fontana Paperbacks, 1980.

Tadd, J. L. 1899. *New Methods in Education: Art, real manual training, nature study, explaining processes whereby hand, eye and mind are educated by means that conserve vitality and develop a union of thought and action*. New York: Orange Judd Co. [entered 1898 in Library of Congress].

Taylor, E. 1984. *William James on Exceptional Mental States: The 1896 Lowell Lectures*. Amherst: Univ. of Massachusetts Press.

Taylor, J. 1925. Biographical memoir. In *Neurological Fragments of J. Hughlings Jackson*. London: Oxford Univ. Press.

Thornton, E. M. 1976. *Hypnotism, Hysteria, and Epilepsy: An Historical Synthesis*. London: William Heinemann Medical Books, Ltd.

Thurman, J. 1866. On the weight of the brain and the circumstances affecting it. *Journal of Mental Science* 12:1–43.

Tizard, B. 1959. Theories of brain localization from Flourens to Lashley. *Medical History* 3:132–44.

de la Tourette, G. 1900. La localisation cérébrale des troubles hystériques. *Revue Neurologique* 8:225–27.

Tuke, D. H. 1878–79. Metalloscopy and expectant attention. *The Journal of Mental Science* 24:598–609.

Tuke, D. H. 1892. Physiognomy of the insane. In *A Dictionary of Psychological Medicine*, edited by D. H. Tuke. Vol. 2:947–50. London: J. & A. Churchill.

Turner, F. M. 1974. *Between Science and Religion: The Reaction to Scientific Naturalism in Late Victorian England*. New Haven and London: Yale Univ. Press.

Turner, J. 1892. Asymmetrical conditions met with in the faces of the insane; with some remarks on the dissolution of expression. *Journal of Mental Science* 38:18–29, 199–211.

Tyms, R. 1949. *Doubles in Literary Psychology*. Cambridge: Bowes and Bowes.

Urquhart, A. R. 1879–80a. On the habitual tendency of individuals to direct themselves to the right or left [abstract of communication by M. Delaunay, May 18, 1879]. *Brain* 2:291–92.

Urquhart, A. R. 1879–80b. Duality of cerebral operations [abstract of lecture by M. Luys, May 15, 1879]. *Brain* 2:294–95.

Valenstein, E. S. 1973. *Brain Control: A Critical Examination of Brain Stimulation and Psychosurgery*. New York/London: John Wiley & Sons.

Vallin, E. 1899. Rapport général sur les Prix décernés en 1899 [with a report of Dumontpallier's death]. *Bulletin de l'Académie de Médecine*, 3rd ser., 42:600–601.

Veith, I. 1965. *Hysteria: The History of a Disease*. Chicago and London: Univ. of Chicago Press.

Verity, R. 1870. *Subject and Object; As Connected with Our Double Brain, and a New Theory of Causation*. London: Longmans, Green, Reader, and Dyer.

Vesey, G.N.A., ed. 1964. *Body and Mind: Readings in Philosophy*. London: George Allen and Unwin.

Vigouroux, R. 1878. Sur la théorie physique de la métalloscopie. *Lancette français: Gazette des Hôpitaux* 51:780.

van Vleuten, C. F. 1907. Linksseitige motorische Apraxie: Ein Beitrag zur Physiologie des Balkens. *Allgemeine Zeitschrift für Psychiatrie und Psychisch-Gerichtliche Medizin* 64:203–39.

Watson, H. 1836. What is the use of the double brain? *The Phrenological Journal and Miscellany* (March) 9:608–11.

Watson, J. B. 1913. Psychology as a behaviorist views it. *Psychological Review* 20:158–77.

Watson, R. I. 1975. Prescriptive theory and the social sciences. In *Determinants and Controls of Scientific Development*, edited by K. D. Knorr et al., 11–35. Dordrecht, Holland: Reidel.

Weil, A. T. 1977. The marriage of the sun and moon. In *Alternative States of Consciousness*, edited by N. E. Zinberg, 37–52. New York: The Free Press.

Weinstein, E., and R. Kahn. 1955. *Denial of Illness: Symbolic and Physiological Aspects*. Springfield, Ill.: Charles C Thomas.

Wernicke, C. 1874. *Der Aphasische Symptomencomplex. Eine Psychologische Studie auf Anatomischer Basis*. Breslau: M. Cohn und Weigart. Reprinted in *Wernicke's Works on Aphasia: A Source Book and Review*, translated by G. E. Eggert, 91–144. The Hague, Netherlands: Mouton Publishers, 1977.

Westminster Medical Society. 1844. [Discussion on Wigan (1844a)]. *The Lancet* 1:85.
Whitehead, A. N. 1926. *Science and the Modern World*. Cambridge: Cambridge Univ. Press.
Whiting, A. 1905. Lecture on the diagnosis of functional nervous disease. *Medical Press and Circular* 2:505–508.
Wigan, A. L. 1844a. The duality of the mind, proved by the structure, functions, and diseases of the brain. *The Lancet* 1:39–41.
Wigan, A. L. 1844b. *A New View of Insanity: The Duality of the Mind*. London: Longman, Brown, Green & Longmans.
Wilks, S. 1872. Cases of diseases of the nervous system: A case of aphasia, with remarks on the faculty of language and the duality of the brain. *Guy's Hospital Report* 17:145–71.
Winslow, F. 1849. The unpublished MSS of the late Alfred [*sic*] Wigan, M.D., author of the "Duality of the Mind" &c. *The Journal of Psychological Medicine and Mental Pathology* 2:497–512.
Witelson, S. F. 1976. Sex and the single hemisphere: specialization of the right hemisphere for spatial processing. *Science* 193:425–27.
Wolff, W. 1943. *The Expression of Personality*. New York and London: Harper & Brothers.
Wolff, W. 1952. *The Threshold of the Abnormal: A Basic Survey of Psychopathology*. London: Medical Publications.
Wolpert, I. 1924. Die Simultanagnosie: Störung der Gesamtauffassung. *Zeitschrift für die gesamte Neurologie und Psychiatrie* 93:397–415.
Word, R. C. 1888. Duality of the brain—A theory of mind-reading and slate-writing. *The Southern Medical Record: A Monthly Journal of Practical Medicine* 18:81–89.
Wynter, A. 1875. *The Borderlands of Insanity*. New York: G. P. Putnam's Sons.
Young, R. 1967. Animal soul. In *The Encyclopedia of Philosophy*. Vol. 1:122–27. New York and London: Macmillan Publishing Co. and The Free Press.
Young, R. 1970. *Mind, Brain, and Adaptation in the Nineteenth Century*. Oxford: Clarendon Press, 1970.
Zangwill, O. 1963. The completion effect in hemianopia and its relation to anosognosia. In *Problems of Dynamic Neurology*, edited by L. Halpern. Jerusalem: Jerusalem Post Press.
Zangwill, O. 1974. Consciousness and the cerebral hemispheres. In *Hemisphere Function in the Human Brain*, edited by S. Dimond and J. G. Beaumont. New York: John Wiley & Sons.

Zeldin, T. 1981. *France 1848–1945: Anxiety and Hypocrisy*. Oxford: Oxford Univ. Press.

Zilboorg, G. 1959. The dynamic process in psychiatry. In *The Historical Development of Physiological Thought*, edited by C. Brooks and P.P.F. Cranefield, 137–47. New York: Hafner Publishing Co.

Index

Library of Congress Cataloging-in-Publication Data

Harrington, Anne, 1960–
Medicine, mind, and the double brain.

Bibliography: p.
Includes index.
1. Cerebral dominance—History—19th century. 2. Laterality—History—
19th century. I. Title. [DNLM: 1. History of Medicine, 19th Cent. 2. Laterality—
history. 3. Neurology—history. WL 11.1 H299m]
QP385.5.H37 1987 152'.09 87–2462
ISBN 0–691–08465–3